BIOREMEDIATION
Science and Applications

Related Society Publications

Defining Soil Quality for a Sustainable Environment
SSSA Special Publication Number 35

Interactions of Soil Minerals with Natural Organics and Microbes
SSSA Special Publication Number 17

Pesticides in the Soil Environment: Processes, Impacts, and Modeling
SSSA Book Series: 2

Rates of Soil Chemical Processes
SSSA Special Publication Number 27

Reactions and Movement of Organic Chemicals in Soils
SSSA Special Publication Number 22

Soil and Water Science: Key to Understanding Our Global Environment
SSSA Special Publication Number 41

Sorption and Degradation of Pesticides and Organic Chemicals in Soil
SSSA Special Publication Number 32

BIOREMEDIATION
Science and Applications

Editors

H.D. Skipper and R.F. Turco

Editorial Committee

J.S. Angle, T.B. Moorman, J. Novak and I.L. Pepper

Organizing Committee

H.D. Skipper, R.F. Turco, M.D. Jawson, D. Linn, R.M. Miller, A. Wollum

Editor-in-Chief SSSA

Jerry M. Bigham

Managing Editor

Jon M. Bartels

SSSA Special Publication Number 43

Soil Science Society of America, Inc.
American Society of Agronomy, Inc.
Crop Science Society of America, Inc.
Madison, Wisconsin, USA

1995

Cover design: P.J. Scullion

American Society of Agronomy, Inc.
Soil Science Society of America, Inc.
Crop Science Society of America, Inc.
677 South Segoe Road, Madison WI 53711 USA

Library of Congress Cataloging-in-Publication Data

C.I.P. data in preparation

Printed in the United States of America

CONTENTS

FOREWORD

Human activity, especially during the twentieth century, has led to contamination of many sites with organic and inorganic compounds that pose a threat to human, animal, or plant health. In the first half of the twentieth century, concern about these contaminated sites was minimal because many of the sites were located in remote areas with sparse populations, and because our understanding of potential toxicities and biogeochemical cycles operating at these sites was limited. During the second half of the twentieth century, however, our knowledge about toxicities, biogeochemical processes, and their consequences has increased substantially. Vastly improved public communication has made this information widely available to the public, resulting in considerable public debate about these issues.

Our increased awareness of the finite resources of planet Earth and our desire to preserve them for future generations has increased the search of technologies to remediate contaminated sites. One of these technologies involves bioremediation, which uses biological activity to mineralize organic contaminants, or to remove inorganic contaminants from affected sites. This technology is for the most part in its infancy, but has a promising future. The information presented in this book reflects the state-of-the-art of the science and technology of bioremediation, and highlights areas that should be understood by successful practitioners. In addition, the ideas and applications discussed should provide guidance and inspiration to all those interested in advancing the field of bioremediation to recover and preserve two of our most precious resources: soil and water.

DAVID E. KISSEL, *president*
Soil Science Society of America

JOHN E. FOSS, *president*
American Society of Agronomy

ROBERT C. SHEARMAN, *president*
Crop Science Society of America

PREFACE

A wide diversity of anthropogenic and naturally occurring hazardous chemicals have been applied, purposely or accidentally, to soils. However, citizens around the world recognize the urgent need to do something, to restore the environment to a safe and healthy condition, and to protect the inhabitants from irreparable harm.

One cleanup approach gaining favor in the last 10 years is bioremediation. Bioremediation is a multidisciplinary strategy or process that uses microorganisms, plants, or enzymes from microorganisms or plants to detoxify pollutants in the soil and water environments. Bioremediation embraces biodegradation which is often defined as the transformation/detoxification of pollutants using microorganisms and plants. To be effective, those using bioremediation technologies must embrace holistic approaches to reclaim and to restore soils to a healthy and sustainable state.

In 1991, the Executive Committee of Division S-3, Soil Microbiology and Biochemistry (since renamed Soil Biology and Biochemistry), formed a S-3 Working Group on Bioremediation with the following general objectives: 1) promote symposia, workshops, and outreach programs on bioremediation; 2) project long term needs to advance bioremediation from a divisional perspective; and 3) "build bridges" with other groups interested in bioremediation. From the S-3 Working Group, a S-3 Bioremediation Organizing Committee of Mike Jawson, EPA, Ada, OK; Dan Linn, DuPont, Newark, DE; Raina Miller, University of Arizona; Ron Turco, Purdue University; Art Wollum, North Carolina State University; and Horace Skipper, Chair, Clemson University was formed and initiated plans in 1992 for the first bioremediation symposium by the Soil Science Society of America. The goal of the symposium and now this book is to show how the disciplines of sciences, engineering, and humanities can work together to achieve maximum progress in this new and important arena.

The S-3 Bioremediation Organizing Committee selected 21 speakers to "build bridges" between basic and applied sciences and engineering, professional societies and private industries, governmental agencies and consultants, as well as environmental and economical concerns. A symposium was developed to address a broad range of potential limiting factors to bioremediation: biological, chemical, physical, engineering, regulator, political, public perceptions, educational. The symposium, "Bioremediation: Science and Applications" was conducted November 6-7, 1993 at the annual meetings of SSSA/ASA/CSSA in Cincinnati, OH with approximately 200 "bioremediologists" in attendance. In addition to Division S-3, the symposium was cosponsored by Divisions A-5, Environmental Quality; S-1, Soil Physics; S-2, Soil Chemistry; A-2, Military Land Use and Management; and A-8, Integrated Agricultural Systems.

When this symposium was organized, the S-3 Bioremediation Organizing Committee also recognized the need for a written document to summarize the presentations. A request was submitted to and approved by the SSSA Executive Committee to publish the symposium as a SSSA Special Publication. President Darrell Nelson appointed an Editorial Committee of Scott Angle, University of Maryland; Tom Moorman, USDA-ARS, Ames, IA; Jeff Novak, USDA-ARS, Florence, SC; Ian Pepper, University of Arizona; and Editors, Horace Skipper, Clemson University and Ron Turco, Purdue University. Of the 21 presentations given in Cincinnati, 18 were prepared for publication in this book.

Topics include: Environmental/Soil Contamination; Sorption/Desorption of Xenobiotics; Enhancing Bioavailability of Xenobiotics; Redox and Energy Aspects; Microflora; molecular Tools; Lignin-degraders; Rhizosphere Degradation; Phytoremediation; Hazardous Waste Mixtures; Heavy Metals; Valdez and Prall's Island Oil Spills; Triazine Contaminated Soils; Landfarming; Engineering Aspects; Soil Science Skills; Science Needs for Bioremediation; and Educational Needs.

We deeply appreciate the authors who made time in their busy schedules to complete their chapters. We also are indebted to other members of the S-3 Bioremediation Organizing Committee who made major contributions to the symposium and to other members of the S-3 Bioremediation Editorial Committee and reviewers who helped make this publication a reality.

We trust the information contained in this book will be useful to students of bioremediation, the scientific community, and to those practitioners who are challenged to make bioremediation successful.

HORACE SKIPPER, *editor*
Clemson University
Clemson, South Carolina

RONALD TURCO, *editor*
Purdue University
West Lafayette, Indiana

CONTRIBUTORS

Todd A. Anderson
Pesticide Toxicology Laboratory, Iowa State University, 115 Insectary, Ames, IA 50011-3140, PHONE: 515-294-8667, FAX: 515-294-9823
EMAIL: etox@iastate.edu

S. M. Arnold
Department of Soil Science, University of Wisconsin, Madison, WI 53706-1299, PHONE: 608-263-5691

Jean-Marc Bollag
Laboratory of Soil Biochemistry, The Pennsylvania State University, University Park, PA 16802, PHONE: 814-863-0843, FAX: 814-865-7836

W. B. Bollag
Laboratory of Soil Biochemistry, The Pennsylvania State University, University Park, PA 16802

Tom Brouns
Pacific Northwest Laboratories, PO Box 999 Mailstop K6-77, Richland, WA 99352, PHONE: 509-376-7855, FAX: 509-376-5368
EMAIL: tm_brouns@pnl.gov

A. W. Bourquin
Camp Dresser & McKee Inc., Ten Cambridge Center, Cambridge, MA 02142

Joel Coats
Pesticide Toxicology Lab, Iowa State University, 116 Insectary, Ames, IA 50011-3140, PHONE: 515-294-4776, FAX: 515-294-9823
EMAIL: jcoats@iastate.edu

Scott D. Cunningham
DuPont Company, Glasgow Site-301, PO Box 6101, Newark, DE 19714-6101, PHONE: 302-451-9940, FAX: 302-451-9138
EMAIL: cunninsd@esvax.dnet.dupont.com

E. Kudjo Dzantor
TVA Mail Center, Tennessee Valley Authority Reservation/NFE BA, CEB-1C, Muscle Shoals, AL 35660, PHONE: 205-386-3593, FAX: 205-386-2191

Allan S. Felsot
Food & Environmental Quality Lab, 100 Sprout Road, Washington State University, Richland, WA 99352, PHONE: 509-375-9365, FAX: 509-375-9460
EMAIL: afelsot@beta.tricity.wsu.edu

Hugh Finklea
CIBA-GEIGY Corporation, PO Box 11, St. Gabriel, LA 70776, PHONE: 504-642-1371, FAX: 504-642-1794

Martin F. Fontenot, Jr.
CIBA-GEIGY Corporation, PO Box 11, St. Gabriel, LA 70776, PHONE: 504-642-1308, FAX: 504-642-1653

William T. Frankenberger, Jr. Department of Soil & Env. Sciences, University of California, Riverside, CA 92521, PHONE: 909-787-3405, FAX: 909-787-4328 EMAIL: williamf@orange.ucr.edu

John A. Glaser USEPA, Risk Reduction Engineering Lab, 26 West Martin Luther King Dr., Cincinnati, OH 45268, PHONE: 513-569-7568, FAX: 513-569-7105

C.P. Leslie Grady, Jr. Environmental Systems Engineering, Clemson University, 342 Computer Court, Anderson, SC 29625, PHONE: 803-656-5570, FAX: 803-656-0672

Robin F. Harris Department of Soil Science, University of Wisconsin, Madison, WI 53706-1299, PHONE: 608-263-5691, FAX: 608-265-2595

Stephen M. Hinton Exxon Corporate Research Company, Route 22 East, Annandale, NJ 08801, PHONE: 908-730-2125, FAX: 908-730-3042 EMAIL: hinton@erenj.com

Michael D. Jawson U.S. Environmental Protection Agency, Robert S. Kerr Environmental Research Laboratory, PO Box 1198, Ada, OK 74820, PHONE: 405-436-8560, FAX: 405-436-8529 EMAIL: jawson@ad3100.ada.epa.gov

K. Jaychandran USDA-ARS, National Soil Tilth Laboratory, 2150 Pammel Drive, Ames, IA 50011, PHONE: 515-294-2308, FAX: 515-294-8125

L. G. Keim Exxon Corporate Research Company, Route 22 East, Annandale, NJ 08801

Richard T. Lamar USDA-Forest Service, Forest Product Lab, Madison, WI 53706, PHONE: 608-231-9469, FAX: 608-231-9262

C. Richard Lee Waterway Experimental Station, Corps of Engineers, PO Box 631, Vicksburg, MS 39180, PHONE: 601-634-3585, FAX: 601-634-4298

Robert J. Lenhard Pacific Northwest Laboratories, PO Box 999 Mailstop K6-77, Richland, WA 99352, PHONE: 509-376-5960, FAX: 509-376-4428 EMAIL: rj_lenhard@pnl.gov

M. E. Losi Department of Soil & Env. Sciences, University of California, Riverside, CA 92521, PHONE: 909-787-5218, FAX: 909-787-4328 EMAIL: markl@orange.ucr.edu

M. S. Mather Petro Environmental Technology, Cincinnati, OH

Raina Miller Department of Soil & Water Science, University of Arizona, Tucson, AZ 85721, PHONE: 520-621-7231, FAX: 520-621-1647 EMAIL: rmmiller@ccit.arizona.edu

V. Minak-Bernero
Exxon Corporate Research Company, Route 22 East, Annandale, NJ 08801

J. Kent Mitchell
University of Illinois, Department of Agricultural Engineering, 332-P Agric. Eng. Sci. Bldg., 1304 West Pennsylvania Avenue, Urbana, IL 61801, PHONE: 217-333-4913, FAX: 217-244-0323
EMAIL: jkm@age2.age.uiuc.edu

Thomas B. Moorman
USDA-ARS, National Soil Tilth Laboratory, 2150 Pammel Drive, Ames, IA 50011, PHONE: 515-294-2308, FAX: 515-294-8125

Jeffrey M. Novak
USDA-ARS, Coastal Plains SWPC Res. Ctr., PO Box 3039, Florence, SC 29502-3039, PHONE: 803-669-5203 Ext. 110, FAX: 803-669-6970
EMAIL: oujnvk@clust1.clemson.edu

Tom A. Pedersen
Camp Dresser & McKee Inc., Ten Cambridge Center, Cambridge, MA 02142, PHONE: 617-252-8000, FAX: 617-621-2565

Michael J. Sadowsky
Department of Soil Science, University of Minnesota, St. Paul, MN 55108, PHONE: 612-625-9734, FAX: 612-625-2208

R. S. Skeen
Pacific Northwest Laboratories, PO Box 999 Mailstop K6-77, Richland, WA 99352

Horace D. Skipper
Department of Agronomy and Soils, Clemson University, PO Box 340359, Clemson, SC 29634-0359, PHONE: 803-656-3525, FAX: 803-656-3443
EMAIL: skipper@clust1.clemson.edu

Ron Turco
Environmental Soil Science, Laboratory for Soil Microbiology, Purdue University, West Lafayette, IN 47907-1150, PHONE: 317-494-8077, FAX: 317-496-1368
EMAIL: rturco@dept.agry.purdue.edu

Jerry B.Weber
Department of Crop Science, PO Box 7620, North Carolina State University, Raleigh, NC 27695-7620, PHONE: 919-515-2511, FAX: 919-515-7959

John Wilson
U.S. Environmental Protection Agency, Robert S. Kerr Environmental Research Laboratory, PO Box 1198, Ada, OK 74820, PHONE: 405-436-8532, FAX: 405-436-8529

1 Soil Contamination and the Feasibility of Biological Remediation

J.-M. Bollag & W.B. Bollag
Laboratory of Soil Biochemistry
Center for Bioremediation and Detoxification
The Pennsylvania State University
University Park, Pennsylvania

Over the past decade, the prevention of soil pollution and cleanup of contaminated soils have become a worldwide environmental priority. In order to develop suitable biotechnological methods to remediate these pollutants, it is necessary to understand the hazards they present and the subsequent fate of these materials in the environment. Research efforts in bioremediation technology have expanded rapidly over the past several years, but basic research is still needed to better understand the biological, chemical, and physical factors affecting biotransformation pathways and reaction rates. Interdisciplinary cooperation is required to solve many of the complex problems involved in biologically remediating contaminated soils.

Society has entered a technological age in which manmade and -modified chemicals are used to produce clothing and containers, to power vehicles and homes, to increase food harvests, and to control potentially harmful pests. Although the benefits of such chemicals are obvious, their widespread use has also created a critical problem--how to dispose of and/or remove these products after use or accidental spillage. Traditional methods of handling waste, such as landfilling or incineration, often exchange one problem for another. For instance, landfilling merely confines the pollution while doing little to remove it--certain garbage items are known to survive for decades in a landfill. Incineration removes the wastes but disposal of ash or residues remains, and new concerns of air pollution are created. In addition, certain wastes, such as contaminated ground water, cannot be burned. The sheer enormity of the problem of removal becomes apparent when one considers the amount of wastes generated and released to the environment. In the true spirit of the age, society has turned to technology to devise better methods of disposal. One such method is bioremediation, in which microorganisms or plants are used to degrade, transform, or in other ways detoxify wastes. Although bioremediation has been successfully used for decades to treat wastewater, the application of this technique to soil and groundwater pollution control is fairly new.

The intensive development and improvement of these alternative uses has made the bioremediation industry a rapidly growing business. It is our intention to examine the degree of contamination that presently exists in the soils of the USA and to discuss bioremediation techniques that can be used todetoxify polluting chemicals. Advantages and disadvantages of these techniques will be addressed, as will possible ways of overcoming associated pitfalls. Finally, potential technologies and improvements for the future will be examined.

GENERATION OF WASTE

This society generates an enormous amount of waste annually. In fact, the USA produces 300 million metric tons of hazardous waste each year, or approximately 1.3 metric tons per capita. These wastes arise from all facets of society: households generate about 4 pounds per person per day; various industries, including the food, textile, lumber, petroleum, chemical, and transportation industry produce millions of pounds more, despite the fact that U.S. industries invest billions of dollars a year in pollution abatement capital expenditures. The chemical industry alone generates over 2500 million pounds of waste, and the amount of chemicals discharged to surface water, injected underground or released to land totals more than one billion pounds (USEPA, 1993). The agricultural industry also produces tons of hazardous chemicals in the form of pesticides -- in 1992 over 400 million pounds of these chemicals, including fungicides, insecticides, and herbicides, were employed in order to increase crop yields (USDA, 1992). Although pesticides and fertilizers have contributed to achieving the goal of larger harvests, they have also contributed to the problems of soil and water pollution. In fact, agriculture accounts for 40% of the pollution entering U.S. rivers (Council on Environmental Quality, 1993) (Fig.1-1). Thus, the sheer volume of chemicals released into the environment represents a potentially serious disposal problem.

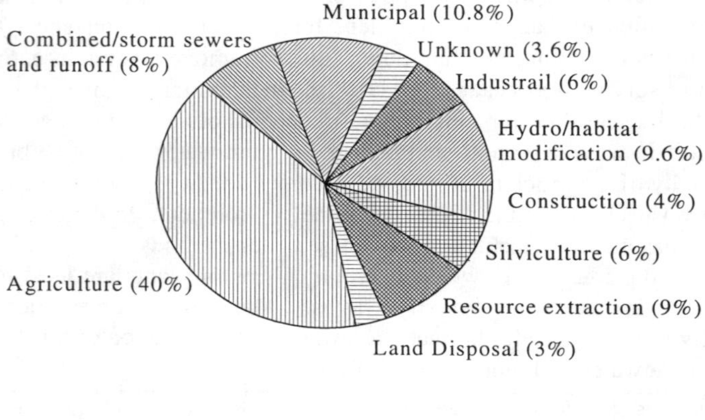

Fig. 1-1. Sources of pollution in U.S. rivers (Council Environ. Quality, 1993).

The manufactured chemicals now present in our environment also represent possible health hazards. The United States Environmental Protection Agency (USEPA) categorizes chemicals by their known or suspected effects as carcinogens, mutagens or as developmental, reproductive, acute, chronic or neurological toxins, and many chemicals fall into more than one of these categories. Over 10 000 pounds of carcinogens, agents known to increase cancer risks, have been released to land in more than 50% of the states in the U.S. (Fig. 1-2). In addition, the average adult ingests approximately 0.02 μg of pesticides such as DDT, dieldrin, and toxaphene per pound of body weight per day (State of the Environment, 1987). Obviously, exposure to these and other chemicals released into the environment can have both acute and cumulative effects on health.

The desire to minimize health effects has led to large expenditures to combat pollution. As previously mentioned, industries themselves spend large sums of money to reduce pollutant output. Pollution control and abatement consume a large percentage of the approximate 20 billion dollars of the U.S. federal gross spending for natural resources and the environment per year (Council on Environmental Quality, 1993).

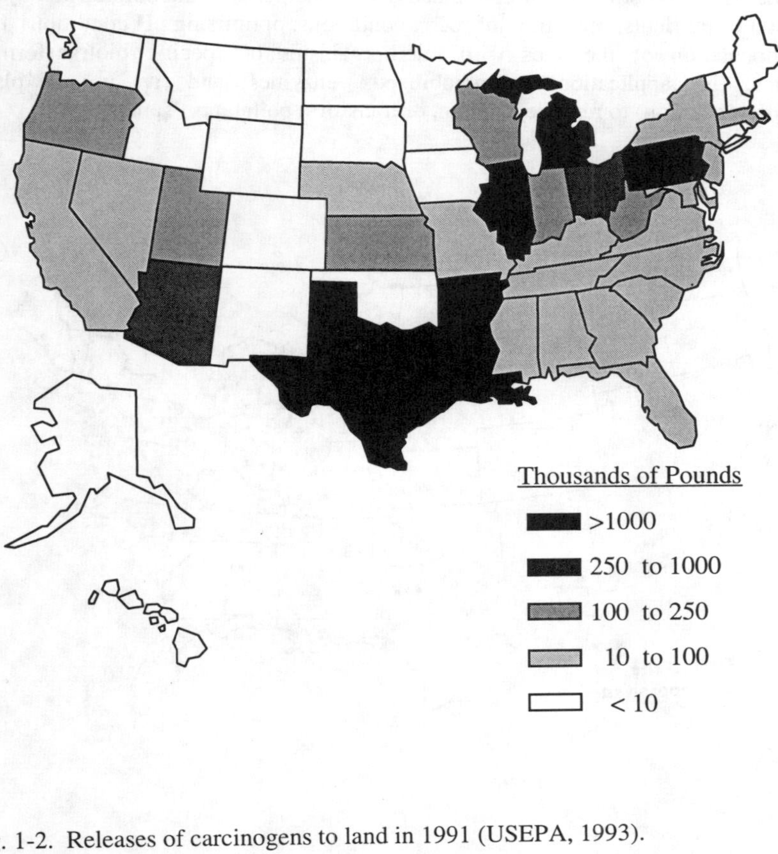

Fig. 1-2. Releases of carcinogens to land in 1991 (USEPA, 1993).

Various agencies in the U.S. government (e.g., the Departments of Energy, Defense, Agriculture, Interior, and Transportation) are budgeting 9 billion dollars for cleanup of federal facilities, and another 2 billion is spent for the enforcement of federal environmental regulations (Executive Office of the President, 1992).

Perhaps the greatest investment in pollution control has been in the identification of Superfund sites. These are areas located throughout the USA which have been determined to possess large amounts of hazardous chemicals requiring cleanup. Over 1 200 such areas have been designated (Fig. 1-3), and the estimated cost for decontamination of these Superfund sites is upwards of 300 billion dollars. This huge monetary investment has led to an interest in new and more cost-effective methods for removal of the hazardous wastes. One technology which shows great promise in terms of cost-effective pollution abatement is bioremediation.

BIOREMEDIATION

Bioremediation is the use of microorganisms or plants to detoxify an environment, mostly by transforming or degrading pollutants. Four basic techniques may be used: (i) stimulation of the activity of indigenous microorganisms by the addition of nutrients, regulation of redox conditions, optimizing pH conditions, etc.; (ii) inoculation of the sites with microorganisms of specific biotransforming abilities; (iii) application of immobilized enzymes; and (iv) use of plants (phytoremediation) to remove, contain, or transform pollutants (Table 1-1).

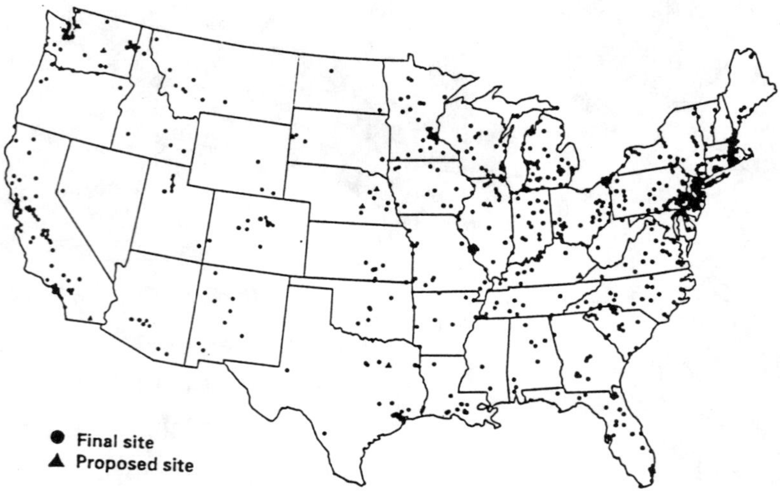

Fig. 1-3. National Priority List Sites, total of 1235 sites (USEPA, 1993).

Table 1-1. Techniques of in situ bioremediation.

Method	Process
Stimulation	To enhance activities of indigenous microorganisms by addition of nutrients, regulation of redox conditions, pH change, or removal of other limiting conditions.
Inoculation	Inoculation of microorganisms of specific biotransforming abilities into contaminated environments.
Enzyme Treatment	Application of immobilized enzymes to transform or degrade specific pollutants.
Phytoremediation	Use of plants to remove, contain, or transform pollutants.

The utility of bioremediation in the degradation of pollutants in the environment has been successfully demonstrated for several chemical compounds. In the 1970s, Raymond (U.S. Patent) showed that bioremediation could reduce the amount of gasoline in contaminated ground waters. Remediation using biological agents has also been achieved for kerosene-, pentachlorophenol- and oil-contaminated soil, oil- and para-nitrophenol-infiltrated aquatic habitats and naphthalene-, phenanthrene- and chlorinated ethene-contaminated ground waters (Table 1-2). Although relatively few contaminating compounds have been successfully bioremediated to date, research is continuing to identify organisms and methods capable of degrading a variety of other pollutants,and bioremediation remains an innovative and attractive pollution abatement technology (Madsen, 1991).

Table 1-2. Classes of chemicals treated by bioremediation.

Class	Representative Chemical	Example
Fuel hydrocarbons	Benzene, Toluene	Hoeppel et al., 1991
PAH's (Polynuclear romatic hydrocarbons)	Creosote	Mueller et al., 1989
PCB's (Polychlorinated biphenyls)	Aroclor	Nies & Vogel, 1990
Chlorinated solvents	TCE†	Nelson et al., 1990
Chlorinated aromatic compounds	Chlorobenzene	Kampbell & Wilson, 1993
Chlorophenols	Pentachlorophenol	Glaser & Lamar, 1995
Nonhalogenated phenolics	2-Methylphenol	
Pesticides	2,4-D, Atrazine	Finklea & Fontenot, 1995; Felsot et al., 1995
Explosives	TNT†	Walker & Kaplan, 1992
Nitrogen heterocyclics	Pyridine	
Radionuclide	Plutonium	
Anions	Nitrate	
Metals	Lead	Cunningham & Berti, 1993; Cunningham & Lee, 1995.

† TCE= Trichloroethylene; TNT=2,4,6- Trinitrotoluene

Several specific methods have been and are being developed for bioremediation of contaminated soil and water. These include composting, landfarming, above-ground bioreactors, solid-phase, and *in situ* treatment. Composting, bioremediation in which biodegradable materials are combined in a pile suitable for the growth of thermophilic microorganisms, has been widely used by farmers as a means of removing natural organic wastes while generating soil organic matter. Recently, this technique has also been used to decrease the concentrations of explosives [TNT (2,4,6-trinitrotoluene), RDX (hexahydro-1,3,5-trinitro-1,3,5-triazine), HMX (octahydro-1,3,5,7-tetranitro-1,3,5,7-tetraazocine) and Tetryl (N-methyl-1,2,4,6-tetranitroaniline)] to less than 10% of their starting levels within 80 days (Walker & Kaplan, 1992). A related method is solid-phase treatment in which contaminated soil or treatment lagoon sludge is placed in a lined treatment bed and amended with nutrients (USEPA, 1991). Alternatively, mounds of soil may be placed in an enclosure or on top of a liner and water with nutrients percolated throughout. This method has been successfully used to remove hydrocarbons from contaminated sludge (Bossert et al., 1984).

Landfarming is an on-site treatment method which is frequently utilized by the petroleum industry. In landfarming, the degradation of pollutants is enhanced by supplementing the soil with nutrients and oxygen, then tilling and irrigating to create an optimal environment for microbial activity and to increase the chance of contact between contaminants and microorganisms. In this technique special plots of land are necessary in which a thick clay layer (natural or constructed) must lie between the landfarm area and a deep water table to prevent groundwater contamination by the leachate. Contaminated wastes are then loaded into the land plot with special consideration given to the rate of loading, moisture content, nutrient availability, etc. Under appropriate conditions of landfarming, it has been possible to demonstrate removal of approximately 70% of the aromatic and 90% of the aliphatic compounds from oil-contaminated wastes within a 35-d period (Litchfield, 1991).

Bioremediation in above-ground bioreactors involves the use of a containment vessel within which microorganisms in suspension or on a fixed solid support are incubated with contaminated wastes in a liquid or slurry form. This technology is particularly useful, as extensive control over the entire process is possible, and ideal conditions for microbial transformation of the chemical can be created. Thus, such factors as nutrient availability, degree of interaction between the organisms and the pollutants, water content and rate of influx and efflux of contaminated and detoxified solutions, respectively, can be tightly regulated to ensure maximal bioremediation. In addition, an inoculum can be chosen that provides extensive activity on the compound to be removed. The degrading population may be present in the contaminated materials or may be added in the form of pure cultures or precise mixtures of microbial species. The inoculum may even include genetically engineered microorganisms, as their containment allowed by the enclosed above-ground bioreactors' design will ameliorate most of the public's and regulatory agencies' misapprehensions about the application of such genetic engineering technologies. Finally, above-ground bioreactors can be operated in an aerobic or an anaerobic mode, or perhaps in series, to facilitate complete biodegradation of contaminants. Using these reactors, the levels of such compounds as pentachlorophenol, methyl ethyl ketone, isopropanol, ethylene glycol, methyl isobutyl ketone, phenol, benzoate, chlorobenzoic acid, organochlorines, and organophosphate pesticides have all been successfully decreased (Litchfield, 1991).

In situ bioremediation involves the use of organisms to remove pollutants at the site of contamination. Often, these organisms are indigenous to the area and may even be adapted for growth on the chemical contaminant in that particular environment. The adapted microbes will transform the compound until one nutrient reaches a limiting concentration. In enhanced *in situ* bioremediation, then, the limiting nutrient(s) is (are) added to the environment to allow the organisms to continue degrading the pollutant. For example, following the Exxon Valdez oil spill in Alaska, the application of nutrients to the oil-soaked beaches stimulated the rate of biodegradation (USEPA, 1991; Hinton 1995). Enhanced *in situ* bioremediation has also been used to remediate halogenated aliphatics, chlorinated pesticides, polychlorinated biphenyls, nitroaromatics, and polycyclic aromatics at contaminated locations (Lee et al., 1988; Morgan & Watkinson, 1989; Stroo, 1992).

An alternative to the enhancement of bioremediation by indigenous microorganisms is the use of an inoculum of an appropriate pure or mixed culture of degrading microorganisms to effect removal of the undesired compound(s) (Gibson & Sayler, 1992). Usually, in this case as well, nutrient amendment is necessary for maximal removal of the pollutants. For each type of bioremediation technique discussed above, successful removal of contaminants has been demonstrated, indicating the utility of these methods for pollution control.

ADVANTAGES AND DISADVANTAGES OF BIOREMEDIATION

Bioremediation is also attractive because of several advantages that its application has over more conventional technology. For instance, bioremediation techniques are typically more economical than more traditional methods -- it is estimated that bioremediation would cost one-third to one-half that of incineration. In addition, in some cases pollutants can be treated at the site, thus reducing the risk of exposure to cleanup personnel and potentially wider exposure as a result of transportation accidents (Gabriel, 1991). Furthermore, the techniques usually can be expected to have minimal environmental impact, since bioremediation is a natural process. Also, waste products are usually not generated, as bioremediation often results in complete degradation of the contaminants. These characteristics make bioremediation techniques potentially ideal for detoxification of chemical pollutants.

However, there are a number of problems which are encountered with bioremediation as well. First, in order for organisms to successfully reduce pollutant levels, their population must increase. For this, their growth conditions must be determined and maintained at the contaminated sites. Controlling these conditions may prove difficult, particularly as conditions may vary so greatly among sites. In addition, the labor required to maintain optimal conditions may be extensive. Even in an ideal environment an organism may prefer to metabolize other more readily available nutrients within a contaminated area, or the pollutant may be completely or partially inaccessible to the degrading organism. Finally, the environment may contain substances (e.g., heavy metals) or organisms that inhibit growth of the degrading population.

Nevertheless, many of these problems associated with bioremediation are already in the process of being overcome. Microorganisms capable of degrading recalcitrant compounds are being continually discovered using methods such as batch enrichment and chemostat selection. Degrading organisms may also be artificially generated by genetic engineering techniques. However, only when the introduced modified organisms has a selective advantage that is linked with the growth on the substrate of concern will they have the chance and advantage to compete successfully with the existing microflora in the contaminated soil (Alexander, 1984). To circumvent the problem of insufficient organismal growth, the application of isolated enzymes, rather than the entire organisms, has been advocated (Munnecke, 1981; Nannipieri & Bollag, 1991). Methods for improving the efficacy and stability of isolated enzymes for field treatments are currently being examined, as are methods for ensuring microbial degradation of the compound of interest in the contaminated environment. Although bioremediation technologies are by no means problem-free, current research is designed to allow bioremediation to reach its full potential as a pollution control technique.

Other potential, but likely solvable, barriers to the use of bioremediation as a pollution abatement technology involve engineering and regulatory questions. In terms of regulatory pitfalls, only recently (in the 1980s) has *in situ* bioremediation gained favor with the regulatory agencies which tend to hamper bioremediation efforts at both the state and federal levels. Lawyers have as much influence on decisions regarding bioremediation as scientists and engineers, leading to such regulations as those which strictly control the use of genetically engineered microorganisms (Stroo, 1992). Nevertheless, the USEPA is sponsoring demonstration sites and educational programs in an attempt to promote bioremediation technologies (USEPA, 1991; Sims et al., 1992), but issues of safety, public health, and environmental liability are still raised. For example, bioremediation methods may effectively diminish concentrations of contaminants; yet, because of the tendency for biological systems to be inhibited by diminishing substrate levels, stringent clean-up goals may still not be achieved. In addition, bioremediation may succeed in completely removing a given pollutant only to generate another possibly toxic by-product; for instance, conversion of TNT to amines or azoxy dimers during composting or conversion of TCE to vinyl chloride by anaerobic bacteria in aquifers. Obviously, use of bioremediation requires careful monitoring by both the users and the regulators of this innovative, but not yet generally recognized methodology.

Other factors limiting the utility of bioremediation concern engineering design and application. First, the chemical and site requiring decontamination must be assessed to determine suitability for one of the bioremediation techniques. Most importantly, it must be decided if, in fact, bioremediation is the option of choice for removal of the pollutant from that particular contaminated area, as alternative methodologies might be better suited to certain scenarios. For bioremediation to be feasible, the compound must be biodegradable and the environment appropriate, although in some cases, inappropriate conditions can be remedied by adjustments involving nutrient amendment, pH or water content alterations, reduction of toxins and/or elimination of predators. In addition, organisms capable of degrading the pollutant must be available and capable of transforming the compound under field conditions. Herein often lie difficulties in moving from bench- and pilot-scale studies to full-scale field operations (Gabriel, 1991). In addition, numerous development costs may inhibit application of the methodologies before success has been adequately demonstrated (Litchfield, 1991). For this reason, multidisciplinary development teams are ideal and should consist of experts in various aspects of biology including microbiology, biochemistry, physiology, ecology and genetics, as well as in engineering, hydrogeology, soil science, and environmental chemistry. The goal of the biologists is to find contaminant-degrading organisms, determine the optimal conditions for growth of the organisms and biotransformation of the pollutant within this environment, and monitor for unwanted by-products in a small-scale investigation. The goal of the engineer, hydrogeologist, and soil scientist is to design a system possessing these optimal growth conditions on a large scale and working as cheaply as possible. Only by thorough coordination and integration of efforts by these diverse staff members can efficient and safe bioremediation be realized.

RESEARCH NEEDS

The limited knowledge available on the nature of bioremediation warrants additional research into its application. Investigation into ways to improve process performance are necessary but are restricted by a paucity of information concerning the basic principles underlying the phenomenon (Gibson & Sayler, 1992). For instance, it is unclear how to sustain growth and existence of the degrading organisms in preference to others in the environment, particularly if the pollutant does not itself support growth. Additionally, the effect of multicomponent mixtures on bioremediation is largely unknown, making a prediction of the outcome of multicomponent biodegradation efforts quite difficult. Research into molecular improvements of degradative capacity are also desirable, as it has become possible to engineer microorganisms possessing expanded ranges of performance and using alternative pathways which avoid the formation of toxic by-products (Gibson & Sayler, 1992). Finally, novel and innovative technologies for bioremediation must be explored. As mentioned previously, one such method may be the use of isolated and immobilized enzymes rather than entire organisms for decontamination of polluted locations.

In addition, the idea has been proposed that binding of chemicals to solid matter via synthetic reactions catalyzed by microbial enzymes should result in detoxification of the compounds with only limited release subsequently (Bollag, 1992). The use of plants for bioremediation (i.e., phytoremediation) has also been advocated (Anderson & Coats, 1995; Walton & Anderson, 1992; Bell, 1992). Although it is thought that this removal is due to soil microorganisms associated with plant roots, the ability of plants to support microbial growth by providing alternative carbon sources may make them potentially useful for new bioremediation methods. Obviously, such novel approaches to the problem of pollution abatement merit further consideration and exploration.

CONCLUSIONS

Investigations into the use of bioremediation as a means of pollution control have yielded promising results, in addition to emphasizing some potential problems and pitfalls. Many of these problems are likely to be solved once additional research provides a deeper understanding of the phenomenon of bioremediation. Other barriers, in particular those that deal not with technical details but with policy decisions, may prove more difficult to overcome. Nevertheless, it is apparent that in order to realize the full potential of this technology, more information must be obtained and novel approaches must be explored. With the cooperation of soil scientists, engineers, microbiologists, hydrogeologists, and others, it should be possible to safely, economically, and efficiently reduce pollutant concentrations at contaminated sites to acceptable levels. Such successful bioremediation efforts should improve the quality of life and reinforce society's confidence in research and technology.

REFERENCES

Anderson T.A. and J.R. Coats. 1995. An overview of microbial degradation in the rhizosphere and its implications for bioremediation. p. 133-141. *In* H.D. Skipper and R.F. Turco (ed.) Bioremediation: Science and applications. SSSA Spec. Publ. 43. ASA,CSSA and SSSA, Madison WI.

Alexander, M. 1984. Ecological constraints on genetic engineering. p.151-168. *In* G. S. Omenn and A. Hollaender (ed.) Basic life science: Genetic Control of Environmental Pollutants. Plenum Press, New York.

Bell, R. M. 1992. Higher plant accumulation of organic pollutants from soils. Project Summary, U.S. Gov. Print. Office, Washington D.C. USEPA/600/SR-92/138

Federal Coordination Council of Science, Engineering and Technology Committee on Life Sciences and Health. 1992. Biotechnology for the 21st century. FCCSET, U.S. Gov. Print. Office, Washington, D.C.

Bollag, J.-M. 1992. Decontaminating soil with enzymes. Environ. Sci. Technol. 26: 1876-1881.

Bossert, I., W.M. Kachel, and R. Bartha. 1984. Fate of hydrocarbons during oily sludge disposal in soil. Appl. Environ. Microbiol. 47: 763-767.

Cunningham, S.C., and W.R. Berti. 1993. Remediation of contaminated soils with green plants : An overview. *In Vitro* Cell. Dev. Biol. 29P: 207-212.

Cunningham, S.C. and C.R. Lee. 1995. Phytoremedition: Plant-Based Remediation of Contaminated Soils and Sediments. p.142-153. *In* H.D. Skipper and R.F. Turco (ed.) Bioremediation: Science and applications. SSSA Spec. Publ. 43. ASA,CSSA and SSSA, Madison WI.

Council on Environmental Quality. Environmental quality 23rd Ann. rep. U.S. Gov. Print. Office, Washington, D. C.

Executive Office of the President, Office of Management and Budget, Washington, D.C., 1992.

Felsot, A.S., J.K. Mitchell, and E.K. Dzantor. 1995. Remediation of herbicide-contaminated soil by combinations of landfarming and biostimulation. p. 232-252. *In* H.D. Skipper and R.F. Turco (ed.) Bioremediation: Science and applications. SSSA Spec. Publ. 43. ASA,CSSA and SSSA, Madison WI.

Finklea, H.C., and M.F. Fontenot, Jr. 1995. Accelerated bioremediation of triazine contaminated soils -- a practical case study. p217-232. *In* H.D. Skipper and R.F. Turco (ed.) Bioremediation: Science and applications. SSSA Spec. Publ. 43. ASA,CSSA and SSSA, Madison WI.

Gabriel, P. F. 1991. Innovative technologies for contaminated site remediation: Focus on bioremediation. J. Air Waste Manage. Assoc. 41: 1657-1660.

Gibson, D. T., and G. S. Sayler. 1992. Scientific foundations of bioremediation - current status and future needs. The Amer. Acad. Microb., Washington, DC.

Glaser J.A. and R.T. Lamar. 1995. Lignin-Degrading Fungi as Degraders of Pentachlorophenol and Creosote in Soil. p.116-132 *In* H.D. Skipper and R.F. Turco (ed.) Bioremediation: Science and applications. SSSA Spec. Publ. 43. ASA,CSSA and SSSA, Madison WI.

Hoeppel, R.E., R.E. Hinchee, and M.F. Arthur. 1991. Bioventing soils contaminated with petroleum hydrocarbons. J. Indust. Microbiol. 8:141-147.

Hinton, S.M. 1995. Bioremediation of the Valdez and Prall's Island oil spills. p. 208-216. *In* H.D. Skipper and R.F. Turco (ed.) Bioremediation: Science and applications. SSSA Spec. Publ. 43. ASA,CSSA and SSSA, Madison WI.

Kampbell, D.H., and B.H. Wilson. 1994. Bioremediation of chlorinated solvents in the vadose zone. p.255-258. *In:* R.E. Hichee, A. Leeson, L. Semprini and S.K. Ong (ed.) Bioremediation of chlorinated and polycyclic aromatic hydrocarbon compounds. Lewis Publishers, Boca Raton FL.

Lee, M.D., J.M. Thomas, R.C. Borden, P.B. Bedient, C.H. Ward, and J.T. Wilson. 1988. Biorestoration of aquifers contaminated with organic compounds. CRC Crit. Rev. Environ. Control. 18:29-89.

Litchfield, C.D. 1991. Practices, potential, and pitfalls in the application of biotechnology to environmental problems. p.147-157. *In* G.S. Sayler, R. Fox; and J. W. Blackburn (ed.) Environmental biotechnology for Waste Treatment; Plenum Press, New York.

Madsen, E.L. 1991. Determining *in situ* biodegradation: Facts and challenges. Environ. Sci. Technol. 25: 1663-1673.

Morgan, P., and R. J. Watkinson. 1989. Microbiological methods for the cleanup of soil and ground water contaminated with halogenated organic compounds. FEMS Microbiol. Rev. 63: 277-300.

Mueller, J.G., P.J. Chapman, and P.H. Pritchard. 1989. Creosote contaminated sites. Environ. Sci. Technol. 23:1197-2101.

Munnecke, D.M. 1981. The use of microbial enzymes for pesticide detoxification. *In* T. Leisinger et al. (ed.) Microbial degradation of xenobiotics and recalcitrant compounds. FEMS Symp. 12. Acad. Press, London.

Nannipieri, P., and J.-M. Bollag. 1991. Use of enzymes to detoxify pesticide-contaminated soils and waters. J. Environ. Qual. 20: 510-517.

Nelson, M.J., J.V. Kinsella, and T. Montoya. 1990. *In situ* biodegradation of TCE contaminated groundwater. Environ. Prog. 9:190-196.

Nies, L., and T.M. Vogel. 1990. Effects of organic substrates on dechlorination of Aroclor 1242 in anaerobic sediments. Appl. Environ. Microbiol. 56:2612-2617.

Raymond, R.L. 1974. Reclamation of hydrocarbon contaminated ground water. U.S. Patent 3 846 290. Date issued 5 November.

Sims, J.L., R.C. Sims, and J.E. Matthews. 1990. Approach to bioremediation of contaminated soil. Hazard Waste Hazard Mater. 8: 117-149.

State of the Environment. 1987. A view toward the nineties. The Conservation Foundation. Washington D.C.

Stroo, H.F. 1992. Biotechnology and hazardous waste treatment. J. Environ. Qual. 21: 167-175.

U.S. Department of Agriculture, Economic Research Service, Agricultural Resources: Inputs, outlook and situation report, AR-28; Washington, DC; October 1992.

U.S. Environmental Protection Agency. 1991. Understanding Bioremediation. U.S. Gov. Print. Office, Washington D.C. USEPA/540/2-91/002.

U.S. Environmental Protection Agency. 1993. Toxics Release Inventory. U.S. Gov. Print. Office, Washington D.C. USEPA/745/R-93/003.

Walker, J.E., and D.L. Kaplan. 1992. Biological degradation of explosives and chemical agents. Biodegradation 3:369-385.

Walton, B.T., and T.A. Anderson. 1992. Plant-microbe treatment systems for toxic waste. Curr. Opin. Biotechnol. 3:267-270.

2 Sorption and Binding of Organic Compounds in Soils and Their Relation to Bioavailability

J.M. Novak
USDA-ARS, Coastal Plains SWPCRC
Florence, South Carolina

K. Jayachandran and T.B. Moorman
USDA-ARS, National Soil Tilth Laboratory
Ames, Iowa

J.B. Weber
North Carolina State University
Raleigh, North Carolina

Widespread pollution of soils and groundwater environments occurs due to the improper disposal, misuse, and accidental release of organic chemicals into the environment. Numerous strategies are available for remediation of contaminated areas which typically involve a physicochemical or biological treatment process (U.S. EPA, 1988). A problem with many of the physicochemical treatment techniques is that the organic chemical contaminant is not degraded. Rather, the contaminant is usually sorbed to a modified matrix and transferred to a repository. Binding of the chemical to a sorbent is costly and does not reduce contaminant toxicity. Utilization of remediation methods that involve microbial decontamination have certain advantages when compared to physicochemical treatments. Bioremediation, through which microorganisms are used to transform organic chemicals to less toxic chemical forms, can provide an economical method to reduce contaminant toxicity. The development, application, and ecological significance of bioremediation to treat organic chemical contaminants have been extensively reviewed (Thomas & Ward, 1989; Sims et al., 1990; Skladany & Metting, 1993). This review will focus primarily on the challenges and potential problems of using bioremediation techniques on pesticide residues in soil. An estimated 341 million kg of pesticides are used globally for agricultural uses (Pimental & Levitan, 1986). Bioremediation is a potential technology for removing these compounds from soil following spills at manufacturing plants, mixing and distribution sites, and farmsteads. Pesticides offer challenges to the bioremediation industry. Many of these compounds are toxic to microorganisms at high concentrations (Moorman, 1989) and their complex chemical structures render them more recalcitrant than many components in typical oil and gas spills. Also, their potential hazard to human health require bioremediation processes to be highly effective in detoxifying trace amounts of these materials.

Before discussion of implementation techniques used for bioremediation, an understanding of factors which regulate bioavailability of pesticides will be discussed. A pesticide is often sorbed to soil as soon as it comes in contact with the surfaces of minerals and organic matter through a variety of chemical or physical processes. Pesticide sorption to soil will affect its availability for microbial degradation and influence the effectiveness of bioremediation to clean up contaminated soil. Sorption is therefore an important process affecting pesticide persistence in soil (Bailey & White, 1964; Hamaker & Thompson, 1972; Helling & Dragun, 1980; Weber et al., 1993).

The bioavailability of adsorbed pesticides in soil is dependent upon the extent of the adsorption reaction and rate of desorption. Chemicals that are readily desorbed are in equilibrium with the solution phase concentration and appear to be available for biodegradation. In contrast, some pesticides appear to enter sorption sites where desorption occurs at very slow rates or the chemicals become bound and nondesorbable (Weber et al., 1993). These processes are critical to water quality because they may limit pesticide bioavailability and regulate degradation (Khan & Iverson, 1982; Calderbank, 1989; Moorman, 1993). This chapter provides an understanding of the implications of sorption and binding of pesticides to soil minerals and organic matter with respect to bioavailability. The effects of sorption and binding of organic compounds has been extensively reviewed. Our intention is to identify gaps in our current understanding of bioavailability and identify areas of future research.

ADSORPTION AND DESORPTION PROCESSES

The soil organic and inorganic fractions are primarily responsible for sorption of pesticides (Hance, 1988; Calderbank, 1989). Both fractions have the capability to sorb pesticides because of their large surface area and diverse functional groups. Numerous reviews explain the nature of soil colloids and describe mechanisms of pesticide sorption (Calvert, 1980; Hance, 1988; Calderbank, 1989; Koskinen & Harper, 1990; Sims et al., 1990; Weber et al., 1993). The intermolecular forces which can attract pesticides to functional groups and subsequently sorb them have been classified as to sorption mechanisms (Bailey & White, 1970; Stevenson, 1982). The nature of the sorptive process is only partly understood, but a number of mechanisms have been identified, such as physical binding through van der Waals forces and chemical binding through dipole-dipole interactions, cation or water bridging, H-bonding, ion exchange, covalent bonding, or ligand exchange, (Khan, 1982; Koskinen & Harper, 1990). Several mechanisms may simultaneously contribute to pesticide sorption and the strength of each binding mechanism probably will vary as a function of time.

The sorption potential of a pesticide to soil is partially controlled by its chemical and physical properties. Depending upon chemical structure and the soil environment pesticides may behave as acidic, basic or nonionic compounds. Pesticides which are positively charged (diquat and paraquat), or can become positively charged by protonation (e.g., *s*-triazines, amitrole, pyrimidines, amines, and other weakly basic pesticides) are all capable of electrostatic attraction to negative charges on soil colloids. This form of electrostatic binding is almost always supplemented by H-bonding or charge-transfer bonding (Hance, 1969).

Nonionic pesticides have been reported to be sorbed to soil organic matter by H-bonding, van der Waals forces and charge-transfer complexes (Stevenson, 1982). Weber et al. (1993) provided a comprehensive review of the sorption potential and bioactivity of many types of pesticides.

Experimental evidence from pesticide sorption studies has contributed to the conceptualization of a site sequential-equilibria model (Wauchope & Meyers, 1985) which partitions the sorption process into specific sites (Fig. 2-1). During equilibration with soil, a portion of a pesticide will partition between the soil solution and the instantaneous equilibrium sites. Pesticide sorption/desorption kinetic studies have shown that this equilibration is usually complete within a few hours. Pesticides in the instantaneous equilibrium and labile sites are available for microbial and chemical degradation. Bioremediation techniques would be applicable to pesticides sorbed at these sites.

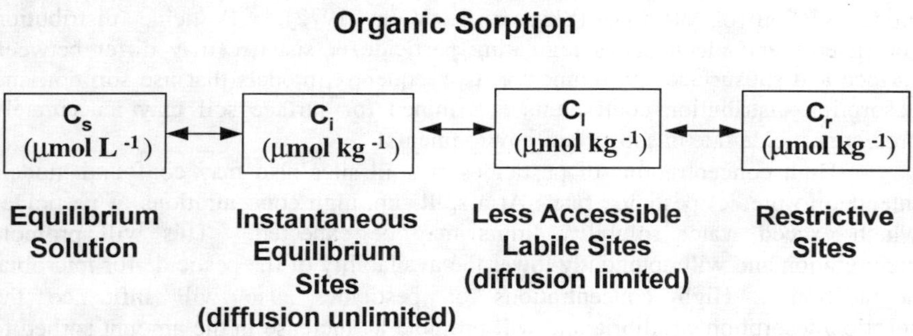

Fig. 2-1. Site sequential equilibria model conceptualizing pesticide sorption by soil (modified from Locke, 1992)

Sorption partition coefficients (K_d, K_{oc}) describe the distribution of pesticide between soil or organic matter and water at or near equilibrium conditions. Typically, pesticide K_d values are determined by batch equilibration and the K_d normalized for organic carbon content which results in the K_{oc} value (Table 2-1). Expressing sorption as K_{oc} has been adopted because of the strong relationship between organic carbon and the sorption of most nonionic pesticides. Pesticide K_{oc} values and organic carbon contents can be used by researchers to estimate the sorption of the chemical to a range of soils and then model potential movement (Helling & Dragun, 1980; McCall et al., 1980). For instance, DDT has a much higher K_{oc} value when compared to other chemicals in Table 2-1 which suggests a much lower leaching potential.

Sorption coefficients are typically determined using water or $CaCl_2$ solutions and agricultural surface soil. Consequently, extrapolation of pesticide behavior in agricultural surface soils, may not accurately predict behavior and bioavailability in a spill site or deep subsurface environment. Surface soil typically has greater pesticide sorption coefficients and degradation rates because of greater organic matter contents and microbial populations than subsurface soil (Mueller et al., 1992; Grundl & Small, 1993). However, it should be noted that organic matter content and

Table 2-1. K_{OC}-- distribution coefficients for several pesticides†

Chemical	K_{OC}
Dicamba	0
Monuron	83
Atrazine	170
Lindane	1,300
Trifluralin	3,900
Chloropyrifos	14,000
DDT	240,000

†After McCall et al., 1980.

microbial populations across three surface soils did not directly relate to biodegradation of atrazine (Skipper & Volk, 1972). Pesticide distribution coefficients and mechanisms regulating pesticide persistence may differ between surface and subsurface environments. Consequently, models that use sorption and desorption distribution coefficients determined for surface soil may inaccurately predict pesticide fate in subsurface environments.

High concentrations of pesticides at spill sites also may confound models intended to predict pesticide fate. At a spill site, high concentrations of pesticides which exceed water solubility limits may be expected. This will promote precipitation and will potentially lower the availability of the pesticide for microbial degradation. High concentrations of pesticides also will influence the sorption/desorption equilibria and will promote an increase in the amount sorbed to soil.

Investigators also should be concerned with the presence of organic cosolvents added to pesticide mixtures as formulating agents. Pesticides have been reported to be more soluble in organic cosolvents relative to water (Miller et al., 1988; Zachara et al., 1988). Kookana et al. (1990) reported that the adsorption/desorption of linuron and simazine by soil varied as a function of a methanol/water mixture. The pesticides were more soluble and more easily desorable in the presence of an organic cosolvent which implies a greater mobility of pesticides in solution and movement to groundwater.

This indicates that the leaching potential of pesticides at spill sites may be greater than predicted by models that use distribution coefficients determined from surface soil and with water as a solvent. Accurate prediction of pesticide fate at a spill site will require determination of pesticide sorption/desorption and precipitation behavior using soil and pore water which closely resemble the environment of the contaminated area.

FORMATION OF BOUND RESIDUES

It is well recognized in pesticide sorption studies that a portion of the sorbed chemical is desorbable, and after exhaustive solvent extraction a portion may not be recoverable (Kaufman et al., 1976; Khan, 1982; 1991). This unextractable fraction is referred to as the bound fraction or residue and is defined as that fraction of pesticide not recoverable by methods commonly used in residue analysis (Khan, 1988). The extent of pesticide residue formation in soils can be quite high: in fact, Khan (1982) reported that between 7 to 90% (expressed as a percentage of applied chemical) of applied pesticides may associate with the bound fraction of soil.

Diffusional processes promote the transfer of sorbed pesticides in the less accessible labile sites into more restrictive sites which eventually become irreversibly bound (Fig. 2-1). The restrictive sites occur in the internal sites of small soil pores, within clays, and inside the organic matrix of humic substances. Migration of pesticides into the restrictive pools may take from several days to a few weeks. McCall and Agin (1985) conducted a desorption kinetic study using picloram and reported that desorption from internal sites of soil required more than 300 h to reach equilibrium.

Both biological and chemical reactions are known to influence the formation of bound pesticide residue. It has been shown that microbial phenoloxidases, laccases, and peroxidases react with various pesticides or their metabolites with model organic matter components to form pesticide residue-organic complexes. Bollag et al. (1980) reported that an extracellular laccase enzyme from *Rhizoctonia praticola* catalyzed reactions between 2,4-dichlorophenol, orcinol, syringic, and vanillic acids to produce various polymeric products. Liu et al. (1987) also showed that soil or laccase enzyme catalyses the reaction of syringic acid with 2,6-diethylaniline. Autoclaved soils formed no products from these reactants, but gamma irradiated soils partially retained the capacity for reaction. Winklemann and Klaine (1991) reported the binding of ^{14}C-atrazine occurred in irradiated soil from Tennessee, which suggested binding by chemical processes.

The pesticide residue-organic complex can react further to become incorporated within soil organic matter by reacting with specific classes of humic substances (fulvic, humic acid, and humin). Bartha (1971) reported that the majority of bound propanol occurred in the humic acid fraction to form a humus-3, 4-dichloroaniline complex. Golab and Amundson (1975) also reported that a triamine derivative of trifuralin was bound to soil humic substances. More recently, Capriel et al. (1985) demonstrated that under outdoor conditions 9 years after atrazine application, the soil contained about 83% of the initially applied radioactivity. The bound ^{14}C-parent and degradates was distributed among the various humic fractions.

With longer residence time in soil, bound pesticide residues tend to lose biological activity and become even more resistant to degradation or extraction (Calderbank, 1989). It has been suggested that the resistance of bound pesticide to degradation could be due to an increase in stronger chemical bonds, physical exclusion of microbes and incorporation into the humic polymers (Khan, 1988; Calderbank, 1989). This phenomenon has been referred to as "aging" of residues.

The use of radiolabeled compounds has facilitated the study of bound residue formation by determining the radioactivity associated with the bound pesticide residue. This is accomplished by extraction and combustion of a sample containing the bound pesticide (and ^{14}C-degradates) to produce $^{14}CO_2$. This provides quantitative measurement of unrecoverable pesticide and avoids determination of the bound residue fraction by difference between initial and final concentration of extractable chemical. Khan (1982) developed a method whereby the thermal profiles of ^{14}C bound residues in soil were obtained by heating the bound residue and monitoring $^{14}CO_2$ evolution associated with specific temperature regimes (200 to 550°C). Using the thermal profile, it was shown that the ^{14}C-labeled pesticide was sorbed by specific soil humic fractions.

EFFECTS OF SORPTION ON BIODEGRADATION

Biodegradation of organic chemicals in the soil is one of the critical processes that determines the fate and behavior of xenobiotics in the environment. Soil microorganisms play a major role in determining the fate of these chemicals in soil. Although abiotic processes play a major role in detoxifying some of these compounds, without microbial activity such organic compounds would accumulate to a greater extent and pollute our soil and groundwater environment. These microorganisms typically occupy a soil volume of <0.1% and exhibit a significant diversity (Torstensson, 1988). The bacterial populations are as high as 10^9 cells g^{-1} of soil and fungi produce greater than 100 000 m hyphal length g^{-1} soil. Algae, protozoa, and microfauna are also a part of living component of the soil; however, their role in biodegradation of xenobiotics has received less attention. Several reviews have focused on pesticide biodegradation emphasizing mechanisms, kinetics, enzymes involved, and the significance of biodegradation (Goring et al., 1975; Hill & Wright, 1978; Bollag, 1982; Kearney & Karns, 1987; Alexander & Scow, 1989; Moorman, 1993). Both microbial extracellular and intracellular enzymes mediate the biodegradation process (Bollag, 1982; Johnson, 1983; Kearney & Karns, 1987). The intracellular enzymes such as amidases, dehalogenases, glutathione-s-transferase, hydroxlases, and oxygenases may not have significant effects on biodegradation of sorbed compounds. However, the extracellular enzymes such as amidases, esterases, and phosphatases with pesticide-degrading ability could be significant in degrading sorbed molecules (Burns & Edwards, 1980; Bollag, 1982).

Sorption decreases the degradation rate of some organic compounds by reducing their availability to microorganisms. Supporting evidence is provided by kinetic studies in model systems containing organic sorbates, sorbents and microorganisms, and studies relating degradation rates in soils to changes in sorption strength. Implicit in both of these approaches is the idea that degradation is limited by the inaccessibility of sorbed molecules to microorganisms. This is ultimately due to rates of desorption and diffusion to microbial cells or to their extracellular enzymes.

Desorption may become limiting to degradation rates in soils containing microbial populations that rapidly degrade available compounds. Desorption processes may be critical to bioremediation as they control the organic chemical availability for degradation. Several model systems have been developed and studied for the availability of sorbed compounds. In one model system, Ogram et al. (1985) showed that 2,4-D sorbed to soil organic matter was completely protected from microbial degradation, and only 2,4-D in soil solution was available for degradation by *Flavobacterium*. In addition, sorbed bacteria also were able to degrade 2,4-D that was in the soil solution. Three soils with a wide range of organic carbon and montmorillonite clay contents were used. Three models were proposed and used to explain their results. Since 2,4-D is metabolized intracellularly, microorganisms unable to contact the molecules were unable to degrade the herbicide. However, the study was conducted with a single strain of 2,4-D degrading bacteria under controlled conditions and may not be applicable to field conditions where greater microbial diversity may be responsible for degrading the herbicide for longer periods time than the 5-h incubation period used in the experiment.

Another model system showed that diquat, a dipyridylium cationic herbicide, was available to microorganisms when sorbed to the external exchange sites of a nonexpanding clay mineral (kaolinite) and unavailable when sorbed in between the silicate sheets of montmorillonite (Weber & Coble, 1968). These dipyridylium herbicides ionize completely in aqueous solutions to yield organic cations which react very strongly with clay minerals, especially montmorillonite which has an expanding lattice structure and high cation exchange capacity. Paraquat, another dipyridylium cationic herbicide, was protected in the interlayers of montmorillonite and not available to plants (Weber et al., 1969a) or microorganisms (Burns & Audus, 1970; Damanakis et al., 1970) but when they were sorbed to organic matter, the compound were available for degradation. In a similar study, a portion (15%) of entrapped diquat was released to plants by displacement with a nontoxic organic cation but none was released with inorganic cations (Weber et al., 1969b). Quinoline, an N-containing heterocyclic contaminant was studied for the effect of sorption on biodegradation (Smith et al., 1992). The quinoline-smectite clay complex was unavailable for bacteria that degrade quinoline which reduced the rates of quinoline mineralization (Fig. 2-2). Quinoline is sorbed strongly to smectite clay minerals by cation exchange reaction, and desorption is a rate-limiting step for biodegradation. These studies indicate that desorption is a rate-limiting step to biodegradation for compounds that are strongly sorbed to clay minerals, particularly between the inner layers of the expanding clay minerals via cation exchange reaction. Recently, a desorption rate-limiting process was described in soil by Shelton and Parkin (1991). A methylcarbamate insecticide, carbofuran, was incubated in soils at water contents of -1.5 Mpa to -0.04 Mpa, and the rates of metabolism of solution and sorbed ^{14}C-carbofuran and that mineralized to $^{14}CO_2$ were measured. At higher soil water contents (-0.04 & -0.08 Mpa) desorption became rate-limiting on biodegradation of soluble carbofuran. However, at lower water contents levels (-0.7 & -1.5 Mpa) sorption did not affect degradation, due to the loss of biological activity and biodegradation. This study suggests that soil moisture can affect the interaction between sorption and degradation processes.

The relationship between soil moisture, sorption, and degradation may not hold for microorganisms that excrete extracellular enzymes that can easily contact and degrade sorbed molecules. Other microorganisms may exhibit chemotaxicity towards some compounds and obtain physical accessibility. The model systems described in the previous paragraphs were studied in laboratory solution cultures and may not represent the soil environment *in situ*.

Aged residues are generally considered to be more protected from microbial degradation. As the age of the chemical in soil increases, strong chemical binding or diffusional barriers limit the desorption process (Hamaker & Goring, 1976; Pignatello, 1989), thereby minimizing microbial degradation. For example, picloram became less available for microbes during 100 d of incubation (McCall & Agin, 1985). In a field study, the bioavailability of aged simazine residues (20 y) vs. freshly added ^{14}C-simazine was compared (Scribner et al., 1991). They reported that aged simazine residue became desorption limited and unavailable for microbial

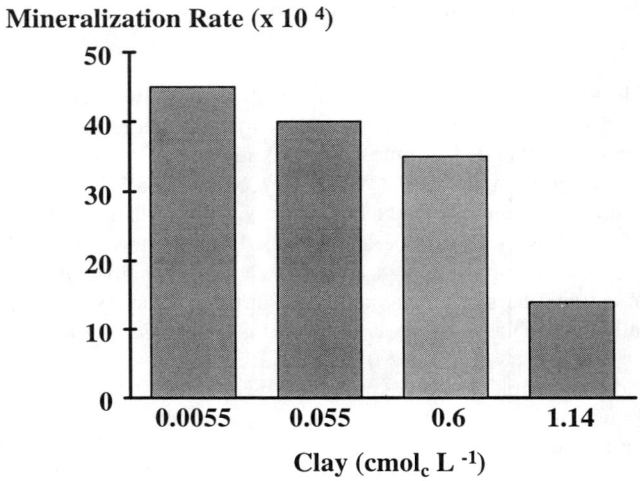

Fig. 2-2. The effect of clay concentration on the mineralization of ^{14}C by subsurface bacterium (Strain 2N2A). The rate constants are for desorption-limited reactions obtained after initial metabolism of solution-phase quinoline (t > 30 min) described by Smith et al., 1992.

degradation. The authors suggested that simazine residues could be partitioned within the organic matter matrix or entrapped in soil micropores that are normally inaccessible to microorganisms. It was suggested that aged residue can desorb slowly from these sites and potentially contaminate groundwater. Recently, significant levels of atrazine found in subsurface tile-drainage from a clay loam soil four years after initial atrazine application suggested that aged residues can become potential contaminants of groundwater (Buhler et al., 1993).

Determination of biodegradation rates are important for predicting the fate and behavior of organic chemicals in soils. There are several kinetic models, including monod, zero-order, first-order, and second-order which serve as a diagnostic tool for determining rates of biodegradation. These kinetic models were developed based on parameters that were essential for biodegradation processes. These models may not be useful in predicting the rate of biodegradation of sorbed chemicals, although first-order kinetics was tested by a few workers (Rijnaarts et al., 1990; Scow & Hutson, 1992).

In the adsorption-desorption and degradation processes, diffusion into soil aggregates plays a major role in controlling microbial utilization of organic chemicals (Brusseau & Rao, 1989; Pignatello, 1989; Brusseau et al., 1991; Scow & Alexander, 1992; Scow & Hutson, 1992). To determine the effects of sorption on spherical aggregates and diffusion into the aggregates on biodegradation processes, a deterministic model called the diffusion-sorption-biodegradation (DSB) model was developed by Scow & Hutson (1992). The DSB model predicted that the larger the aggregate size, the greater the sorption and slower degradation would occur as a result of low availability of chemicals to microorganisms. This model did not consider the possible distribution of microorganisms inside the pore spaces of aggregates where biodegradation may occur. However, another model developed by Dhawan et al. (1991) incorporated an intra-aggregate microbial component for biodegradation. They suggested that strongly sorbed chemicals may be degraded faster than the time required to diffuse out of the aggregates. Similarly Rijnaarts et al. (1990) applied a sorption-retarded radial diffusion model (RDM) to study the effect of intraparticle diffusion and desorption on the biodegradation of α-hexachlorocyclohexane in a soil slurry containing aggregates. Desorption and intraparticle diffusion limited the rate of biodegradation of the chemical.

The bioavailability of sorbed organic compounds has been examined for a wide variety of compounds. However, many of these studies utilized prokaryotic organisms in the experimental systems. The role of fungi in the biodegradation of sorbed compounds has received less attention. Fungi may be able to degrade sorbed compounds and contribute significantly to large scale (*in situ*) bioremediation. Fungi secrete a wide variety of both intra-and extracellular enzymes, and have the advantage of physical accessibility for sorbed compounds.

As seen in this section, sorption plays a major role inhibiting the microbial degradation of organic pollutants. Remediation of contaminated soils by adding montmorillonite clay or activated carbon may reduce the immediate threat of further leaching or volatilization, but offers only a temporary solution to soil decontamination of many compounds. An understanding of the ecology of microorganisms, their degradative capabilities, edaphic factors, physico-chemical properties of the compounds, and sorption-desorption mechanisms is needed to provide a permanent solution to remediation of contaminated sites.

BIODEGRADATION OF BOUND RESIDUES

There is a growing interest concerning the biodegradation of bound residues in soils. Bound residues include parent and degradation products and can occur in soils, plants, or in other biological material (Kaufman et al., 1976; Klein & Scheunert, 1982; Khan, 1982; 1991; Fuhr, 1987; Calderbank, 1989). Bound residue formation originating from microbial enzymes and plant root exudates can not be discounted.

The availability of bound residues to microorganisms and the fate of these bound residues with relevance to bioremediation has not been fully investigated. The availability of bound residues to plants is also an important process. Plant roots are able to extract the bound residue from soil by changing the soil chemical environment. Once the residue is taken up by the root, it may be translocated elsewhere within the plant. The availability of bound residues to plants has been extensively reviewed elsewhere (Duke et al., 1991; Khan, 1991; Weber et al., 1993).

Microorganisms may use organic matter as a nutrient source and release the bound residues into the soil solution, where they may undergo biodegradation. Microbial mineralization of soil bound ^{14}C-3,4-dichloroaniline was demonstrated by Hsu and Bartha (1974) by incubating the bound ^{14}C-dichloroaniline with fresh soil. Saxena and Bartha (1983) confirmed the microbial involvement in the release and mineralization of a 3,4-dichloroaniline-humic acid complex.

While microbial involvement in the release and degradation of bound residues has been implicated, mechanisms of release from organic matter complexes are uncertain. The addition of soil amendments has been shown to stimulate microbial activity and increase the release and degradation of bound residues. When ^{14}C-parathion residue was mixed with fresh soil, 17% of the bound residue was mineralized (Racke & Lichtenstein, 1985). Addition of glucose or cow manure stimulated the microbial population and activity and caused a significant increase in mineralization of ^{14}C-bound parathion residue. Addition of chloramphenicol (bactericide) or captafol (fungicide) to the soil decreased the evolution of $^{14}CO_2$ from bound residues. Another study, also suggested that addition of growth substrates (glucose) enhanced the release of soil-bound ^{14}C-atrazine due to increased microbial activity (Khan & Behki, 1990). Organic matter itself can serve as a nutrient source to stimulate the microbial activity in order to release bound residues (Khan & Ivarson, 1982; Khan & Behki, 1990). This could be considered as a cometabolic process, where microbial growth substrates serves as a carbon and energy source for the microorganisms to release the bound residues from organic polymer complexes. Once the compounds are released into the soil solution, they undergo biotic and abiotic transformations and may be mineralized.

Soil microbial communities are genetically and physiologically diverse and the rate of bound residue release and subsequent biodegradation will vary depending on the associated species, activity, and enzyme secretion. A significant amount of bound ^{14}C-prometryn and its metabolites were released from an organic soil by a mixed population of microorganisms (Khan & Ivarson, 1981). In a subsequent study, it was shown that various physiological groups of soil microorganisms (cellulolytic, proteolytic, lipolytic, and lignolytic) did not differ in their ability to release and degrade the bound residues of ^{14}C-prometryn (Khan & Ivarson, 1982). However, in subsequent studies, Khan & Behki (1990) showed that inoculated bacterial species efficiently released and metabolized more ^{14}C-bound atrazine residue than the native microorganisms. The inoculated bacteria (*Pseudomonas* sp Strain 192 & 194) were previously isolated from an enrichment culture utilizing atrazine as a carbon source while producing dechlorinated and dealkylated atrazine (Behki & Khan, 1986). Apparently, the inoculated bacterial species utilized organic matter as growth substrate and then released the bound ^{14}C from the soil. The free residues were then metabolized by the same species.

Fungal populations are often associated with organic matter and they may play an important role in the release of bound residues from soil. With their extensive hyphal network, fungi can penetrate the organic matrix or inner layers of clay minerals and release bound residues into the soil solution, where bacteria and fungi can degrade them. Plant roots may take up the released compounds and further metabolize them. Recently, a study showed that two vesicular-arbuscular mycorrhizal (VAM) fungi increased the bioavailability of bound ^{14}C-residue to onion plants from an organic soil treated with ^{14}C-fonofos (Nelson & Khan, 1990). The VAM hyphae were capable of removing ^{14}C-residues from soil and translocating the residue to the onion plants. These results suggest that under field conditions VAM associations may greatly increase the bioavailability of soil-bound residues to plants. However, pesticides have varying effects on VAM (Trappe et al., 1984).

The white-rot fungi may be a potential source for enhancing the release of bound residues from soil. These fungi secrete an extracellular ligninase enzyme that cleaves lignin polymers into soluble intermediates. Lignin is a complex and highly resistant polymer of phenolic subunits that becomes associated within soil organic matter. The white-rot and other fungi that degrade lignin also may degrade soil organic matter and release the bound residues. The enzymatic systems of these fungi also cleave a variety of compounds such as chloranilines, pentachlorophenol, DDT, and polychlorinated biphenyls (Morgan & Watkinson, 1989).

Bound pesticide residues are biodegradable, but the process is very slow compared to biodegradation processes of highly soluble residues. The aging and environmental significance of bound residues have received less attention than soluble residues. Due to slow degradation, bound pesticide residues can continue to accumulate with continuous pesticide application. As indicated by Calderbank (1989), bound residue formation should plateau at some point because the amount being degraded should be in equilibrium with the amount being added. However, this equilibrium concentration for most pesticides in unknown.

IMPLICATIONS OF BIOAVAILABILITY TO BIOREMEDIATION

Pesticide contaminated surface soils can be treated by landfarming and composting. These techniques may involve excavation, removal, and spreading of the contaminated soil on a site equipped with a liner or some other barrier to prevent leaching. The soil may be farmed, composted, or manipulated at the decontamination site to increase biological activity through the use of soil amendments, microbial inoculant, tillage, to increase aeration, and irrigation, to maintain optimal water contents for microbiological activity. The use of pesticide-tolerant plants to activate rhizosphere microorganisms also is being examined for bioremediation applications (Anderson, 1993; Anderson & Coats, 1995). After the remediation is complete, the soil can be returned to the original site. A variation of this process is land-spreading in which pesticide contaminated soils are applied to crops at rates equivalent to conventional agronomic application rates (Felsot et al., 1995).

In highly contaminated soils where the pesticide concentration exceeds the solubility limits, the initial rate of biodegradation may depend solely upon the maximum degradation rate of the indigenous or inoculated microorganisms. However, the high concentration of pesticide will increase the total amount of chemical sorbed to the soil and will probably enhance chemically mediated processes, such as diffusion into the soil organic matter. As the concentration decreases, or in soils that are less contaminated, the fraction of chemical in sorption sites inaccessible to microorganisms may eventually limit biodegradation. Dzantor et al. (1993) reported that aged residues, applied as a soil-containing mixture of alachlor, trifluralin, and atrazine, degraded more slowly during landfarming than an equivalent quantity of herbicide sprayed on uncontaminated soil. The aged residues probably contained a mixture of strongly sorbed and bound residues.

Organic materials are often added at landfarming sites and during composting (Ziegenfuss et al., 1991). Other minerals such as inorganic nitrogen and phosphorus also are added to stimulate microbial activity. The addition of organic and inorganic materials is particularly important for pesticides that are biodegraded by cometabolic means (McCormick & Hiltbold, 1966; Moorman, 1993). However, organic matter additions to stimulate biological activity may have the unintended effect of increasing sorption. Waste-activated carbon, digested sewage sludge, and manure applied at 0.5 to 2.1 t ha^{-1} applied to soil increased the sorption of alachlor and/or reduced volatilization losses, with the waste-activated carbon having the greatest effect (Guo et al., 1993). Alachlor remaining after the 30-d experiment in the waste-activated carbon amended soil ranged from 80 to 100% compared to 14 to 26% in the unamended soil. The other amendments increased degradation. Dzantor et al. (1993) reported that addition of sewage sludge which contained considerable quantities of stabilized organic matter increased microbial activity and subsequent alachlor and atrazine degradation, but prolonged the persistence of trifluralin. Trifluralin has a much higher soil partition coefficient than atrazine or alachlor. Addition of sewage sludge may have increased the adsorbed and bound fractions of trifluralin slowing degradation. In the same experiment, corn meal addition did not affect trifluralin degradation. Manure addition also has been reported to increase the degradation of a variety of polynuclear aromatic hydrocarbons (PAH) compounds in soil (Sims et al., 1990). Amendments may vary substantially in effectiveness to increase biodegradation. The ultimate usefulness of a specific amendment may depend on its relative ability to stimulate microorganisms or adsorb and bind pesticides.

Some reports have indicated that pesticides form associations with dissolved organic carbon (DOC), which can increase movement (Carter & Suffet, 1982; Madhun et al., 1986; Lee & Farmer, 1989). In other studies no effects on movement were found (Spurlock & Biggar, 1990). These interactions are complex, in that the DOC may increase the soluble fraction of pesticides, but also increase microbial activity. The net effect appears to depend upon the hydrophobic nature of the pesticide (hydrophobicity increases binding to DOC) and the nature of DOC. Lee and Farmer (1989) reported that the association constant (K_{doc}) between pesticides and dissolved humic and fulvic acids was greatest for DDT and least for bromacil, but napropamide has a greater K_{doc} value than lindane. The waste-activated carbon, digested sewage sludge and manure discussed previously (Guo et al., 1993) all reduced alachlor leaching in comparison to leaching in an unamended sandy soil. The potential effects of DOC may be beneficial, in that pesticide solubilized by DOC may have greater long-term bioavailability. In bioremediation operations where leaching is prevented by a liner or other means, the potential effects of organic matter additions and DOC mediated transport are of no concern.

Slurry phase or pump and treat technologies also are affected by degradation-sorption interactions. These technologies have been used for remediation of a variety of organic contaminants in surface and subsurface soils. Slurried soils within bioreactors can rely upon indigenous or inoculated microorganisms to degrade desorbable materials. Depending upon the nature of the sorbed material and rate of biodegradation in the reactor, the desorption rate may become limiting at some point in the process. Burbaker & Stroo (1992) showed that highly sorbed PAHs remained sorbed in both slurry and soil columns. High initial concentrations of the PAHs were removed by biodegradation, but the remaining fraction remained adsorbed to the soil. Tightly bound pesticide residues may be retained on the residual sediment that is removed periodically from the bioreactors or clarifiers. Pump and treat operations make a similar use of bioreactors, but in this case adsorption and desorption reactions control the movement of organics from the zone of contamination to the extraction wells (Morgan & Watkinson, 1989).

While extractable pesticide residues can be assessed with conventional solvent extraction and analytical chemistry, bound (nonextractable) fraction represents a source of residues that are difficult to measure and can potentially enter into the environment. The existing experimental evidence is derived from experiments that were designed to assess the impact of bound residues in agricultural contexts. The high concentrations of pesticides at spill sites in which bioremediation technologies might be employed will require a different assessment of bound residues. Duke et al. (1991) suggested that 20 to 70% of applied herbicide becomes bound residue, but less than 3% (of the original amount) becomes bioavailable, as indicated by plant uptake. If the initial herbicide concentration at the spill site were 50 to 100 kg ha^{-1} and proportionally similar quantities of bound residues were formed, a small percentage of the bound residues released during the landfarming operation may cause phytotoxic or other environmental effects. Many herbicide metabolites have some herbicidal activity (Duke et al., 1991). Unfortunately, little conclusive data exist to assess either the quantity of bound residues formed during bioremediation of pesticides wastes, or their implications. The assessment of the impact of bound residues is complicated by the fact that analytical methodology for measurement of bound residues is not widely tested or available.

SUMMARY

The sorption and binding of organic chemicals plays a critical role in determining their fate in soil. In general, sorbed chemicals in labile sites are available for microbial transformations and bioremediation techniques should be applicable. Chemicals sorbed in more restrictive and bound sites, out of reach of microorganisms, will persist and thereby lengthen the bioremediation process. Bioremediation strategies should account for the potential soil effects on these sorbed chemicals.

Microorganisms are capable of degrading both sorbed and bound residues. The degradation, however, may require an extensive time period to remediate soil contaminants to safe levels. An understanding of the diversity, survival, and degradative capabilities of specific microorganisms and microbial complexes are essential to achieve remediation of organic chemicals in soil.

Bioremediation was effective for the transformation of some organic chemicals (BTEX, oils) to less toxic forms. The technology, however, is influenced by a number of poorly studied factors that will affect pesticide degradation. Possible areas of future research to improve the application and efficacy of bioremediation include:

- long-term environmental fate of a bound residue;
- factors regulating pesticide bioavailability;
- identification of microorganisms responsible for bound residue degradation;
- evaluation of environmental conditions necessary to insure that bioremediation will occur.

REFERENCES

Alexander, M., and K.M. Scow. 1989. Kinetics of biodegradation in soil. p. 243-269. *In* B.L. Sawhney and K. Brown (ed.) Reactions and movement of organic chemicals in soils. SSSA Spec. Publ. 22. ASA and SSSA, Madison, WI.

Anderson, T.A., E.A. Guthrie, and B.T. Walton. 1993. Bioremediation in the rhizosphere:plant roots and associated microbes clean contaminant soil. Environ. Sci. Technol. 27:2630-2636.

Anderson T.A. and J.R. Coats. 1995. An overview of microbial degradation in the rhizosphere and its implications for bioremediation. p. 133-141. *In* H.D. Skipper and R.F. Turco (ed.) Bioremediation: Science and applications. SSSA Spec. Publ. 43. ASA,CSSA and SSSA, Madison WI.

Bailey, G.W., and J.L. White. 1964. Review of adsorption and desorption of organic pesticides by soil colloids, with implications concerning pesticide bioactivity. J. Agric. Food Chem. 12:324-332.

Bailey, G.W., and J.L. White. 1970. Factors influencing the adsorption, desorption and movement of pesticides in soil. Residue Rev. 32:29-92.

Bartha, R. 1971. Fate of herbicide-derived chloroanalines in soil. J. Agric. Food Chem. 19:385-390.

Behki, R.M., and S.U. Khan. 1986. Degradation of atrazine by Pseudomonas: N-dealkylation and dehalogenation of atrazine and its metabolites. J. Agric. Food Chem. 34:746-749.

Bollag, J.M. 1982. Microbial metabolism of pesticides. p. 125-168. *In* J.P. Rosazza (ed.) Microbial transformations of bioactive compounds. C.R.C. Press, Boca Raton, FL.

Bollag, J.-M., S.-Y. Liu, and R.D. Minnard. 1980. Cross-coupling of phenolic humus constituents and 2,4-dichlorophenol. Soil Sci. Soc. Am. J. 44:52-56.

Brusseau, M.L., R.E. Jessup, and P.S.C. Rao. 1991. Nonequilibrium sorption of organic chemicals: Elucidation of rate-limiting processes. Environ. Sci. Technol. 25:134-142.

Brusseau, M.L., and P.S.C. Rao. 1989. The influence of sorbate-organic matter interactions on sorption nonequilibrium. Chemosphere 18:1691-1706.

Buhler, D.D., G.W. Randall, W.C. Koskinen, and D.L. Wyse. 1993. Atrazine and alachlor losses from subsurface tile drainage of a clay loam soil. J. Environ. Qual. 22:583-588.

Burbaker, G.R., and H.F. Stroo. 1992. In situ bioremediation of aquifers containing polyaromatic hydrocarbons. J. Hazardous Materials. 32:163-177.

Burns, R.G., and L.J. Audus. 1970. Distribution and breakdown of paraquat in soil. Weed Res. 10:49-58.

Burns, R.G., and J.A. Edwards. 1980. Pesticide breakdown by soil enzymes. Pestic. Sci. 2:506-512.

Calderbank, A. 1989. The occurrence and significance of bound pesticide residue in soil. Residue Rev. 108:71-103.

Calvert, R. 1980. Absorption-desorption phenomena. p. 1-30. *In* R.J. Hance (ed.) Interactions between herbicides and the soil. Academic Press, New York.

Capriel, P., A. Haisch and S.U. Khan. 1985. Distribution and nature of bound (nonextractable) residues of atrazine in a mineral soil nine years after herbicide application. J. Agric. Food Chem. 33:567-569.

Carter, C.W., and I.H. Suffet. 1982. Binding of DDT to dissolved humic substances. Environ. Sci. Technol. 16:753-740.

Damanakis, M., D.S.H. Drennan, J.D. Fryer, and K. Holly. 1970. The adsorption and mobility of paraquat on different soils and soil constituents. Weed Res. 19:264-277.

Dhawan, S., L.T. Fan, L.E. Erickson, and P. Tuitemwong. 1991. Modeling, analysis, and simulation of bioremediation of soil aggregates. Environ. Prog. 10:251-260.

Duke, S.O., T.B. Moorman, and C.T. Bryson. 1991. Phytotoxicity of pesticide degradation products. p. 188-204. *In* L. Somasundaram and J.R. Coats (ed.) Pesticide transformation products: Fate and significance in the environment. ACS symp. ser. no. 459. Washington, D.C.

Dzantor, E.K., A.S. Felsot, and M.J. Beck. 1993. Bioremediating herbicide contamination soils. Appl. Biochem. Biotechnol. 39/40:621-630.

Felsot, A.S., J.K. Mitchell, and E.K. Dzantor. 1995. Remediation of herbicide-contaminated soil by combinations of landfarming and biostimulation. p.232-252. *In* H.D. Skipper and R.F. Turco (ed.) Bioremediation: Science and applications. SSSA Spec. Publ. 43. ASA,CSSA and SSSA, Madison WI.

Fuhr, F. 1987. Non-extractable pesticide residues in soil. p. 381-389. *In* R. Greenhalgh and T.R. Roberts (ed.) Pesticides science and biotechnology. Proc. 6th Int. Congr. Pestic. Chem. IUPAC. Blackwell Scientific Publications, London.

Golab, T., and M.E. Amundson. 1975. Degradation of trifluralin, oxzalin, and isopropalin in soil. Environ. Qual. Safety Supp. III. p. 258-271.

Goring, C.A.I., D.A. Laskowski, J.W. Hamaker, and R.W. Meikle. 1975. Principles of pesticide degradation in soil. p. 135-172. *In* R. Haque and V.H. Freed (ed.) Environmental dynamics of pesticides. Plenum Publishing Corp., NY.

Grundl, T., and G. Small. 1993. Mineral contributions to atrazine and alachlor sorption in soil mixtures of variable organic carbon and clay content. J. Contaminant. Hydrol. 14:117-128.

Guo, L., T.J. Bicki, A.S. Felsot, and T.D. Hinesly. 1993. Sorption and movement of alachlor in soil modified by carbon-rich wastes. J. Environ. Qual. 22:186-194.

Hamaker, J.W., C.A.I. Goring. 1976. Turnover of pesticide residues in soil. p. 219-243. *In* D.D. Kaufmann, G.G. Still, G.D. Paulson, and S.K. Bandal (ed.) Bound and conjugated pesticide residues. ACS Symp. Ser. No. 29, Amer. Chem. Soc., Washington, D.C.

Hamaker, J.W., and J.M. Thompson. 1972. Adsorption. p. 49-143. *In* C.A.I. Goring and J.W. Hamaker (ed.) Organic chemicals in the soil environment. Marcel Dekker, New York, NY.

Hance, R.J. 1969. Influence of pH, exchangeable cation and the presence of organic matter on the adsorption of some herbicides by montmorillonite. Can. J. Soil Sci. 49:357-364.

Hance, R.J. 1988. Adsorption and bioavailability. p. 2-19. *In* R. Grover (ed.) Environmental chemistry of herbicides. Vol 1. CRC Press, Inc. Boca Raton, FL.

Helling, C.S., and J. Dragun. 1980. Soil leaching tests for toxic organic chemicals. p. 43-88. *In* Test protocols for environmental fate and movement of toxicants. Proc. Symp. Assoc. Off. Anal. Chem. Meet. 21-22 Oct. 1980. Washington, D.C.

Hill, I.R., and S.J.L. Wright. 1978. Microbiological aspects of pesticide behavior in the environment. *In* Pesticide microbiology, Academic Press, London.

Hsu, T.S., and H. Bartha. 1974. Biodegradation of chloroaniline humus complexes in soil and in culture solution. Soil Sci. 118:213-219.

Johnson, L.M. 1983. Enzymes associated with the microbial metabolism of insecticides. p. 131-145. *In* R. Lal (ed.) Insecticide microbiology. Springer-Verlag, New York.

Kaufman, D.D., G.G. Still, G.D. Paulson, and S.K. Brandal. 1976. Bound and conjugated pesticide residues. Am. Chem. Soc. Symp. Ser., 29. 382 p.

Kearney, P.C., and J.S. Karns. 1987. Microbial metabolism. p. 93-101. *In* J.W. Biggar and J.N. Seiber (ed.) Fate of pesticides in the environment. Proc. Agri. Expt. Stn., Univ. of California, CA.

Khan, S.U. 1982. Bound pesticide residues in soils and plants. Residue Rev. 84:1-25.

Khan, S.U. 1988. Bound residues. p. 266-279. *In* R. Grover and J.J. Cessna (ed.) Environmental chemistry of herbicides. Vol. II. CRC Press, Inc. Boca Raton, FL.

Khan, S.U. 1991. Bound (nonextractable) pesticide degradation products in soils. p. 108-121. *In* L. Somasundaram and J.R. Coats (ed) Pesticide transformation products-fate and significance in the environment. ACS Symp. Ser. No. 459. Washington, D.C.

Khan, S.U., and R.M. Behki. 1990. Effects of *Pseudomonas* species on the release of bound ^{14}C residues from soil treated with ^{14}C-atrazine. J. Agric. Food Chem. 38:2090-2093.

Khan, S.U., and K.C. Ivarson. 1981. Microbial release of unextracted (bound) residues from an organic soil treated with prometryn. J. Agric. Food Chem. 29:1301-1303.

Khan, S.U., and K.C. Ivarson. 1982. Release of soil bound (nonextractable) residues by various physiological groups of microorganisms. J. Environ. Sci. Health B. 17:737-749.

Klein, W., and I. Scheunert. 1982. Bound pesticide residues in soil, plants and food with particular emphasis on the application of nuclear techniques in agrochemicals: Fate in food and the environment, International Atomic Energy Agency, Vienna, p. 177-205, SM-263-/38.

Kookana, R.J., R.G. Gerritse, and L.A.G. Aylmore. 1990. Effect of organic cosolvent on adsorption and desorption of linuron and simazine in soil. Aust. J. Soil Res. 28:717-725.

Koskinen, W.C., and S.S. Harper. 1990. The retention process: mechanisms. p. 51-77. *In* H.H. Chang (ed.) Pesticides in the soil environment:processes, impacts, and modeling. Soil Science Society of America, Madison, WI.

Lee, D.Y., and W.J. Farmer. 1989. Dissolved organic matter interaction with napropamide and four other nonionic pesticides. J. Environ. Qual. 18:468-474.

Liu, S.-Y., R.D. Minard and J.-M. Bollag. 1987. Soil-catalyzed complexation of the pollutant 2,6-diethylaniline with syringic acid. J. Environ. Qual. 16:48-53.

Locke, M.A. 1992. Sorption-desorption kinetics of alachlor in surface soil from two soybean tillage systems. J. Environ. Qual. 21:558-566.

Madhun, Y.A., J.L. Young, and V.H. Freed. 1986. Binding of herbicides by water-soluble organic materials from soil. J. Environ. Qual. 15:64-68.

McCall, P.J., and G.L. Agin. 1985. Desorption kinetics of picloram as affected by residence time in the soil. Environ. Toxicol. Chem. 4:37-44.

McCall, P.J., D.A. Laskowski, R.L. Swann, and H.J. Dishburger. 1980. p. 89-109. Measurement of sorption coefficients of organic chemicals and their use in environmental fate analysis. *In* Test protocols for environmental fate and movement of toxicants. Proc. Symp. Ass. Off. Anal. Chem. Meet. 21-22 Oct. 1980. Washington, D.C.

McCormick, L. L., and A. E. Hiltbold. 1966. Microbiological decomposition of atrazine and diuron in soil. Weeds 14:77-82.

Meuller, T.C., T.B. Moorman, and C.E. Snipes. 1992. Effect of concentration, sorption, and microbial biomass on degradation of the herbicide fluometuron in surface and subsurface soils.J. Agric. Food Chem. 40:2517-2522.

Miller, N.A., D.C. Wolf and H.D. Scott. 1988. Influence of methanol and hexane on soil adsorption of simazine. Water, Air, Soil Pollut. 39:1012-112.

Moorman, T.B. 1989. A review of pesticide effects on soil microorganisms and microbial processes related to soil fertility. J. Prod. Agric. 1:14-23.

Moorman, T.B. 1993. Pesticide degradation by soil microorganisms: environmental, ecological, and management effects. p. 121-165. *In* J.L. Hatfield and B.A. Stewart (ed.) Soil Biology: Effects on Soil Quality, Adv. Soil Sci., Lewis Publ. Boca Raton, FL.

Morgan, P., and R.J. Watkinson. 1989. Microbiological methods for the cleanup of soil and ground water contaminated with halogenated organic compounds. FEMS Microbiol. Rev. 63:277-300.

Nelson, S.D., and S.U. Khan. 1990. Effect of endomycorrhizae on the bioavailability of bound ^{14}C residues to onion plants from an organic soil treated with ^{14}C-fonofos. J. Agric. Food Chem. 38:894-898.

Ogram, A.V., R.E. Jessup, L.T. Ou, and P.S.C. Rao. 1985. Effects of sorption on biological degradation rates of (2,4-dichlorophenoxy) acetic acid in soils. Appl. Environ. Microbiol. 49:582-587.

Pignatello, J.J. 1989. Sorption dynamics of organic compounds in soils. p. 45-80. *In* B.L. Sawhney and K. Brown (ed.) Reactions and movement of organic chemicals in soils. SSSA Spec. Publ. 22. ASA and SSSA, Madison, WI.

Pimental, D., and L. Levitan. 1986. Pesticides: amounts applied and amounts reaching pests. Biosci. 36:86-91.

Racke, D.D., and E.P. Lichtenstein. 1985. Effects of soil microorganisms on the release of bound ^{14}C residues from soils previously treated with ^{14}C-parathion. J. Agric. Food Chem. 33:938-943.

Rijnaarts, H.H.M., A. Bachmann, J.C. Jumelel, and A.J.B. Zehnder. 1990. Effect of desorption and intraparticle mass transfer on the aerobic biomineralization of alpha-hexachlorocyclohexane in a contaminated calcareous soil. Environ. Sci. Technol. 24:1349-1354.

Saxena, A., and R. Bartha. 1983. Microbial mineralization of humic acid-3,4-dichloroaniline complexes. Soil Biol. Biochem. 15:59-62.

Scow, K.M., and M. Alexander. 1992. Effect of diffusion on the kinetics of biodegradation: Experimental results with synthetic aggregates. Soil Sci. Soc. Am. J. 56:128-134.

Scow, K.M., and J. Hutson. 1992. Effect of diffusion and sorption on the kinetics of biodegradation: Theoretical considerations. Soil Sci. Soc. Am. J. 56:119-127.

Scribner, S.L., T.R. Benzing, S. Sun, and S.A. Boyd. 1991. Desorption and bioavailability of aged simazine residues in soil from a continuous corn field. J. Environ. Qual. 21:115-120.

Shelton, D.R., and T.B. Parkin. 1991. Effect of moisture on sorption and biodegradation of carbofuran in soil. J. Agric. Food Chem. 39:2063-2068.

Sims, J.L., R.C. Sims, and J.E. Matthews. 1990. Approach to bioremediation of contaminated soil. Hazard. Waste Hazard. Mater. 7:117-149.

Skipper, H. D., and V. V. Volk. 1972. Biological and chemical degradation of atrazine in three Oregon soils. Weed Sci. 20:344-347.

Skladany, G.J., and F.B. Metting, Jr. 1993. Bioremediation of contaminated soil. p. 483-513. *In* Soil Microbial Ecology. Marcel Dekker, Inc. New York, NY.

Smith, S.C., C.C. Ainsworth, S.J. Traina, and R.J. Hicks. 1992. Effect of sorption on the biodegradation of quinoline. Soil Sci. Soc. Am. J. 56:737-746.

Spurlock, F.C., and J.W. Biggar. 1990. Effect of naturally occurring soluble organic matter on the adsorption and movement of simazine [2-chloro-4,6-bis (ethylamino)-s-triazine] in Hanford sand loam. Environ. Sci. Technol. 24:736-741.

Stevenson, F.J. 1982. Humus Chemistry. John Wiley & Sons, New York, NY.

Thomas, J.M., and C.H. Ward. 1989. *In* situ biorestoration of organic contaminants in the subsurface. Environ. Sci. Technol. 23:760-766.

Torstensson, L. 1988. Microbial decomposition of herbicides in the soil. Outlook on Agric. 17:120-124.

Trappe, J.M., R. Molina, and M. Castellano. 1984. Reactions of mycorrhizal fungi and mycorrhiza formation to pesticides. Annu. Rev. Phytopathol. 22:331-359.

U.S. Environmental Protection Agency. 1988. Technology screening guide for treatment of CERCLA soils and sludges. Publ. EPA/5402/2-88.

Wauchope, R.D., and R.S. Meyers. 1985. Adsorption-desorption kinetics of atrazine and linuron in freshwater-sediment aqueous slurries. J. Environ. Qual. 14:132-136.

Weber, J.B., J.A. Best and J.U. Gonese. 1993. Bioavailability of sorbed agricultural chemicals. p. 153-196. *In* D.M. Linn and T. Carski (ed.) Sorption and degradation of agricultural chemicals in soils. Spec. Pub. 32. Soil Science Society of America, Inc. Madison, WI.

Weber, J.B., and H.D. Coble. 1968. Microbial decomposition of diquat adsorbed on montmorillonite and kaolinite clays. J. Agric. Food Chem. 16:475-478.

Weber, J.B., R.C. Meek and S.B. Weed. 1969a. The effect of cation exchange capacity on the retention of diquat and paraquat by three-layer type clay minerals: II. Plant availability of paraquat. Soil Sci. Soc. Am. Proc. 33:382-385.

Weber, J.B., S.B. Weed and J.A. Best. 1969b. Displacement of diquat from clay and its phytotoxicity. J. Agric. Food Chem. 17:1075-1076.

Winkelmann, D.A., and S.J. Klaine. 1991. Degradation and bound residue formation of atrazine in a western Tennessee soil. Environ. Toxicol. Chem. 10:335-345.

Zachara, J.M. C.C. Ainsworth, R.L. Schmidt, and C.T. Resch. 1988. Influence of cosolvents on quinoline sorption by subsurface materials and clays. J. Contam. Hydrol. 2:343-364.

Ziegenfuss, P.S., R.T. Williams, and C.A. Myler. 1991. Hazardous materials composting. J. Hazardous Materials. 28:91-99.

3 Surfactant-Enhanced Bioavailability of Slightly Soluble Organic Compounds[1]

R.M. Miller
Department of Soil and Water Science
University of Arizona
Tucson, Arizona

Rising remediation costs have increased interest in the development and application of innovative remediation technologies. Specifically, development of new bioremediation technologies is attractive because of the potential for both successful and economical *in situ* treatment of contaminated sites. A common constraint to efficient bioremediation of contaminated sites is the low aqueous solubility of many organic contaminants. These include a variety of fuel-related compounds (e.g., aliphatics, aromatics, and heterocyclics), and higher molecular weight chlorinated organic compounds. Limited contaminant solubility has several consequences for remediation. For slightly soluble biodegradable contaminants both degradation rates and levels are reduced. For slightly soluble recalcitrant contaminants, removal by flushing is retarded because of extensive sorption by soil. Development of efficient bioremediation technologies will require that contaminant availability issues be addressed. To accomplish this, the mechanisms underlying microbial solubilization and biodegradation of slightly soluble organic contaminants must first be understood.

MICROBIAL BIODEGRADATION OF SLIGHTLY SOLUBLE ORGANIC COMPOUNDS

Organic compounds can exist in the liquid, solid, or in the vapor phase at ambient temperature. Liquid hydrocarbons will form two-phase systems with water. The hydrocarbon layer will rest on top of the aqueous phase if less dense than water(fuel hydrocarbons) or settled below the aqueous phase if more dense than water (chlorinated solvents). Four types of uptake by microorganisms have been described for liquid hydrocarbons as shown in Fig. 3-1 (Nakahara et al., 1977; Boultan & Ratledge, 1984; Zhang & Miller, 1992).

[1]This research was supported by the U.S. Department of Energy (DOE) Subsurface Science Program, Office of Health and Environmental Research (OHER) Grant No: DE-FG03-93ER61526.

Biodegradation of dissolved hydrocarbon is the major type of utilization for liquid hydrocarbons with relatively high aqueous solubilities, for example, aliphatics of chain-length < C_{10} (Table 3-1). In contrast, biodegradation of liquid hydrocarbons with extremely low solubilities (e.g. hydrocarbons with chain length > C_{10}), results from direct contact of cells with both large oil drops and with submicron droplets (Fogel et al., 1981). The relative contributions of large drops and submicron droplets to hydrocarbon degradation depends on the type of microorganism, the hydrocarbon, and environmental conditions (Nakahara et al., 1977). For example, in aquatic ecosystems, hydrocarbons form heterogeneous emulsions. These emulsions contain small droplets (diameter < 1μm) of pseudosolubilized oil (Gutnick & Rosenberg, 1977), as well as larger aggregates. The pseudosolubilized oil is more susceptible to microbial degradation while the larger aggregates may persist for long periods of time due to limited surface area available for contact with microorganisms (Atlas, 1981).

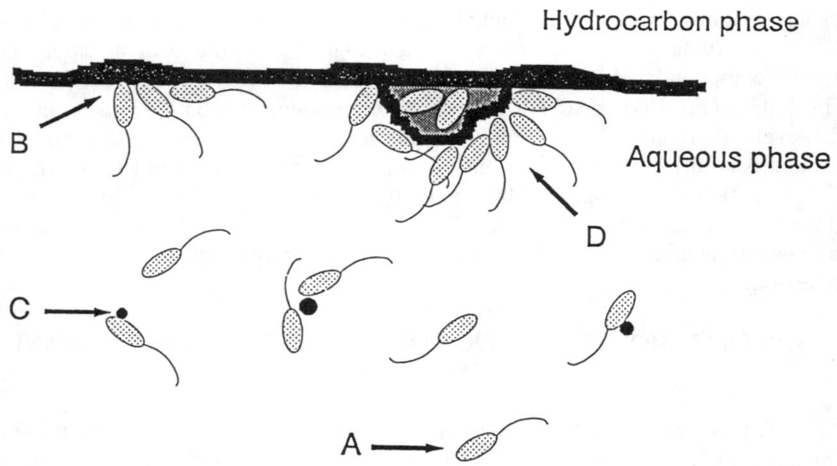

Fig. 3-1. Modes of microbial uptake of hydrocarbons: A, uptake of hydrocarbons dissolved in the aqueous phase surrounding degrading cells; B, uptake via direct contact of degrading cells at the aqueous-hydrocarbon interface of large oil drops in water; C, uptake through direct contact of degrading cells with fine or submicron size oil droplets dispersed in the aqueous phase; D. enhanced uptake as a result of production of biosurfactants or emulsifiers which effectively increase the apparent aqueous solubility of the hydrocarbon.

Table 3-1. Representative solubilities of common organic contaminants.

	Aqueous Solubility	
Compound	(mg L^{-1})	Source
Butane (C_4H_{10})	61.4	Singer &Finnerty, 1984.
Decane ($C_{10}H_{22}$)	0.052	"
Hexadecane ($C_{16}H_{34}$)	0.0063	"
Benzene (C_6H_6)	1,780	Verschueren, 1983.
Naphthalene ($C_{10}H_8$)	30	"
Anthracene ($C_{14}H_{10}$)	1.29	"
Benzo(a)pyrene ($C_{20}H_{12}$)	0.003	"
Biphenyl ($C_{12}C_{10}$)	7.5	Verschueren, 1983.
Arochlor 1221 (mixture, 21% Cl)	0.59	"
Arochlor 1254 (mixture 54% Cl)	0.057	"

Less information is available about the biodegradation of slightly soluble solid-phase organic compounds. However, evidence indicates that biodegradation is restricted to dissolved phase. Studies with naphthalene and bibenzyl (Wodzinski & Bertolini, 1972), phenanthrene (Wodzinski & Coyle, 1974), and n-eicosane (C_{20}) (Chakravarty et al., 1972), all showed that addition of solid substrate to a biodegrading culture already saturated with substrate had no effect on the culture generation time.

Thomas et al. (1986) measured rates of dissolution of naphthalene and 4-chlorobiphenyl and compared these rates with the rate of biodegradation. Results showed that for both compounds the bacteria were obtaining more carbon than was available in the dissolved state. From the measured dissolution rates in sterile media, it was calculated that only 0.044 pg substrate $cell^{-1}$ produced was available as dissolved substrate. Yet, the growth rate suggested that several times as much (0.11 to 0.31 pg substrate $cell^{-1}$) was actually available for use. The authors suggested several possible explanations for their observations. First, rapid microbial removal of dissolved substrate may have increased the dissolution rate. Second, release of bacterial emulsifiers during growth may have increased the dissolution rate. Third, direct contact with available surface area may affect growth rate. There is at least one other explanation for this behavior; vapor phase assimilation. Some solid hydrocarbons, such as naphthalene and 4-chlorobiphenyl (as well as other low MW solid hydrocarbons), are relatively volatile at the cultivation temperature used (30°C) and can support bacterial growth on a minimal medium in the vapor phase (Kiyohara et al., 1982).

Other studies indirectly support the idea that biodegradation of solid hydrocarbons occurs only in the solubilized state. Bossert and Bartha (1986) studied the biodegradation of various polycyclic aromatic hydrocarbons (PAH) in soil. They found a general increase in recalcitrance with increasing ring number and a strong negative correlation between the recalcitrance of a PAH and its water solubility. This finding is supported by a number of different studies (summarized by Cerniglia & Heitkamp, 1989) which compared PAH degradation in both acclimated and nonacclimated systems. Hallas and Vestal (1978) studied the growth of *Mycobacterium convolutum* on solid n-alkanes. Growth on hydrocarbons of longer chain length was characterized by a decrease in cell yield and an increase in the lag phase, e.g., 0 hours for nonadecane (C_{19}) to approximately 15 hours for hexacosane (C_{26}). Utilization of the plant paraffins (C_{27} to C_{35}) is slow but examples have been reported. Hankin and Kolattukudy (1968) showed that *Micrococcus cerificans* was able to grow on n-nonacosane (C_{29}). However, in order to get growth on this substrate, the n-nonacosane was either thinly coated on diatomaceous earth to maximize surface area, or more effectively, was ultrasonically emulsified. In addition, large inocula were used to avoid extremely slow growth. Haines and Alexander (1974) reported a bacterium which grew slowly on tetracontane (C_{40}) and tetratetracontane (C_{44}). Biological oxygen demand (BOD), a relatively insensitive method, was used to measure biodegradability and appreciable O_2 uptake occurred only after 20 days.

The difference in biodegradation of liquid and solid phase slightly soluble organics is emphasized by a study of diphenylmethane biodegradation (Wodzinski & Larocca, 1977). Diphenylmethane has a melting point of 24°C. The rate of biodegradation at 30°C (liquid), was significantly enhanced by further addition of diphenylmethane to a substrate saturated solution. However, at 20°C (solid), the biodegradation rate was unaffected by addition of diphenylmethane to a substrate saturated solution.

In summary, biodegradation rates typically decrease as aqueous solubility decreases. Biodegradation of liquid-phase hydrocarbons occurs through utilization of dissolved substrate and by direct contact at the water-hydrocarbon interface of either large or small hydrocarbon droplets. Thus, biodegradation is rate-limited by both solubility and by the aqueous-organic interfacial area. In contrast, biodegradation of solid-phase hydrocarbons occurs primarily by utilization of dissolved substrate and is therefore rate-limited by hydrocarbon solubility and by the rate of hydrocarbon dissolution.

MICROBIAL ENHANCEMENT OF ORGANIC COMPOUND UPTAKE

Several adaptations can occur during growth on liquid hydrocarbons. These include the development of increased cell affinity for hydrocarbons, the production of extracellular surface-active or emulsifying agents, and intracellular modifications which aid in hydrocarbon metabolism.

Affinity

It has long been observed that hydrocarbon-degrading microorganisms have an affinity for binding to hydrophobic surfaces such as hydrocarbon droplets (Miura et al., 1977). However, many microorganisms adhere to surfaces. For example, Rosenberg et al. (1980) showed that *Staphylococcus aureus* and early phase *Serratia marcescens* adhere to a variety of hydrocarbons but cannot degrade them. In addition, *Acinetobacter* strains that adhere to and utilize liquid hydrocarbons, also adhere to hydrocarbons that they do not degrade. Thus, there are both specific and nonspecific mechanisms by which microorganisms can adhere to hydrocarbons. One example of a nonspecific adherence mechanism is cell surface fimbriae. Rosenberg et al. (1982) showed that the surface of *Acinetobacter calcoaceticus* RAG-l was covered with fine fimbriae. Mutants without fimbriae did not adhere to hydrocarbons or other hydrophobic surfaces. RAG-l cells that were sheared to remove the fimbriae showed a significant reduction in adherence to hydrocarbons.

An example of a specific adherence mechanism that increases microbial affinity for hydrocarbon substrates is the presence of hydrophobic complexes in the cell wall. Käppeli et al. (1978) isolated a polysaccharide-fatty acid complex from the surface of yeast cells. The complex was identified as a mannan containing 4% covalently-linked fatty acids and was found only in cells grown on alkanes. Similar observations have been made for bacterial cells. Kennedy et al. (1975) demonstrated with phase and scanning electron microscopy that a hydrocarbon degrading *Acinetobacter* sp. showed affinity for paraffinic hydrocarbons, while bacteria unable to use the hydrocarbons did not. Zhang and Miller (1994) have observed that cell-surface hydrophobicity of *Pseudomonas* isolates increases during growth on hydrocarbons.

Biosurfactants

A second type of adaptation microorganisms have developed is the capacity to produce and release surfactants. Surfactants are amphipathic molecules that can modify the properties of a liquid medium at a surface or interface by reducing surface tension. Distilled water has a surface tension of 72 dyn cm^{-1}. An effective biosurfactant can lower this to under 30 dyn cm^{-1} (Singer & Finnerty, 1984). At low concentration, surfactants are present as individual molecules. However, as the concentration of the surfactant is increased, a concentration is reached where no further change in interfacial properties takes place. The amount of biosurfactant needed to reach this concentration is called the critical micelle concentration (cmc). At the cmc, surfactant molecules aggregate to form monolayer (micelle) or bilayer (vesicle, lamella) structures that have the ability to encapsulate hydrocarbon molecules resulting in either solubilization or emulsification of the hydrocarbon.

Biosurfactants are produced by many bacterial genera. The chemical structure of biosurfactants varies (Fig. 3-2), but all biosurfactants described thus far in the literature are anionic or nonionic. Biosurfactants can be classified into several broad groups: glycolipids, lipopeptides, lipopolysaccharides, phospholipids, and fatty acids/neutral lipids (Rosenberg, 1986; Lang & Wagner, 1987; Fiechter, 1992). Both biosurfactant yield and composition are affected by growth conditions including carbon source, culture medium nutrients (e.g., nitrogen, phosphate, iron), temperature, pH, and aeration (Syldatk & Wagner, 1987; Hommel & Ratledge, 1993). In addition, there are species level differences in the chemical structure of biosurfactants. For example, the rhamnolipids produced by various *Pseudomonas* spp. differ in; the number of rhamnose molecules (1 to 2), the length of the two lipid moieties, and the presence or absence of a carboxy group. Biosurfactant molecular weights range from approximately 500 to 1,500 although some exceptions exist, for example, *Pseudomonas* strains growing on hexadecane have been reported to produce protein-containing surface-active substances with molecular weights of up to 14,300 (Lang & Wagner, 1987).

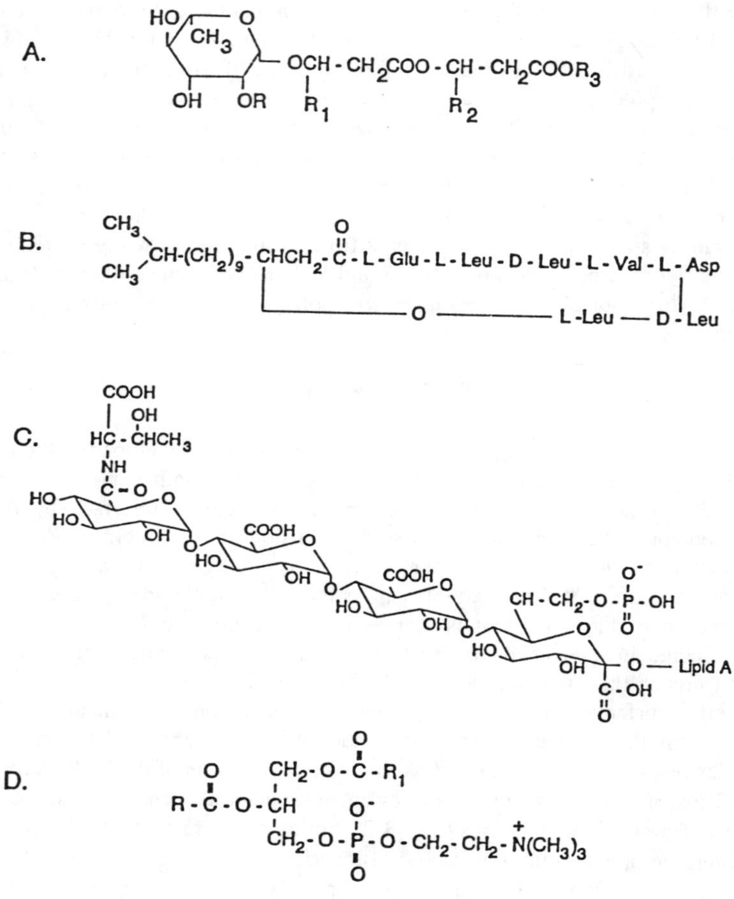

Fig. 3-2. Examples of biosurfactant structures. (A) rhamnolipid, a glycolipid produced by *Pseudomonas aeruginosa*. R = H, rhamnosyl, or CO-CH=CH-$(CH_2)_6$-CH_3; R_1 and R_2 = C_nH_{2n+1}, R_3 = H or CH_3. (B) surfactin, a lipopeptide produced by *Bacillus subtilus*. (C) a lipopolysaccharide from *Rhodopseudomonas sphaeroides*. The lipid A moiety usually consists of a phosphorylated ß-1,6-linked D-glucosamine disaccharide backbone with ester- and amide-linked fatty acids (Cloutier & Kosaric, 1993). (D) phosphatidylcholine, a phospholipid often used in the preparation of liposomes.

Intracellular Modifications

Microorganisms grown on alkanes show a variety of intracellular adaptations to growth on hydrocarbons. Inclusion bodies containing unmetabolized hydrocarbon are frequently seen in bacteria. Inclusion bodies may play a role in reducing the toxicity of large amounts of hydrocarbon that accumulate in the cell. Extensive cytoplasmic membrane systems are present in many hydrocarbon-grown bacteria. These membranes may serve as the site for initial hydrocarbon oxidation by providing a hydrophobic environment to maintain the solubilized state of the substrate. It is also postulated that the enzymes required for initial oxidation may be bound to these membranes (Singer & Finnerty, 1984).

Peroxisomes are commonly seen in yeasts that are grown on hydrocarbons. These organelles function in fatty acid metabolism (Singer & Finnerty, 1984) and like bacterial inclusions may act to reduce toxicity of the hydrocarbon substrate and metabolites to the cell. Inclusion bodies are also seen in yeast cells although not as frequently as in bacterial cells. Pooling of hydrocarbon occurs at the plasma membrane and frequently the rate of membrane synthesis is higher in hydrocarbon-grown yeast (Boultan & Ratledge, 1984).

SOLUBILIZATION OF ORGANIC COMPOUNDS

Theoretical Considerations

In order to increase the solubility of hydrocarbons in water, either their structure must be altered or the environment surrounding the hydrocarbon must be altered to accommodate the hydrocarbon. One of the early studies of solubilization of hydrocarbons in surfactant-hydrocarbon systems used x-ray analysis to show that hydrocarbons were incorporated into the surfactant micelle center (Fig. 3-3.) (Klevens, 1950). This was distinguished from alignment of the hydrocarbon with the surfactant molecules because alignment causes swelling of the micelle which can be seen by an increase in x-ray spacing. Further, long chain alcohols that incorporate into micelles by alignment with surfactant molecules greatly increased the amount of hydrocarbon the surfactant solubilized. As in the case of solubility in water, solubility in these micelles was affected by the size and the degree of unsaturation of the hydrocarbon (Klevens, 1950). In general, the larger the molecule the less soluble it was, while aromaticity tended to enhance solubilization.

For hydrocarbons of three rings or less it was calculated that the number of potassium laurate surfactant molecules per micelle was a constant value of approximately 150, corresponding to a micelle diameter of 6 nm. The number of molecules of hydrocarbon solubilized in each of these micelles was dependent on the hydrocarbon, for example, 118 molecules of benzene were encapsulated while 10 molecules of naphthalene and 2 molecules of phenanthrene were encapsulated. For larger polycyclics, the number of surfactant molecules in each micelle increased significantly. For example, it was estimated that micelles of diameter 12 nm (600 molecules surfactant) would form to solubilize 1 molecule of anthracene, a linear 3-ring aromatic, and micelles of diameter 36 nm (5000 molecules surfactant) would form to solubilize one molecule of naphthacene, a linear 4-ring aromatic hydrocarbon. Hydrocarbon solubilization in micelles also depends on the surfactant type as shown by Almgren et al. (1979). This difference was explained by differentiating between surfactants which solubilize hydrocarbons in the micelle hydrocarbon tail core and surfactants which solubilize in the core as well as through binding of hydrocarbons to surfactant polar head groups.

Surfactants, depending on their chemical structure, can also form bilayer structures (Fig.3-3). In aqueous solution, the polar surfactant heads of a bilayer face the outside while the nonpolar tails are sandwiched between the heads. Thus, the environment both inside and outside a vesicle is hydrophilic (aqueous) while the environment within the bilayer, composed of the nonpolar surfactant tails, is hydrophobic. Bilayers can also exist as flexible sheets or planar bilayers, the largest of the basic surfactant structures. A bilayer sheet is essentially unlimited in size. If the solution on both sides of the bilayer is the same, the properties and behavior of the two bilayer surfaces will be identical. However, if the bilayer is at an interface e.g., air-water or liquid-liquid, the bilayer may develop asymmetric properties.

The incorporation of nonamphipathic lipophilic molecules such as hydrocarbons into vesicles is an area that has not been well studied (Szoka & Papahadjopoulos, 1980). However, phospholipid vesicles (liposomes) have been studied intensively as model membrane systems and for use as drug delivery systems. Nonpolar drugs are often incorporated directly into the lipid bilayer by adding the drug to the organic solvent in which the phospholipid is dissolved when forming vesicles, prior to evaporation of the solvent. Phospholipid solubilization of such lipid-soluble material is often considerable and vesicles are stable for long periods of time (Kimelberg & Mayhew, 1978).

Surfactant Solubilization of Organic Compounds

Both synthetic and microbial surfactants increase the apparent aqueous solubility of aliphatic and aromatic organic compounds. Table 3-2 summarizes recent studies of synthetic surfactant solubilization of a variety of organic compounds with low aqueous solubility. This relationship can be expressed quantitatively in terms of the molar solubilization ratio (MSR). The molar solubilization ratio is defined as the number of moles of organic compound solubilized per mole of surfactant added to the solution. The MSR values in Table 3-2 represent the slope of the organic solubility curve (organic solubility vs. surfactant concentration) taken at surfactant concentrations greater than the CMC. As shown in Table 3-2, the MSR is directly related to the aqueous solubility of the organic compound. If the MSR is normalized with respect to the aqueous solubility of the organic compounds tested, it becomes clear that there is much greater relative enhancement of apparent solubility for compounds with low aqueous solubility, e.g., DDT, than for compounds with higher aqueous solubility, e.g., dichloroethylene.

There are few quantitative reports of biosurfactant solubilization of organic compounds partly because relatively few biosurfactants have been completely purified and characterized. The available data suggest that biosurfactants have great potential for solubilization of organics. For example, the MSR for an octadecane/rhamnolipid system studied by Zhang and Miller (1992) was 0.16. Normalizing the MSR with respect to aqueous octadecane gives a value of 6.7×10^6 which is comparable to values reported for synthetic surfactants in Table 3-2. Cryo-transmission electron microscopy showed that the majority of structures in this

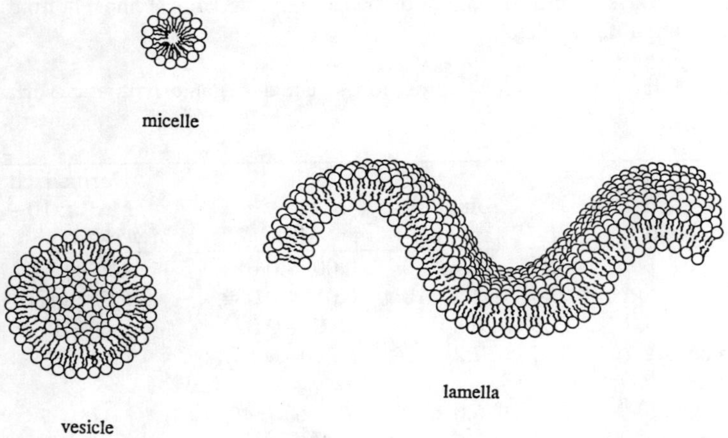

Fig. 3-3. Basic surfactant structures include micelles, vesicles and lamellae.

system (at pH 7.0) are micellar (Champion et al., 1994). Goswami and Singh (1991) found that a *Pseudomonas* glycoprotein (500 mgL^{-1}) enhanced the apparent solubility of hexadecane from 0.006 $mgL^{-1}$to 32 mgL^{-1}, an increase of over 3 orders of magnitude. Similarly, Falatko and Novak (1992) reported increases in apparent solubility of several aromatic compounds of up to 15-fold by an undefined mix of biosurfactants (600 mgL^{-1} TOC).

Vesicle structures also have potential for solubilization of organics. Miller and Bartha (1989) used sonication to solubilize octadecane (C_{18}) and hexatriacontane (C_{36}) into phosphatidylcholine vesicles 30 to 40 nm in diameter. Solubilization of octadecane and hexatriacontane by the vesicles occurred with MSRs of 3.1 and 0.8 mol hydrocarbon mol^{-1} phospholipid respectively. The normalized MSR values are 1.3 x 10^8 and 3.3 x 10^7. In contrast benzo(a)pyrene, a 5-ring aromatic, was solubilized very poorly by phosphatidylcholine vesicles. However, vesicles composed of phosphatidylcholine, phosphatidylserine, and cholesterol (1:1:1) solubilized 0.09 mol benzo(a)pyrene mol^{-1} phospholipid mix (R.M. Miller, unpublished data).

Table 3-2. Molar solubilization ratios for selected organic/synthetic surfactant systems.

Compound	MW	Solubility mol L^{-1}	MSR[†]	Normalized MSR x 10^{-3} (MSR/solubility)
DDT[‡]	355	1.1 x 10^{-8}	0.005 - 0.019	460 - 1730
Octadecane[§]	254	2.4 x 10^{-8}	0.013 - 0.034	540 - 1420
Pyrene[§]	202	6.8 x 10^{-7}	0.04 - 0.07	59 - 103
Phenanthrene[¶]	178	7.2 x 10^{-6}	0.10 - 0.16	14 - 22
Phenanthrene[#]			0.034 - 0.15	5 - 21
Trichlorobenzene[‡]	182	6.6 x 10^{-5}	0.006 - 0.49	0.09 - 7
Naphthalene[¶]	128	2.5 x 10^{-4}	0.320 - 0.37	1.3 - 1.5
Tetrachloroethylene[††]	167	1.4 x 10^{-3}	1.3 - 4.0	0.93 - 2.9
Trichloroethylene[††]	132	1.0 x 10^{-3}	1.2 - 15.2	0.12 - 1.5
1,2-dichloroethylene[††]	97	5.0 x 10^{-2}	5.2 - 14.2	0.10 - 0.28

[†]MSR (Molar Solubilization Ratio) is expressed as the moles of organic compound solubilized/mole of micellar surfactant. MSR values reported the range of results from various surfactants tested for each organic compound.
[‡]Kile and Chiou (1989). Surfactants used: Brij 35, CTAB, SDS, Triton X-100, Triton X-114, Triton X-405.
[§]Thai and Maier (1992). Surfactants used: Triton X-114, Corexit 0600.
[¶]Edwards et al. (1991). Surfactants used: Brij 30, Igepal CA-720, Tergitol NP-10, Triton X-100.
[#]Tiehm (1994). Surfactants used: Arkopal N-300, Brij 35, Genapol X-150, Marlipal 013/90, Pluronic PE6400, Sapogenat T-300, Tegopren 5851, Triton X-102, SDS
[††]West (1992). Surfactants used: Brij-30, Igepal CA-720, Tergitol NP-10, Triton X-100.

These data show that both the biosurfactant structure and the hydrocarbon structure are important in determining surfactant solubilization of hydrocarbons. There are a myriad of surfactant structures produced in nature that form micelle and bilayer structures (Fig. 3-2). As yet, no systematic study of biosurfactant solubilization of organic compounds has been done. Information gained from such studies would be valuable in determining which biosurfactants are most efficient in solubilizing organics.

SURFACTANT-ENHANCED BIODEGRADATION OF ORGANIC COMPOUNDS: SOLUTION STUDIES

The ability of a surfactant to enhance biodegradation of a slightly soluble organic compound depends on the extent to which organic bioavailability is increased by the surfactant. Bioavailability will depend on the relative increase in organic solubilization and perhaps more importantly, on the ability of the degrading cell to take up the surfactant-solubilized organic from solution. Therefore, there must be compatibility between the degrading organism and the surfactant if the organic is solubilized within surfactant structures.

Synthetic Surfactants

Studies show that addition of synthetic surfactants can enhance the biodegradation of slightly soluble hydrocarbons. This effect however, is not consistent or well defined. In many cases surfactant addition has little or no effect and can even inhibit growth. Aiba et al. (1969) and Mimura et al. (1971) found that yeast growth rates decreased in the presence of synthetic surfactants. Microscopic examination showed that surfactant addition, even very low levels, caused interference of the direct interaction between yeast cells and hydrocarbon.

Whitworth et al. (1973) found that additions of Tween 20, Tween 80, Span 20, and Span 80 enhanced growth of *Candida lipolytica* on dodecane (C_{12}). These detergents could also serve as carbon sources for *C. lipolytica*. Tween 20 has also been shown to stimulate the utilization of hexadecane by several strains of *Pseudomonas aeruginosa* (Singer & Finnerty, 1984). Kawai et al. (1977) investigated the biodegradation of polypropylene glycol and found that Triton X-100 inhibited growth of the four bacterial strains tested, but 0.03% Tween 85 significantly increased polymer biodegradation. Thai and Maier (1992) found that addition of Corexit 0600 stimulated octadecane biodegradation above the CMC but had no effect below the CMC. The addition of Brij 35 had no effect on octadecane biodegradation.

Breuil and Kushner (1980) studied the effect of addition of fatty acids, lipids, and detergents on hexadecane-biodegrading bacteria. Long-chain fatty acids were found to enhance biodegradation, while short-chain fatty acids (C_8 to C_{12}) inhibited biodegradation. It was postulated that the longer chain fatty acids induced production of enzymes used to degrade hexadecane, whereas the shorter chain fatty acids inhibited production of these enzymes. Addition of lipids had an effect similar to that of fatty acids. Detergent addition (0.001-0.01% Triton X-100, Brij 35 or F2-70) was found to shorten the lag phase and increase the total amount of growth.

Detergents added to washed cells (free of emulsifying agents produced by the cell) induced the same growth rates as found in normal unwashed cells.

Although most studies have focused on aliphatic hydrocarbons, a recent study by Tiehm (1994) investigated the effect of nine surfactants on degradation of aromatic hydrocarbons. This work showed that some of the surfactants were toxic to the degrading organisms, some served as preferential carbon sources, and some enhanced the degradation of the model aromatic compounds.

These data make it obvious that biodegradation of organic compounds can be enhanced through addition of artificial surfactants. However, no systematic investigation has been undertaken to determine whether enhancement can only be achieved through a unique approach for each organic compound/organism, or whether a generalized approach can be defined to optimize biodegradation of organic compounds by microorganisms. This type of investigation will require delineation of the effects of the following parameters on biodegradation: surfactant type, structure, and concentration, substrate structure and concentration, and microbial type and numbers.

Biosurfactants

Torulopsis bombicola growth on various n-alkanes C_{10}-C_{20}, was enhanced by addition of 0.02% sophorose lipids (Ito & Inoue, 1982). In these experiments, growth on 2% alkane was quantified by absorbance at 650 nm. Surfactant-induced increases in growth (expressed as growth in the presence of surfactant and growth in the absence of surfactant) ranged from slight (1.3) to large (9.4), and in some cases growth only occurred in the presence of surfactant. This effect was observed almost exclusively with *Torulopsis* yeasts, the growth of 12 other alkane-utilizing yeast genera tested being unaffected or inhibited by addition of sophorose lipids. Subsequent work showed that *T. bombicola* produced a sophorose lipid (Cooper & Paddock, 1984), indicating that biosurfactant enhancement of growth may depend on whether the degrader has specificity for the added biosurfactant. Zhang and Miller (1994) have demonstrated a similar effect of rhamnolipid on bacterial growth. In this study of 4 octadecane-degrading *Pseudomonas* isolates, addition of rhamnolipid stimulated octadecane biodegradation up to 4-fold for 2 of the isolates except at very low rhamnolipid concentrations which initially inhibited biodegradation. Octadecane biodegradation by the other 2 isolates was inhibited at both high and low concentrations of rhamnolipid. Degradation was quantified both by protein increase and by measurement of evolved $^{14}CO_2$ (using ^{14}C-octadecane) in these experiments. Falatko and Novak (1992) found that biosurfactants produced during growth on gasoline constituents enhanced biodegradation of gasoline constituents, while biosurfactants produced during growth on glucose and vegetable oil inhibited biodegradation of gasoline. The biosurfactant-induced increase in biodegradation was modest for all four constituents tested (toluene, m-xylene, naphthalene, and 1,2,4-trimethylbenzene). The largest increase was for 1,2,4-trimethylbenzene which degraded in 7 days in the presence of the biosurfactant and in 12 days in the control. In this study, disappearance of substrate was quantified by gas chromatography as a measure of degradation.

Several groups have reported the production of abnormal amounts of phospholipids from microorganisms grown on hydrocarbons. For instance, the yeast

Candida tropicalis produces significantly more phospholipid when grown on n-alkanes instead of glucose (Cooper & Zajic, 1980). Schaeffer and Umbreit (1963) and Jones and Benson (1965) studied the agent produced by *Thiobacillus thiooxidans* to facilitate uptake of elemental sulfur. They identified the major components as phosphatidylinositol and phosphatidylglycerol, although other phospholipids were present as well. The surface tension of the growth medium fell from 72 dyn cm^{-1} to 49 dyn cm^{-1} during incubation of the organism, and it was found that cell-free supernatant from a 7-day old culture retained the ability to wet sulfur. Käppeli and Finnerty (1979) have described an extracellular phospholipid vesicle produced by hexadecane grown *Acinetobacter* spp. HO1-N. The major constituent of the vesicle was phosphatidylethanolamine (67.6%) followed by phosphatidylglycerol (21.0%), triacyllysocardiolipin (5.5%), cardiolipin (3.8%) and acylphosphatidylglycerol (2.1%). The composition of the vesicle, which solubilizes hexadecane, varied significantly from that of the cytoplasmic or outer membranes of the cell. The vesicle was, however, similar to intracellular hexadecane inclusion body membranes found in HOl-N. Experimental data indicate that the vesicles fuse with the cell envelope to allow transfer of the solubilized hexadecane to hydrocarbon inclusion bodies in the cell (Singer & Finnerty, 1984).

A variety of fatty acids and neutral lipids are also found as extracellular products. Makula and Finnerty (1972) showed that the type of fatty acid produced extracellularly was dependent on the n-alkane used as carbon source. Fatty acids are also often found as part of a mixture of biosurfactants in culture broths of microorganisms grown on hydrocarbons.

SURFACTANT-ENHANCED BIODEGRADATION OF ORGANIC COMPOUNDS: SOIL STUDIES

The ability of a surfactant to enhance the biodegradation of organic compounds in soil will depend on: i) the solubilization of the organic contaminant by surfactant monomers and micelle/vesicle structures; ii) the sorption of the organic contaminant by the soil; iii) the effect of the surfactant on soil structure; iv) the biodegradability of the surfactant, i.e., whether the surfactant is an alternate carbon source; v) the biological compatibility of the surfactant in terms of toxicity to degrading microorganisms; and vi) the ability of the degrading microorganisms to assimilate the surfactant-solubilized organic compound.

Synthetic Surfactants

Similar to solution studies of biodegradation, the effects of synthetic surfactants on biodegradation of organics in the presence of soil can be positive or negative. For instance, Graves and Leavitt (1991) found that surfactant amendment (Adsee, 799) inhibited biodegradation of petroleum hydrocarbons by serving as an alternate carbon source. This study was done using a bench-top land-treatment system consisting of 2-3 kg petroleum-contaminated soil contained in polyethylene boxes. In comparison, a study of two nonionic surfactants demonstrated a positive effect of Corexit 0600 on octadecane biodegradation in sand and a negative effect of Brij 35 (Thai, 1993). Aronstein et al. (1991) and Aronstein and Alexander (1992) have shown that in some cases two nonionic surfactants, Alfonic 810-60 and Novel

II 1412-56, increased the rate of biodegradation of biphenyl and phenanthrene in soil and soil slurries. In these studies, surfactant concentrations were 10-fold to 100-fold higher than substrate concentrations and degradation was quantified by measurement of evolved $^{14}CO_2$ Results reported support the hypothesis that enhanced biodegradation is dependent on surfactant type, surfactant concentration, hydrocarbon type, soil type, and moisture conditions. For example, the effect of Alfonic 810-60 on biodegradation of phenanthrene was compared in three soils: an aquifer sand with 0.4% organic matter (9%, 7%, and 3% mineralization in the presence of 100, 10, and 0 μg surfactant g^{-1} soil); a Lima silt loam with 7.6% organic matter (8%, 42%, and 23% mineralization in the presence of 100, 10 and 0 μg surfactant g^{-1} soil); and a marsh soil with 32.9% organic matter (8%, 51%, 7% mineralization in the presence of 100, 10, 0 μg surfactant g^{-1} soil). The studies described above indicate that surfactants can enhance biodegradation of contaminants in soil. However, because of the complexity of the experimental systems used, the data are difficult to interpret. This indicates once again how the lack of knowledge concerning mechanisms of surfactant-enhanced degradation is limiting their application to remediation technologies.

Biosurfactants

Biosurfactant amendments seem to stimulate organic biodegradation in soil. In a study of several glycolipid biosurfactants, Oberbremer et al. (1990) showed that both the rate and extent of hydrocarbon degradation in a soil slurry was enhanced by biosurfactants. A hydrocarbon mixture was used that included tetradecane, pentadecane, hexadecene, 1,2,4-trimethylcyclohexane, pristane, phenyldecane and naphthalene. The model system used in this study contained 10% soil, 1.35% hydrocarbon, and 200 mgL^{-1} surfactant (14.8 g kg^{-1} hydrocarbon mixture). Without surfactant present 81% of the hydrocarbon mixture was degraded in 114 h. However, in the presence of surfactants, up to 99% of the hydrocarbon mixture was degraded in 79h. Similarly, Jain et al. (1992) found that addition of a *Pseudomonas* biosurfactant enhanced the biodegradation of tetradecane, pristane, and hexadecene in a silt loam with 2.1% organic matter. Interestingly, biodegradation of 2-methylnaphthalene, the hydrocarbon with the highest water solubility used in this study, was not stimulated. The soil in each batch test was adjusted to 25% moisture content and amended with 100 μg g^{-1} biosurfactant and 8000 μg g^{-1} hydrocarbon mixture.

The clear advantage of biosurfactants over synthetic surfactants is the possibility for *in situ* production. Unfortunately, there is very little information concerning *in situ* biosurfactant production. Perhaps the strongest evidence for biosurfactant production in soil comes from a study by Oberbremer and Müller-Hurtig (1989) that demonstrated biosurfactant production during degradation of hydrocarbons in a stirred soil reactor (99 g soil dry weight L^{-1}). In this study, a mixture of hydrocarbons was added to the reactor including tetradecane, 1-hexadecene, pristane, trimethylcyclohexane, phenyldecane, and naphthalene. During the first 35 h of operation, hydrocarbon removal was primarily by stripping. Hydrocarbons affected by stripping were trimethylcyclohexane most of which was removed, and naphthalene which was stripped to a much lesser degree. After this lag period, the most water-soluble component remaining, naphthalene, was degraded rapidly (as measured by oxygen uptake and naphthalene disappearance). This was followed by a second lag period which lasted for approximately 13 h. At the end of the second lag period a decrease in surface tension in the solution from 25 to 5 mN m^{-1} was measured which occurred simultaneously with a second period of degradation. During this period, the remaining hydrocarbon components were degraded at similar rates. The sharp decrease in surface tension which accompanied degradation of the hydrocarbons with low water solubility suggests production of surfactant was occurring. This was confirmed by isolation of a glycolipid in supernatant samples which was identified as a trehalose lipid.

This study did not explain how surfactant production was stimulated, it was simply a reported phenomenon. There are reports that surfactant production is a result of nutrient stress. Mulligan et al. (1989b) have shown that rhamnolipid production by *Pseudomonas aeruginosa* is dependent on phosphate metabolism and is stimulated under phosphate-limited conditions. Other known parameters which affect biosurfactant production include nitrogen source, carbon source, multivalent cations, and antibiotics (Syldatk & Wagner, 1987; Mulligan et al., 1989a). This raises the obvious question of whether biosurfactant production in an indigenous population can be stimulated in a controlled way. To answer this question, the mechanism by which surfactant production is induced and the required nutrient and environmental conditions must be understood. A second question to be asked is whether surfactant production can be selectively stimulated.

Surfactant-Enhanced Flushing of Organic Compounds from Soil

Although complete destruction (mineralization) of organic contaminants is preferred, some organic molecules are highly resistant to biodegradation. These contaminants may be amenable to surfactant-facilitated removal by soil washing or flushing. The ability of a surfactant to remove an organic contaminant from soil depends on: i) the solubilization of the organic contaminant by surfactant monomers and micelle/vesicle structures; ii) the sorption of the organic contaminant by the soil; and iii) the sorption of the surfactant by the soil and its effect on soil surface wettability (Liu et al., 1991), and iv) the size and shape of the surfactant/organic structures, which if too large may be filtered out by soil pores.

SYNTHETIC SURFACTANTS

Both synthetic and biological surfactants have potential for use in facilitated flushing to remove contaminants from soil. For example, a series of batch studies by Liu et al. (1991) showed removal of anthracene (up to 60%) and pyrene (up to 80%), by a variety of surfactants. The amount of hydrocarbon removal was dependent both on the surfactant used (some surfactants were ineffective, e.g. sodium dodecylbenzenesulfonate), and on the amount of soil present in the batch experiment. In general, with increasing amounts of soil, there was a decrease in hydrocarbon solubilization. McDermott et al. (1989) used a four-stage bench-scale reactor to extract PCBs from soil. Results showed that a 1% sodium dodecylbenzenesulfonate solution reduced the level of PCBs from 1000 to 40 mg L^{-1} using a laboratory-spiked soil. The surfactant was less effective using a PCB-contaminated field soil reducing the PCB concentration from a range of 100 to 168

$\mu g\ g^{-1}$ to a range of 18 to 24 $\mu g\ g^{-1}$. The authors suggest the reduced effectiveness in the field soil may be due to the lower level of initial contamination or the age of contaminant residue. It is not clear how similar these soils are, but differences in soil properties such as organic matter would also impact PCB removal (Jafvert, 1991).

In batch experiments using the surfactant Surfynol 485 (2%), Peters et al. (1992) found removal efficiencies of up to 97% for a No. 2 diesel fuel-contaminated soil. However, removal of diesel fuel from intact cores was less than 1%. The low recovery of diesel was ascribed to the development of extensive channeling in the soil cores during shipping and handling which resulted in preferential flow of the surfactant solution.

Biosurfactants

Falatko and Novak (1992) studied biosurfactant-facilitated removal of gasoline from a sand. In this study, a layer of gasoline was added in a coarse grain sand to the top of a packed sand column. Biosurfactant solution (600 mg L^{-1} TOC) increased the effluent concentration of 4 gasoline constituents, toluene, m-xylene, 1,2,4- trimethylbenzene, and naphthalene up to 15-fold. The relative increase in solubility was greatest for 1,2,4-trimethylbenzene, the constituent with the lowest aqueous solubility. Van Dyke et al. (1993) surveyed a variety of biosurfactants for removal of hexachlorobiphenyl from soil. Out of 13 biosurfactants tested, 7 removed hexachlorobiphenyl much more efficiently than the water control (compare 13 to 48% removal to 1% removal for the control). The most efficient biosurfactants were produced by two *Pseudomonas aeruginosa* strains and *Acinetobacter calcoaceticus* RAG-1. Harvey et al. (1990) report increased removal of crude oil from gravel by 1% solutions of a *Pseudomonas* glycolipid biosurfactant. The increase in removal was temperature dependent reaching a maximum of over 70% at 70°C. At 30°C, approximately 20% of the oil was removed. Total oil concentrations were not reported.

SUMMARY

Both the structure and physical state of an organic contaminant affect contaminant bioavailability and therefore, biodegradation. In general, bioavailability and biodegradation rates increase with increasing substrate aqueous solubility. However, organic compounds in the liquid and gaseous phases are more bioavailable than solid-phase organic compounds because microorganisms can degrade liquid and gaseous phase organics by direct contact. Microorganisms have several adaptations to facilitate uptake of slightly soluble liquid organics including development of hydrophobic cell wall complexes to increase hydrocarbon affinity, production of extracellular surfactants to enhance hydrocarbon solubilization or wettability, and intracellular modifications to increase hydrocarbon tolerance and biodegradation rates.

Surfactants, both synthetic and biological, can be used to increase the apparent aqueous solubility of organic contaminants. The relative enhancement of apparent organic solubility is much higher for organic compounds with very low aqueous solubility. Both types of surfactants can also enhance the biodegradation of organic contaminants. However, the interaction of surfactants with microorganisms is not well-defined, and in some studies surfactants have been found to inhibit biodegradation.

The application of biosurfactants to remediation of contaminated sites has several obvious advantages over synthetic surfactants. The natural variety in biosurfactant structures is hard to reproduce synthetically. However, as yet, no systematic study of biosurfactants has been done so that the optimal surfactant structure for use in remediation, remains unclear. Information from such studies will be necessary to determine which biosurfactants are most efficient in solubilizing organics, enhancing biodegradation, or facilitating the removal of contaminants from soils by flushing. The second obvious advantage to biosurfactants is the potential for *in situ* production of biosurfactants. However, very little is known about the production of biosurfactants in soils, whether it occurs, and if it occurs how production might be stimulated. Specific research needs for biosurfactant-enhanced remediation include:

1) Structure-function analysis of surfactant solubilization of organics.

2) Investigation of the transport of surfactant-solubilized organics into cells.

3) Investigation of the interaction of biosurfactants with soil.

4) Optimization of biodegradation and/or flushing in surfactant/soil/organic systems

5) Investigation of *in situ* production of biosurfactants.

6) Scale-up and cost analysis for *ex situ* biosurfactant production

REFERENCES

Aiba, S., K.L. Huang, V. Moritz, and J.I. Someya. 1969. Cultivation of yeast cells by using n-alkanes as the sole carbon source: II. an approach to the mechanism of the microbial uptake of n-alkanes. J. Ferm. Tech. 47:211-215.

Almgren, M. F. Grieser, J.R. Powell, and J.K. Thomas. 1979. A correlation between the solubility of aromatic hydrocarbons in water and micellar solutions, with their normal boiling points. J. Chem. Engr. Data 24:285-287.

Aronstein, B.N., and M. Alexander. 1992. Surfactants at low concentrations stimulate biodegradation of sorbed hydrocarbons in samples of aquifer sands and soil slurries. Env. Toxicol. Chem. 11:1227-1233.

Aronstein, B.N., Y.M. Calvillo, and M. Alexander. 1991. Effect of surfactants at low concentration on the desorption and biodegradation of sorbed aromatic compounds in soil. Environ. Sci. Technol. 25:1728-1731.

Atlas, R.M. 1981. Microbial biodegradation of petroleum hydrocarbons: an environmental perspective. Microbiol. Reviews. 45:180-209.

Bossert, I. and R. Bartha. 1986. Structure-biodegradability relationships of polycyclic aromatic hydrocarbons in soil. Bull. Environ. Contam. Toxicol. 37:490-495.

Boultan, C.A. and C. Ratledge. 1984. The physiology of hydrocarbon-utilizing microorganisms. p. 11-77. *In* A. Wiseman (ed.) Topics in Enzyme and Fermentation Biotechnology Vol. 9 Halstead Press, New York.

Breuil, C., and D.J. Kushner. 1980. Effects of lipids, fatty acids, and other detergents on bacterial utilization of hexadecane. Can. J. Microbiol. 26:223-231.

Cerniglia, C.E. and M.A. Heitkamp. 1989. Microbial degradation of polycyclic aromatic hydrocarbons (PAH) in the aquatic environment. *In* U. Varanasi, (ed.) Metabolism of Polycyclic Aromatic Hydrocarbons in the Aquatic Environment CRC Press, Boca Raton, Florida.

Chakravarty, M., Amin, P.M. Singh, H.D., Baruah, J. N. and Iyengar, M.S. 1972. A kinetic model for microbial growth on solid hydrocarbons. Biotech. Bioeng. 14:61-63.

Champion J.T., J.C. Gilkey, H. Lamparski, J. Retterer, and R.M. Miller. 1995. Electron microscopy of rhamnolipid (biosurfactant) morphology: effects of pH, cadmium, and octadecane. J. Colloid and Interface Sci. 170:569-574.

Cloutier, L. and Kosaric, N. 1993. Microbial lipopolysaccharides. p. 305-325. In N. Kosaric, (ed.) Biosurfactants; production, properties, applications Marcel Dekker, New York..

Cooper, D.G. and D.A. Paddock. 1984. Production of a biosurfactant from *Torulopsis bombicola*. Appl. Environ. Microbiol. 47:173-176.

Cooper, D.G. and J.E. Zajic. 1980. Surface-active compounds from microorganisms. Adv. Appl. Microbiol. 26:229-253.

Edwards, D.A., R.G. Luthy, and Z. Liu. 1991. Solubilization of polycyclic aromatic hydrocarbons in micellar nonionic surfactant solutions. Environ. Sci. Technol. 25:127-133.

Falatko, D.M., and J.T. Novak. 1992. Effects of biologically produced surfactants on the mobility and biodegradation of petroleum hydrocarbons. Water Environ. Res. 64:163-169.

Fiechter, A. 1992. Biosurfactants: moving towards industrial application. Trends Biotech. 10:208-217.

Fogel, S., R. Lancione, and A. Sewall. 1981. A literature and laboratory investigation of the influence of water solubility on the biodegradation of organic chemicals. JBF Scientific Corp. Dec. 1991.

Goswami, P. and H.D. Singh. 1991. Different modes of hydrocarbon uptake by two *Pseudomonas* sp. Biotech. Bioeng. 37:1-11.

Graves, D., and M. Leavitt. 1991. Petroleum biodegradation in soil: the effect of direct application of surfactants. Remediation. pp. 147-166.

Gutnick, D.L., and Rosenberg, E. 1977. Oil tankers and pollution: a microbiological approach. Ann. Rev. Microbiol. 31:379-397.

Haines, J.R., and M. Alexander. 1974. Microbial degradation of high-molecular-weight alkanes. Appl. Microbiol. 28:1084-1085.

Hallas, L.E., and J.R. Vestal. 1978. The growth of *Mycobacterium convolutum* on solid n-alkane substrates: effect on cellular lipid composition. Can. J. Microbiol. 24:1197-1203.

Hankin, L., and P.E. Kolattukudy. 1968. Metabolism of a plant wax paraffin (n-nonacosane) by a soil bacterium (*Micrococcus cerificans*). J. Gen. Microbiol. 51:457-463.

Harvey, S., I. Elashvili, J.J. Valdes, D. Kamely, and A.M. Chakrabarty. 1990. Enhanced removal of Exxon Valdez spilled oil from Alaskan gravel by a microbial surfactant. Biotechnology 8:228-230.

Hommel R.F., and C. Ratledge. 1993. Biosynthetic mechanisms of low molecular weight surfactants and their precursor molecules. p. 3-63. In N. Kosaric (ed.) Biosurfactants; production, properties, applications. Marcel Dekker, New York.

Ito, S., and S. Inoue. 1982. Growth of yeasts on alkanes with sophorolipids. Appl. Environ. Microbiol. 43:1278-1283.

Jafvert, C.T. 1991. Sediment- and saturated-soil-associated reactions involving an anionic surfactant (Dodecylsulfate). 2. Partition of PAH compounds among phases. Environ. Sci. Technol. 25:1039-1045.

Jain, D.K., H. Lee, and J.T. Trevors. 1992. Effect of addition of *Pseudomonas aeruginosa* UG2 inocula or biosurfactants on biodegradation of selected hydrocarbons in soil. J. Indus. Microbiol. 10: 87-93.

Jones, G.E., and A.A. Benson. 1965. Phosphatidylglycerol in *Thiobacillus thiooxidans*. J. Bacteriol. 89:260-261.

Käppeli, O., and W.R. Finnerty. 1979. Partition of alkane by an extracellular vesicle derived from hexadecane-grown *Acinetobacter*. J. Bacteriol. 140:707-712.

Käppeli, O., M. Müller, and A. Fiechter. 1978. Chemical and structural alterations at the cell surface of *Candida tropicalis* induced by hydrocarbon substrate. J. Bact. 133:952-958.

Kawai, F., K. Hanada, T. Yoshiki, and K. Ogata. 1977. Bacterial degradation of a water-insoluble polymer (polypropylene glycol). J. Ferm. Tech. 55:89-96.

Kennedy, R.S., W.R. Finnerty, K. Sudarsanan, and R.A. Young. 1975. Microbial assimilation of hydrocarbons. Arch. Microbiol. 102:75-83.

Kile, D.E., and C.T. Chiou. 1989. Water solubility enhancements of DDT and trichlorobenzene by some surfactants below and above the critical micelle concentration. Environ. Sci. Technol. 23:832-838.

Kimelberg, H.K., and E.G. Mayhew. 1978. Properties and biological effects of liposomes and their uses in pharmacology and toxicology. CRC Crit. Rev. Tox. 6:25-79.

Kiyohara, H., K. Nagao, and K. Yana. 1982. Rapid screen for bacteria degrading water-insoluble solid hydrocarbons on agar plates. Appl. Environ. Microbiol. 43:454-457.

Klevens, H. 1950. Solubilization of polycyclic hydrocarbons. J. Phys. Colloid Chem. 54:283-298.

Lang, S., and F. Wagner. 1987. Structure and properties of biosurfactants. In: Biosurfactants and biotechnology (N. Kosaric, W.L. Cairns, N.C.C. Gray, eds). Marcel Dekker, New York. pp. 21-45.

Liu, Z. S. Laha, and R.G. Luthy. 1991. Surfactant solubilization of polycyclic aromatic hydrocarbon compounds in soil-water suspensions. Wat. Sci. Tech. 23:475-485.

Makula, R.A., and W.R. Finnerty. 1972. Microbial assimilation of hydrocarbons: cellular distribution of fatty acids. J. Bacteriol. 112:398-407.

McDermott, J.B., R. Unterman, M.J. Brennan, R.E. Brooks, D.P. Mobley, C.C. Schwartz, and D.K. Dietrich. 1989. Two strategies for PCB soil remediation: biodegradation and surfactant extraction. Environ. Progress 8:46-51.

Miller, R.M., and R. Bartha. 1989. Evidence from liposome encapsulation for transport-limited microbial metabolism of solid alkanes. Appl. Environ. Microbiol. 55:269-274.

Mimura, A., S. Watanabe, and I. Takeda. 1971. Biochemical engineering analysis of hydrocarbon fermentation. J. Ferm. Tech. 49:255-271.

Miura, Y., M. Okazaki, S. Hamada, S. Murakawa, and R. Yugen. 1977. Assimilation of liquid hydrocarbon by microorganisms I. mechanism of hydrocarbon uptake. Biotech. Bioeng. 19:701-714.

Mulligan, C.N., G. Mahmourides, and B.F. Gibbs. 1989a. Biosurfactant production by a chloramphenicaol tolerant strain of *Pseudomonas aeruginosa*. J. Biotechnol. 12:37-44.

Mulligan, C.N., G. Mahmourides, and B.F. Gibbs. 1989b. The influence of phosphate metabolism on biosurfactant production by *Pseudomonas aeruginosa*. J. Biotech. 12:199-210.

Nakahara, T., L.E. Erickson, and J.R. Gutierrez. 1977. Characteristics of hydrocarbon uptake in cultures with two liquid phases. Biotech. Bioeng. 19:9-25.

Oberbremer, A., and R. Müller-Hurtig. 1989. Aerobic stepwise hydrocarbon degradation and formation of biosurfactants by an original soil population in a stirred reactor. Appl. Microbiol. Biotechnol. 31:582-586.

Oberbremer, A., R. Müller-Hurtig, and F. Wagner. 1990. Effect of the addition of microbial surfactants on hydrocarbon degradation in a soil population in a stirred reactor. Appl. Microbiol. Biotechnol. 32:485-489.

Peters, R.W., C.D. Monetemagno, L. Shem, and B-A.G. Lewis. 1992. Surfactant flooding of diesel-fuel-contaminated soil. p 85-98. *In* G.F. Vandegrift, D.T. Reed, and I.R. Tasker (ed.) Environmental Remediation: Removing Organic and Metal Ion Pollutants. ACS Symposium Series No. 509, American Chemical Society..

Rosenberg, E. 1986. Microbial surfactants. Crit. Rev. Biotechnol. 3:109-132.

Rosenberg, E., E.A. Bayer, J. Delarea, and E. Rosenberg. 1982. Role of thin fimbriae in adherence and growth of *Acinetobacter calcoaceticus* RAG-1 on hexadecane. Appl. Environ. Microbiol. 44:929-937.

Rosenberg, E., D.L. Gutnick, and E. Rosenberg. 1980. Adherence of bacteria to hydrocarbons: a simple method for measuring cell-surface hydrophobicity. FEMS Micro. Lett. 9:29-33.

Schaeffer, W.I., and W.W. Umbreit. 1963. Phosphatidylinositol as a wetting agent in sulfur oxidation by *Thiobacillus thiooxidans*. J. Bact. 85:492-493.

Singer, M., and Finnerty, W. 1984. Microbial metabolism of straight-chain and branched alkanes. p 1-59. *In* R. Atlas (ed.) Petroleum Microbiology. Macmillan Pub. Co., New York, NY.

Syldatk, C., and F. Wagner. 1987. Production of biosurfactants. pp. 89-120. *In* N. Kosaric, W.L. Cairns, N.C.C. Gray, (ed.) Biosurfactants and biotechnology Marcel Dekker, New York..

Szoka, F., and Papahadjopoulos, D. 1980. Comparative properties and methods of preparation of lipid vesicles (liposomes). Ann. Rev. Biophys. Bioeng. 9:467-508.

Thai, L.T., and W.J. Maier. 1992. Solubilization and biodegradation of octadecane in the presence of two commercial surfactants. Proceedings of the 47th Annual Purdue University Industrial Waste Conference.

Thai, L.T. 1993. Solubilization and biodegradation of octadecane in the presence of nonionic surfactants. M.S. Thesis, University of Minnesota.

Thomas, J.M., J.R. Yordy, J.A. Amador, and M. Alexander. 1986. Rates of dissolution and biodegradation of water insoluble-organic compounds. Appl. Environ. Microbiol. 52:290-296.

Tiehm, A. 1994. Degradation of polycyclic aromatic hydrocarbons in the presence of synthetic surfactants. Appl. Environ. Microbiol. 60:258-263.

Verschueren, K. 1983. Handbook of environmental data on organic chemicals. Van Nostrand Rheinhold, New York.

Van Dyke, M.I., S.L. Gulley, H. Lee and J.T. Trevors. 1993. Evaluation of microbial surfactants for recovery of hydrophobic pollutants from soil. J. Indust. Microbiol. 11:163-170.

West, C.C. 1992. Surfactant-enhanced solubilization of tetrachloroethylene and degradation products in pump and treat remediation. p.182-191. In D.A. Sabatini and R.C. Knox (ed) Transport and Remediation of subsurface Contaminants: Colloidal, Interfacial and Surfactant Phenomena. ACS Symposium Series No. 491. American Chemical Society, Washington DC.

Whitworth, D.A., M. Moo-Young, and T. Viswanatha. 1973. Hydrocarbon fermentations: oxidation mechanism and nonionic-surfactant effects in a culture of *Candida lipolytica*. Biotech. Bioeng. 15:649-675.

Wodzinski, R.S., and Bertolini, D. 1972. Physical state in which naphthalene and bibenzyl are utilized by bacteria. Applied Microbiol. 23:1077-1081.

Wodzinski, R.S., and Coyle, J.E. 1974. Physical state of phenanthrene for utilization by bacteria. Applied Microbiol. 27:1081-1084.

Wodzinski, R.S., and Larocca, D. 1977. Bacterial growth kinetics on diphenylmethane and naphthalene-heptamethylnonane mixtures. Appl. Environ. Microbiol. 33:660-665.

Zhang, Y., and R.M. Miller. 1992. Enhanced octadecane dispersion and biodegradation by a *Pseudomonas* rhamnolipid surfactant (biosurfactant). Appl. Environ. Microbiol. 58:3276-3282.

Zhang, Y., and R.M. Miller. 1994. Effect of a *Pseudomonas* rhamnolipid biosurfactant on cell hydrophobicity and biodegradation of octadecane. Appl. Environ. Microbiol. 60: 2101-2106.

4 Redox and Energy Aspects of Soil Bioremediation[1]

R.F. Harris
Department of Soil Science
University of Wisconsin-Madison
Madison, Wisconsin

S.M. Arnold
Environmental Toxicology Center,
University of Wisconsin-Madison,
Madison, Wisconsin

Biogeochemical cycling of elements in terrestrial ecosystems is driven by the chlorophyll mediated extraction of low energy electrons from the oxygen in water to release fully oxidized oxygen gas, followed by phototrophic elevation of the energy state of the electrons to an energy level sufficiently high to reduce fully oxidized carbon in the form of carbon dioxide/bicarbonate to a high energy, intermediate oxidation organic carbon state. The process is called oxygenic photosynthesis, and organisms that use it are called phototrophs. Photosynthetic terrestrial plants tend to favor the fully oxidized forms of nitrogen (nitrate), sulfur (sulfate), and phosphorus (phosphate) as macroelement body building nutrient sources (Fig. 4-1). The photosynthetically produced organic carbon supplies high energy electrons to reduce nitrate to ammonium prior to incorporation into the amino and heterocyclic N organic monomer components of proteins, nucleic acids, and lipids and other N containing polymers of the cell. Similarly, sulfate is reduced to the sulfide level in the cysteine and methionine organic components of cell proteins, although phosphorus remains fully oxidized in all the nucleic acid, phospholipids and other organic phosphate-containing monomers and polymers.

Following plant death, the organic residues enter the soil and are attacked by chemotrophic soil organisms as a source of food (Fig. 4-1). The carbohydrate, lignin, protein, nucleic acid, lipid and other polymers are depolymerized to release their component sugar, aromatic, amino acid, purine and pyrimidine, N-base, glycerol, phosphatidic acid and simple fatty acid monomers. These monomers are used by the soil organisms in part as a source of body building material (anabolic, assimilatory, immobilization reactions), and in part as energy sources for growth and maintenance (catabolic, dissimilatory, mineralization reactions). The energy reactions involve controlled catalytic removal of relatively high energy electrons from organic carbon

[1]Supported by the College of Agricultural and Life Sciences Hatch project 3581 and the Wisconsin Water Resources Center, University of Wisconsin-Madison.

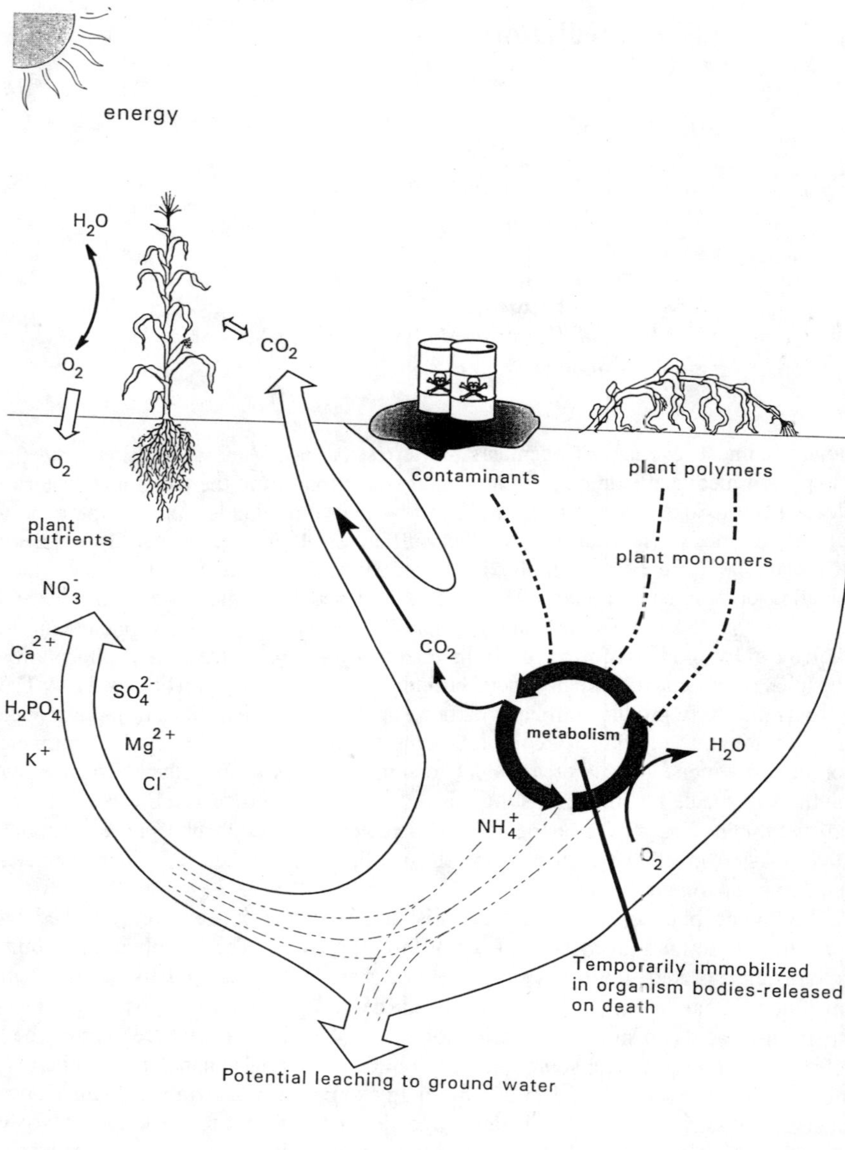

Fig 4-1. Natural nutrient and contaminant cycling in soil.

to release fully oxidized carbon dioxide. Sulfur is released in varying states of oxidation, and nitrogen in the fully reduced ammonium state; organic phosphate is mineralized to release inorganic phosphate in its unchanged fully oxidized state. Reduced inorganic sulfur compounds and ammonium are ultimately used as electron donors for autotrophic growth, with the release of fully oxidized sulfur as sulfate and nitrogen as nitrate. These reactions complete the C, P, S, and N biogeochemical cycles. From a groundwater protection standpoint, it is desirable to couple mineralization release of inorganic nutrients with plant uptake to minimize downward losses of nutrients from the ecosystem (Fig. 4-1).

In aerobic soil, the electron acceptor for all the energy reactions is fully oxidized oxygen in the form of gaseous dioxygen, which is converted to the fully reduced water state, thereby completing the oxygen cycle (Fig. 4-1). In the temporary absence of oxygen as electron acceptor, the oxidized forms of other elements may serve as electron acceptors for energy generation. Microbial choices for alternate assimilatory and energy nutrients are commonly dictated by the sometimes conflicting rules of maximized growth yields and minimized metabolic complexity.

Contaminants enter terrestrial ecosystems in a variety of chemical forms and redox states. The objective of bioremediation is to manipulate the ecosystem in such a way as to promote the soil organism bioreactor to integrate the contaminants into the natural biogeochemical cycles and thereby render them harmless. Basic approaches currently used to promote bioremediation include i) *In situ* activation of the indigenous organisms for accelerated target contaminant degradation by addition of growth limiting dissimilatory (electron donor and/or electron acceptor) and/or assimilatory nutrients, ii) *In situ* inoculation of indigenous populations with selectively enriched organisms of enhanced contaminant degradation capability, with or without supplementary nutrient amendment, and iii) *Ex situ* establishment of nutrient and environment controlled bioreactor populations of genetically improved and/or selectively enriched single organisms or consortia of organisms capable of accelerated degradation of target contaminants. Successful development and implementation of these approaches is facilitated by an understanding of the forces driving ecosystem microbial ecology and related biochemical transformations. A major objective of this chapter is to provide, within the context of conventional redox and energy concepts and terminology (Zumdahl, 1989; Harris, 1982; Chang, 1984; Tables 4-1a,b and 4-2a,b,c), a simplified redox and energy framework for understanding and predicting these driving forces. Emphasis will be placed on presentation and interpretation of the predictive implications of a redox and energy data base for biogeochemical cycling of the elemental components of nutrients and contaminants.

Organizationally, the introduction has addressed the interfacing of nutrient element transformations and contaminant degradation in biogeochemical cycles, with recognition of the range of redox states shown by the components of the immobilization and mineralization processes involved in the cycles. General aspects of the redox nature of assimilatory and dissimilatory reactions, and the competitive driving forces controlling their partitioning and integration to yield a characteristic growth yield and associated substrate consumption and product release, are summarized in Fig. 4-2. Rules for assigning and deriving oxidation states (numbers) of elements (Table 4-1a) and calculating reductance degree, and available electron

Table 4–1a. Redox groundrules and equations: Oxidation numbers (states) of elements in nutrients, biomass and contaminants.

A. Literal interpretation of oxidation number of an element and oxidation number semantics

Literally, the oxidation number of an element is the number of valence electrons assigned to ("possesed" by or "bound" to) the element relative to its base elemental form of assigned oxidation number 0.

1. A gain of 1, 2, 3, or 4 valence e^- by an element (caused by a gain of e^- by ionization to form an anion or by assignment of shared e^- in covalent bonds to the element from a less electronegative element (commonly H)), results in assignment of an oxidation number of –I, –II, –II, or –IV, respectively, to the element.
2. A loss of 1, 2, 3, 4, 5, 6, or 7 valence e^- by an element (caused by a loss of e^- by ionization to form a cation or by assignment of shared e^- in covalent bonds to a more electronegative element (commonly O)), results in assignment of an oxidation number of I, II, III, IV, V, VI, or VII, respectively, to the element.
3. A gain of valence e^- by an element by any mechanism that is counterbalanced exactly by a loss of valence e^- by any mechanism results in assignment of an oxidation number of 0 to the element.

Basic rules

B. By convention, the oxidation number of an element in its elemental form is assigned a value of zero,

Oxidation number of element M in elemental form = 0

Using this equation,

H in H_2 = 0, O in O_2 = 0, N in N_2 = 0, S in elemental S = 0, etc.

C. The ionic charge on a material plus the sum of the weighted oxidation numbers of the elements in the material is zero (where the weighted qualification recognizes the need to include the stoichiometric coefficients of the elements in the calculation). This gives rise to the equation,

Oxidation number of element M in material =
(sum charges – sum weighted oxidation numbers of other elements)/stoichiometric coefficient of M

Using this equation,

1. The oxidation number of an element in its ionically charged mono–element form is: H in H^+ = I, Mn in Mn^{2+} = II, Fe in Fe^{3+} = III, O in O_2^- = –0.5, Cl in Cl^- = –I, S in $S^=$ = –II, etc.
2. The oxidation number of an electron for half reaction balancing purposes is: e in e^- = –I

Rules generally consistent with the relative electronegativity series (Cotton & Wilkinson, 1972):
O (3.44) > Cl (3.16) > N > (3.04) > S (2.58) > Se (2.55) ≥ C (2.55) >
W (2.36 > Pb (2.33) > Re (2.28 > H (2.20) >
P (2.19) > As (2.18) > Mo, Te, Sb, Hg, Sn, Cu, Co, Fe, Cd, Cr, V, Mn (2.16 – 1.55)

D. For most biogeochemicals,

1. The oxidation number of H = I (for exceptions, see Table 4–2a)
2. The oxidation number of O = –II (for exceptions, see Table 4–2a)

E. For most organic metabolites, biopolymers and related biomass,

1. The oxidation number of P = V
2. The oxidation number of N = –III
3. The oxidation number of Cl = –I
4. The average oxidation number of S = –II

F. For other common components of available electron half reactions of biogeochemicals (consistent with B–D, above),

1. The oxidation numbers of H and O in H_2O are: H = I, O = –II
2. The oxidation numbers of H and N in NH_4^+/NH_3 are: H = I, N = –III
3. The oxidation numbers of H, O and P in $H_3PO_4/H_2PO_4^-/HPO_4^=$ are: H = I, O = –II, P = V
4. The oxidation numbers of H and S in $H_2S/HS^-/S^=$: H = I, S = –II

Table 4–1b. Redox groundrules and equations: Reductance degree, available electron content and energy potential of nutrients, biomass and contaminants containing a given target element.

A. Dominant form of a material at biological pH

The mass balance and energetics of chemical reactions are a function of the ionic form of the reaction components. The dominant form at pH 7 is derived from (e.g., Harris, 1982):

Log([protonated]/[deprotonated]) = pKa − 7

pKa = (ΔGf°(deprotonated) − ΔGf°(protonated)/5.7079

where square brackets identify concentration, pKa is −log Ka (H^+ ionization constant), ΔGf° is the standard (unit activity) Gibbs energy of formation of the protonated or deprotonated compound, and 5.7079 is 2.303 (ln to log conversion factor) * R (gas constant) * T (standard 298.15 °K temperature).

B. Available electron content (av e^-) of a material containing a specified target element

The available e^- content (av e^-) of an element M–containing material is defined as the number of M–bound e^- in the material (Harris & Adams, 1979; Harris, 1982)). Options for its derivation are,

1. As a function of the reductance degree of element M in the material,

   Reductance degree of element M in material =
   oxidation number of most oxidized form of M − oxidation number of M in material

   *Available electron content of material = element M–bound electron content of material = reductance degree of element M in material * stoichiometric coefficient of M in material*

2. As a function of the available e^- half reaction of the material (see C, below),

   Available electron content of material =
   stoichiometric coefficient for the e^- in the available e^- half reaction for the material

C. Available electron half reaction of a material containing a specified target element

The available e^- half reaction of an element M–containing material is the oxidation half reaction in which the material as e^- donor is converted to the most oxidized form of M and M is the only element that shows a change in oxidation number in the reactant compared to the product side of the half reaction. To derive the available e^- half reaction (see Table 4–5a, part A for basic components),

1. Set the stoichiometric coefficient for the material (e^- donor) at unity, and mass balance for target element M to obtain the coefficient for the product representing the most oxidized form of M
2. Mass balance for any non redox active elements (other than H and O) to obtain their coefficient(s)
3. Mass balance for O to obtain the coefficient for H_2O
4. Mass balance for H to obtain the coefficient for H^+
5. Balance the ionic charges to obtain the coefficient for e^-
6. Cross check (using Table 4–1a rules) that the target element M is the only element that shows a difference in oxidation number in the reactant as compared to the product side of the available e^- half reaction.

D. Available electron energy potential ($Eh^{07}c$) of a material

Technically, $Eh^{07}c$ is the standard (unit activity) biological (pH 7) reduction potential for the reversal of the available e^- half reaction for the material; i.e., it is the standard pH 7 potential of the oxidized side of the reaction to accept electrons and become converted to the reduced side of the half reaction, relative to the potential of the standard (pH 0) $H^+/½H_2$ couple to accept electrons.

$Eh^{07}c$ is calculated from the ΔGf° and ΔGf^7 H^+ of the components of the available e^- half reaction (remembering that this is an oxidation rather than a conventional reduction half reaction),

$$Eh^{07}c = \{(\text{weighted } \Delta Gf^0 \text{ products} + \text{weighted } \Delta Gf^7\, H^+) - (\text{weighted } \Delta Gf^0 \text{ reactants})\}/nF$$

where n = number of electrons in the half reaction (av e^- content of material), F = Faraday (96.487 kJ•V^{-1}•mole), and the weighted qualification recognizes the need to include the stoichiometric coefficients of the half reaction components in the calculation. By convention, ΔGf° e^- = 0 for energy calculations.

Table 4–2a. Redox and energy properties of biogeochemicals: Representative simple (H and/or O) hydrogen and oxygen monomers.

Symbol	Target element (M) — Target element–containing material†		pH 7 dominant form						Ref. §§
	Name (most reduced to most oxidized forms of element M)	Formula of proton–ionizable forms	Formula weight (g/mole)	State‡	Gibbs energy of formation ΔGf^0 (kJ/mole)	Redox, available electron content and energy properties: Mass and charge balanced half reaction identifying oxidation no. of elements above symbol §, available electron content of material ¶, and conc. (brackets) of reaction components#	Reductance degree of M †† (e^-/M)	Available electron energy ‡‡ $Eh^{07}c$ (e^-V)	
H	Hydride	H^-	2.016	aq	217.15	$\underset{[1\,M]}{\overset{(-I)}{H^-}} = \underset{[10^{-7}M]}{\overset{(I)}{H^+}} + \overset{(-I)}{2e^-}$	2	−1.3323	1
	Hydrogen	H_2	2.016	g	0	$\underset{[1\,Atm]}{\overset{0}{H_2}} = \underset{[1\,M]}{\overset{(I)}{2H^+}} + \overset{(-I)}{2e^-}$	1	0.0000	2
			2.016	g	0	$\underset{[1\,Atm]}{\overset{0}{H_2}} = \underset{[10^{-7}M]}{\overset{(I)}{2H^+}} + \overset{(-I)}{2e^-}$	1	−0.4141	2
	Proton	H^+	1.008	aq	0	$\underset{[1\,M]}{\overset{(I)}{H^+}}$	0	–	2
			1.008	aq	−39.956 ¶¶	$\underset{[10^{-7}M]}{\overset{(I)}{H^+}}$	0	–	2

– continued –

Table 4–2a. Continued.

Symbol	Target element (M) — Target element–containing material — Name (most reduced to most oxidized forms of element M)	Formula of proton–ionizable forms	pH 7 dominant form † — Formula weight (g/ mole)	State ǂ	Gibbs energy of formation ΔGf^0 (kJ/mole)	Redox, available electron content and energy properties — Mass and charge balanced half reaction identifying oxidation no. of elements above symbol §, available electron content of material ¶, and conc. (brackets) of reaction components #	Reductance degree of M †† (e^-/M)	Available electron energy ǂǂ $Eh^{07}c$ (e^-V)	Ref. §§
O	Water	H_2O	18.015	l	–237.19	$\overset{I\ (-II)}{H_2O} = 0.5\overset{0}{O_2} + 2\overset{(I)}{H^+} + 2\overset{(-I)}{e^-}$	2	0.8150	2
	Hydrogen peroxide	$H_2O_2/HO_2^-/O_2^=$	34.015	aq	–134.10	$\overset{I\ (-I)}{H_2O_2} = \overset{0}{O_2} + 2\overset{(I)}{H^+} + 2\ \overset{(-I)}{e^-}$	1	0.2808	4
	Superoxide	HO_2/O_2^-	31.999	aq	28.87	$\overset{(-0.5)}{O_2^-} = \overset{0}{O_2} + \overset{(-I)}{e^-}$	0.5	–0.2992	4
	Dioxygen	O_2	31.999	g	0.00	$\overset{0}{O_2}$	0	–	2

† ¶ †† ǂǂ Derived in accordance with Table 4–1b.

ǂ Standard abbreviations for physical state: aq = aqueous, c = crystalline, g = gas, l = liquid, s = solid.

§ For hydrogen: oxidation number of H in H_2 = 0, from rule 1, Table 4–1a; oxidation numbers in parentheses are derived using rule 2, Table 4–1a.
For oxygen: oxidation number of O in O_2 = 0 from rule 1, Table 4–1a; oxidation number of H in H_2O = I, from rule 3, Table 4–1a, oxidation numbers in parentheses are derived using rule 2, Table 4–1a.

Standard concentrations (aqueous, moles/liter; gas, atmospheres) except for pH 7, or noted otherwise.

§§ References for ΔGf^0: 1, Latimer (1952); 2, Harris (1982); 3, Sadiq and Lindsay (1979); 4, Thauer et al. (1977); 5, Wagman et al. (1968). Data expressed in kcals were converted to kJoules using the standard 4.184 conversion factor.

¶¶ Derived from the standard concentration dependence of ΔGf: $\Delta Gf^7 H^+ = \Delta Gf^0 H^+ + RT\ln[10^{-7}] = 0 + 5.7079 \times \log[10^{-7}]$

Table 4–2b. Redox and energy properties of biogeochemicals: Representative simple (element and/or H or O) monomers of biologically redox active elements.

Symbol	Name (most reduced to most oxidized forms of target element)	Formulae of major proton–ionizable forms	Formula weight (g/mole)	State‡	Gibbs energy of formation ΔGf^0 (kJ/mole)	Mass and charge balanced half reaction identifying oxidation numbers of elements (above symbol) §, available electron content of material ¶, and concentration of reaction components #	Reductance degree of M †† (e^-/M)	Available electron energy ‡‡ $Eh^{07}c$ (e^-V)	Ref.§§
		Target element (M) –containing material; pH 7 dominant form †							
As	Arsine	AsH_3	77.95	g	68.91	$\overset{(-III)}{As}\overset{I}{H_3} + 4\overset{I}{H_2}\overset{-II}{O} = \overset{I}{H_2}\overset{(V)}{As}\overset{-II}{O_4^-} + 9\overset{I}{H^+} + 8\overset{-I}{e^-}$	8	−0.3019	5
	Arsenic	As	74.92	c	0.00	$\overset{0}{As} + 4\overset{I}{H_2}\overset{-II}{O} = \overset{I}{H_2}\overset{(V)}{As}\overset{-II}{O_4^-} + 6\overset{I}{H^+} + 5\overset{-I}{e^-}$	5	−0.0917	5
	Arsenite	$H_3AsO_3/H_2AsO_3^-/HAsO_3^=$	125.94	aq	−639.90	$\overset{I}{H_3}\overset{(III)}{As}\overset{-II}{O_3} + \overset{I}{H_2}\overset{-II}{O} = \overset{I}{H_2}\overset{(V)}{As}\overset{-II}{O_4^-} + 3\overset{I}{H^+} + 2\overset{-I}{e^-}$	1	0.0204	5
	Arsenate	$H_3AsO_4/H_2AsO_4^-/HAsO_4^=$	140.94	aq	−753.29	$\overset{I}{H_2}\overset{(V)}{As}\overset{-II}{O_4^-}$	0	–	5
C	Methane	CH_4	16.04	g	−50.75	$\overset{(-IV)}{C}\overset{I}{H_4} + 2\overset{I}{H_2}\overset{-II}{O} = \overset{(IV)}{C}\overset{-II}{O_2} + 8\overset{I}{H^+} + 8\overset{-I}{e^-}$	8	−0.2386	4
						$\overset{(-IV)}{C}\overset{I}{H_4} + 3\overset{I}{H_2}\overset{-II}{O} = \overset{I}{H}\overset{(IV)}{C}\overset{-II}{O_3^-} + 9\overset{I}{H^+} + 8\overset{-I}{e^-}$	8	−0.2447	
	Ethane	C_2H_6	30.07	g	−32.89	$\overset{(-III)}{C_2}\overset{I}{H_6} + 4\overset{I}{H_2}\overset{-II}{O} = 2\overset{(IV)}{C}\overset{-II}{O_2} + 14\overset{I}{H^+} + 14\overset{-I}{e^-}$	7	−0.2713	4
	Formaldehyde	CH_2O	30.03	aq	−130.54	$C\overset{I}{H_2}\overset{-II}{O} + \overset{I}{H_2}\overset{-II}{O} = \overset{(IV)}{C}\overset{-II}{O_2} + 4\overset{I}{H^+} + 4\overset{-I}{e^-}$	4	−0.4831	4
	Formate	$H_2CO_2/HCO_2^-/CO_2^=$	45.02	aq	−351.04	$\overset{I}{H}\overset{(II)}{C}\overset{-II}{O_2^-} = \overset{(IV)}{C}\overset{-II}{O_2} + \overset{I}{H^+} + 2\overset{-I}{e^-}$	2	−0.4316	4
	Bicarbonate	$H_2CO_3/HCO_3^-/CO_3^=$	91.13	aq	−586.89	$\overset{I}{H}\overset{(IV)}{C}\overset{-II}{O_3^-}$	0	–	2
	Carbon dioxide	CO_2	44.01	g	−394.38	$\overset{(IV)}{C}\overset{-II}{O_2}$	0	–	2
Cd	Cadmium	Cd	112.40	c	0.00	$\overset{0}{Cd} = \overset{(II)}{Cd^{2+}} + 2\overset{-I}{e^-}$	2	−0.4035	3
	Cadmium^{2+}	Cd^{2+}	112.40	aq	−77.86	$\overset{(II)}{Cd^{2+}}$	0	–	3

– continued –

Table 4–2b. Continued.

Symbol	Name (most reduced to most oxidized forms of target element)	Formulae of major proton–ionizable forms	Formula weight (g/ mole)	State ‡	Gibbs energy of formation ΔGf^0 (kJ/mole)	Mass and charge balanced half reaction identifying oxidation numbers of elements (above symbol) §, available electron content of material ¶, and concentration of reaction components #	Reduc– tance degree of M †† (e^-/M)	Available electron energy‡‡ $Eh^{07}c$ (e^-V)	Ref.§§
			Target element (M) –containing material			pH 7 dominant form†			
Cl	Chloride	HCl/Cl^-	35.45	aq	−131.26	$\overset{(-I)}{Cl^-} + 4\,\overset{I\ -II}{H_2O} = \overset{(VII)\ -II}{Cl\,O_4^-} + 8\,\overset{I}{H^+} + 8\,\overset{-I}{e^-}$	8	0.9739	3
	Chlorine	Cl_2	70.91	g	0.00	$\overset{0}{Cl_2} + 8\,\overset{I\ -II}{H_2O} = 2\,\overset{(VII)\ -II}{Cl\,O_4^-} + 16\,\overset{I}{H^+} + 14\,\overset{-I}{e^-}$	7	0.9187	3
	Chlorate	$HClO_4/ClO_4^-$	99.45	aq	−8.62	$\overset{(VII)\ -II}{Cl\,O_4^-}$	0	–	5
Co	Cobalt	Co	58.93	c	0.00	$\overset{0}{Co} = \overset{(III)}{Co^{3+}} + 3\,\overset{-I}{e^-}$	3	0.4625	5
	Cobalt^{2+}	Co^{2+}	58.93	aq	−54.39	$\overset{(II)}{Co^{2+}} = \overset{(III)}{Co^{3+}} + \overset{-I}{e^-}$	1	1.9513	5
	Cobalt^{3+}	Co^{3+}	58.93	aq	133.89	$\overset{(III)}{Co^{3+}}$	0	–	5
Cr	Chromium	Cr	52.00	c	0.00	$\overset{0}{Cr} + 4\,\overset{I\ -II}{H_2O} = \overset{(VI)\ -II}{Cr\,O_4^=} + 8\,\overset{I}{H^+} + 6\,\overset{-I}{e^-}$	6	−0.1705	5
	Chromium^{2+}	Cr^{2+}	52.00	aq	−176.15	$\overset{(II)}{Cr^{2+}} + 4\,\overset{I\ -II}{H_2O} = \overset{(VI)\ -II}{Cr\,O_4^=} + 8\,\overset{I}{H^+} + 4\,\overset{-I}{e^-}$	4	0.2006	1
	Chromium^{3+}	Cr^{3+}	52.00	aq	−215.48	$\overset{(III)}{Cr^{3+}} + 4\,\overset{I\ -II}{H_2O} = \overset{(VI)\ -II}{Cr\,O_4^=} + 8\,\overset{I}{H^+} + 3\,\overset{-I}{e^-}$	3	0.4033	1
	Chromate$^=$	$Cr\,O_4^=$	116.00	aq	−727.85	$\overset{(VI)\ -II}{Cr\,O_4^=}$	0	–	5
Cu	Copper	Cu	63.55	c	0.00	$\overset{0}{Cu} = \overset{(II)}{Cu^{2+}} + 2\,\overset{-I}{e^-}$	2	0.3398	3
	Cuprous	Cu^+	63.55	aq	50.63	$\overset{(I)}{Cu^+} = \overset{(II)}{Cu^{2+}} + \overset{-I}{e^-}$	1	0.1548	3
	Cupric	Cu^{2+}	63.55	aq	65.56	$\overset{(II)}{Cu^{2+}}$	0	–	3

– continued –

Table 4–2b. Continued.

Symbol	Name (most reduced to most oxidized forms of target element)	Formulae of major proton–ionizable forms	Formula weight (g/ mole) mole)	State‡	Gibbs energy of formation ΔGf^0 (kJ/mole) (kJ/mole)	Mass and charge balanced half reaction identifying oxidation numbers of elements (above symbol) §, available electron content of material ¶, and concentration of reaction components #	Reduc– tance degree of M †† (e^-/M) (e^-/M)	Available electron energy‡‡ $Eh^{07}c$ (e^-V) (e^-V)	Ref. §§
						Target element (M) / Target element (M) –containing material / pH 7 dominant form †			
Fe	Iron	Fe	55.85	c	0.00	$\overset{0}{Fe} + 3\,\overset{I}{H_2}\overset{-II}{O} = \overset{(III)}{Fe}(\overset{-II}{O}\overset{I}{H})_3 + 3\,\overset{I}{H^+} + 3\,\overset{-I}{e^-}$	3	−0.4023	3
	Ferrous	Fe^{2+}	55.85	aq	−91.21	$\overset{(II)}{Fe^{2+}} + 3\,\overset{I}{H_2}\overset{-II}{O} = \overset{(III)}{Fe}(\overset{-II}{O}\overset{I}{H})_3 + \overset{I}{H^+} + \overset{-I}{e^-}$	1	−0.2614	3
						$\overset{(II)}{Fe^{2+}} = \overset{(III)}{Fe^{3+}} + \overset{-I}{e^-}$	1	0.7710	
	Ferric hydroxide	$Fe(OH)_3$	106.87	c	−708.14	$\overset{(III)}{Fe}(\overset{-II}{O}\overset{I}{H})_3$	0	–	3
	Ferric	Fe^{3+}	55.85	aq	−16.82	$\overset{(III)}{Fe^{3+}}$	0	–	3
Hg	Mercury	Hg	200.59	c	0.00	$\overset{0}{Hg} = \overset{(II)}{Hg^{2+}} + 2\,e^-$	2	0.8535	3
	Dimercurous	Hg_2^{2+}	401.18	aq	153.61	$\overset{(I)}{Hg_2^{2+}} = 2\,\overset{(II)}{Hg^{2+}} + 2\,e^-$	1	0.9110	3
	Mercuric	Hg^{2+}	200.59	aq	164.70	$\overset{(II)}{Hg^{2+}}$	0	–	3
Mn	Manganese	Mn	54.94	c	0.00	$\overset{0}{Mn} + 4\,\overset{I}{H_2}\overset{-II}{O} = \overset{(VII)}{Mn}\overset{-II}{O_4^-} + 8\,\overset{I}{H^+} + 7\,\overset{-I}{e^-}$	7		3
	Manganous	Mn^{2+}	54.94	aq	−230.58	$\overset{(II)}{Mn^{2+}} + 4\,\overset{I}{H_2}\overset{-II}{O} = \overset{(VII)}{Mn}\overset{-II}{O_4^-} + 8\,\overset{I}{H^+} + 5\,\overset{-I}{e^-}$	5	0.8483	3
	ß Mn dioxide	MnO_2	86.94	c	−465.85	$\overset{(IV)}{Mn}\overset{-II}{O_2} + 2\,\overset{I}{H_2}\overset{-II}{O} = \overset{(VII)}{Mn}\overset{-II}{O_4^-} + 4\,\overset{I}{H^+} + 3\,\overset{-I}{e^-}$	3	1.1399	3
	Permanganate	$HMnO_4/MnO_4^-$	118.94	aq	−450.45	$\overset{(VII)}{Mn}\overset{-II}{O_4^-}$	0	–	3

– continued –

Table 4–2b. Continued.

Symbol	Name (most reduced to most oxidized forms of target element)	Formulae of major proton–ionizable forms	Formula weight (g/mole)	State‡	Gibbs energy of formation ΔGf^0 (kJ/mole)	Mass and charge balanced half reaction identifying oxidation numbers of elements (above symbol) §, available electron content of material ¶, and concentration of reaction components #	Reduc–tance degree of M †† (e^-/M)	Available electron energy‡‡ $Eh^{07}c$ (e^-V)	Ref.§§
	Target element (M) –containing material; pH 7 dominant form †								
Mo	Molybdenum	Mo	95.94	c	0.00	$\overset{0}{Mo} + 4\,\overset{I}{H}_2\overset{-II}{O} = \overset{(VI)}{Mo}\overset{-II}{O}_4^{=} + 8\,\overset{I}{H}^+ + 6\,\overset{-I}{e}^-$	6	–0.3576	3
	Molybdate	$H_2MoO_4/HMoO_4^-/MoO_4^=$	159.94	aq	–836.13	$\overset{(VI)}{Mo}\overset{-II}{O}_4^{=}$	0	–	3
N	Ammonium	NH_4^+/NH_3	18.04	aq	–79.45	$\overset{(-III)}{N}\overset{I}{H}_4^+ + 3\,\overset{I}{H}_2\overset{-II}{O} = \overset{(V)}{N}\overset{-II}{O}_3^- + 10\,\overset{I}{H}^+ + 8\,\overset{-I}{e}^-$	8	0.3627	2
	Dinitrogen	N_2	28.01	g	0.00	$\overset{0}{N}_2 + 6\,\overset{I}{H}_2\overset{-II}{O} = 2\,\overset{(V)}{N}\overset{-II}{O}_3^- + 12\,\overset{I}{H}^+ + 10\,\overset{-I}{e}^-$	10	0.7470	2
	Nitrite	HNO_2/NO_2^-	46.01	aq	–37.70	$\overset{(III)}{N}\overset{-II}{O}_2^- + \overset{I}{H}_2\overset{-II}{O} = \overset{(V)}{N}\overset{-II}{O}_3^- + 2\,\overset{I}{H}^+ + 2\,\overset{-I}{e}^-$	2	0.4328	2
	Nitrate	HNO_3/NO_3^-	62.01	aq	–111.46	$\overset{(V)}{N}\overset{-II}{O}_3^-$	0	–	2
P	Phosphine	PH_3	33.40	g	13.39	$\overset{(-III)}{P}\overset{I}{H}_3 + 4\,\overset{I}{H}_2\overset{-II}{O} = \overset{I}{H}_2\overset{(V)}{P}\overset{-II}{O}_4^- + 9\,\overset{I}{H}^+ + 8\,\overset{-I}{e}^-$	8	–0.7276	3
	Phosphorus	P	30.97	c	0.00	$\overset{0}{P} + 4\,\overset{I}{H}_2\overset{-II}{O} = \overset{I}{H}_2\overset{(V)}{P}\overset{-II}{O}_4^- + 6\,\overset{I}{H}^+ + 5\,\overset{-I}{e}^-$	5	–0.8880	3
	Phosphite	$H_3PO_3/H_2PO_3^-/HPO_3^=$	79.98	aq	–799.44	$\overset{I}{H}\overset{(III)}{P}\overset{-II}{O}_3^{=} + \overset{I}{H}_2\overset{-II}{O} = \overset{I}{H}_2\overset{(V)}{P}\overset{-II}{O}_4^- + \overset{I}{H}^+ + 2\,\overset{-I}{e}^-$	2	–0.7294	3
	Phosphate	$H_3PO_4/H_2PO_4^-/HPO_4^=$	96.99	aq	–1137.42	$\overset{I}{H}_2\overset{(V)}{P}\overset{-II}{O}_4^-$	0	–	3
Pb	Lead	Pb	207.20	c	0.00	$\overset{0}{Pb} + 2\,\overset{I}{H}_2\overset{-II}{O} = \overset{(IV)}{Pb}\overset{-II}{O}_2 + 4\,\overset{I}{H}^+ + 4\,\overset{-I}{e}^-$	4	0.2566	3
	Plumbous	Pb^{2+}	207.20	aq	–24.69	$\overset{(II)}{Pb}{}^{2+} + 2\,\overset{I}{H}_2\overset{-II}{O} = \overset{(IV)}{Pb}\overset{-II}{O}_2 + 4\,\overset{I}{H}^+ + 2\,\overset{-I}{e}^-$	2	0.6412	3
	Lead dioxide	PbO_2	239.20	c	–215.52	$\overset{(IV)}{Pb}\overset{-II}{O}_2$	0	–	3

– continued –

Table 4–2b. Continued.

Symbol	Name (most reduced to most oxidized forms of target element)	Formulae of major proton–ionizable forms	Formula weight (g/mole)	State‡	Gibbs energy of formation $\Delta G f^0$ (kJ/mole)	Mass and charge balanced half reaction identifying oxidation numbers of elements (above symbol) §, available electron content of material ¶, and concentration of reaction components #	Reductance degree of M †† (e^-/M)	Available electron energy ‡‡ $Eh^{07}c$ (e^-V)	Ref.§§
		Target element (M) – containing material; pH 7 dominant form †							
R	Rhenium	Re	186.21	c	0.00	$\overset{0}{Re} + 4\,\overset{I}{H_2}\overset{-II}{O} = \overset{(VII)}{Re}\overset{-II}{O_4^-} + 8\,\overset{I}{H^+} + 7\,\overset{-I}{e^-}$	7	−0.0969	5
	Rhenium dioxide	ReO_2	218.21	c	−368.19	$\overset{(IV)}{Re}\overset{-II}{O_2} + 2\,\overset{I}{H_2}\overset{-II}{O} = \overset{(VII)}{Re}\overset{-II}{O_4^-} + 4\,\overset{I}{H^+} + 3\,\overset{-I}{e^-}$	3	−0.0407	5
	Rhenate	$HReSO_4/ReO_4^-$	250.20	aq	−694.54	$\overset{(VII)}{Re}\overset{-II}{O_4^-}$	0	–	5
S	Sulfide	$H_2S/HS^-/S^=$	34.08	aq	−27.91	$\overset{I}{H_2}\overset{(-II)}{S} + 4\,\overset{I}{H_2}\overset{-II}{O} = \overset{(VI)}{S}\overset{-II}{O_4^=} + 10\,\overset{I}{H^+} + 8\,\overset{-I}{e^-}$	8	−0.2169	3
	Sulfur	S	32.06	c	0.00	$\overset{0}{S} + 4\,\overset{I}{H_2}\overset{-II}{O} = \overset{(VI)}{S}\overset{-II}{O_4^=} + 8\,\overset{I}{H^+} + 6\,\overset{-I}{e^-}$	6	−0.1994	3
	Sulfite	$H_2SO_3/HSO_3^-/SO_3^=$	81.07	aq	−527.06	$\overset{I}{H}\overset{(IV)}{S}\overset{-II}{O_3^-} + \overset{I}{H_2}\overset{-II}{O} = \overset{(VI)}{S}\overset{-II}{O_4^=} + 3\,\overset{I}{H^+} + 2\,\overset{-I}{e^-}$	2	−0.5190	3
	Sulfate	$H_2SO_4/HSO_4^-/SO_4^=$	96.06	aq	−744.54	$\overset{(VI)}{S}\overset{-II}{O_4^=}$	0	–	3
Sb	Stibine	SbH_3	124.77	g	147.74	$\overset{(-III)}{Sb}\overset{I}{H_3} + 1.5\,\overset{I}{H_2}\overset{-II}{O} = 0.5\,\overset{(V)}{Sb_2}\overset{-II}{O_5} + 8\,\overset{I}{H^+} + 8\,\overset{-I}{e^-}$	8	−0.3745	5
	Antimony	Sb	121.75	c	0.00	$\overset{0}{Sb} + 2.5\,\overset{I}{H_2}\overset{-II}{O} = 0.5\,\overset{(V)}{Sb_2}\overset{-II}{O_5} + 5\,\overset{I}{H^+} + 5\,\overset{-I}{e^-}$	5	−0.0444	5
	Antim. hydroxide	$Sb(OH)_3$	172.77	c	−684.00	$\overset{(III)}{Sb}(\overset{-II}{O}\overset{I}{H})_3 - 0.5\,\overset{I}{H_2}\overset{II}{O} = 0.5\,\overset{(V)}{Sb_2}\overset{II}{O_5} + 5\,\overset{I}{H^+} + 2\,\overset{I}{e^-}$	2	0.3672	5
	Antimony oxide	Sb_2O_5	323.50	c	−829.27	$\overset{(V)}{Sb_2}\overset{-II}{O_5}$	0	–	5
Se	Selenide	$H_2Se/HSe^-/Se^=$	79.97	aq	43.93	$\overset{I}{H}\overset{(-II)}{Se^-} + 4\,\overset{I}{H_2}\overset{-II}{O} = \overset{(VI)}{Se}\overset{-II}{O_4^=} + 9\,\overset{I}{H^+} + 8\,\overset{-I}{e^-}$	8	0.1345	5
	Selenium	Se	78.96	c	0.00	$\overset{0}{Se} + 4\,\overset{I}{H_2}\overset{-II}{O} = \overset{(VI)}{Se}\overset{-II}{O_4^=} + 8\,\overset{I}{H^+} + 6\,\overset{-I}{e^-}$	6	0.3242	5

– continued –

Table 4–2b. Continued.

Symbol	Name (most reduced to most oxidized forms of target element)	Formulae of major proton–ionizable forms	Formula weight (g/mole)	State ‡	Gibbs energy of formation ΔGf^0 (kJ/mole)	Mass and charge balanced half reaction identifying oxidation numbers of elements (above symbol) §, available electron content of material ¶, and concentration of reaction components #	Reductance degree of M †† (e^-/M)	Available electron energy ‡‡ $Eh^{07}c$ (e^-V)	Ref. §§
		Target element (M) –containing material / pH 7 dominant form †							
	Selenite	$H_2SeO_3/HSeO_3^-/SeO_3^{=}$	126.96	aq	−369.87	$\overset{(IV)}{\mathrm{Se}}\,\overset{-II}{\mathrm{O}_3^{=}} + \overset{I}{\mathrm{H}_2}\overset{-II}{\mathrm{O}} = \overset{(VI)}{\mathrm{Se}}\,\overset{-II}{\mathrm{O}_4^{=}} + 2\,\overset{I}{\mathrm{H}^+} + 2\,\overset{-I}{e^-}$	2	0.4443	5
	Selenate	$H_2SeO_4/HSeO_4^-/SeO_4^{=}$	142.96	aq	−441.41	$\overset{(VI)}{\mathrm{Se}}\,\overset{-II}{\mathrm{O}_4^{=}}$	0	–	5
Sn	Tin	Sn	118.69	c	0.00	$\overset{0}{\mathrm{Sn}} + 2\,\overset{I}{\mathrm{H}_2}\overset{-II}{\mathrm{O}} = \overset{(IV)}{\mathrm{Sn}}\,\overset{-II}{\mathrm{O}_2} + 4\,\overset{I}{\mathrm{H}^+} + 4\,\overset{-I}{e^-}$	4	−0.5314	5
	Stannous oxide	SnO	134.69	c	−256.90	$\overset{(II)}{\mathrm{Sn}}\,\overset{-II}{\mathrm{O}} + \overset{I}{\mathrm{H}_2}\overset{-II}{\mathrm{O}} = \overset{(IV)}{\mathrm{Sn}}\,\overset{-II}{\mathrm{O}_2} + 2\,\overset{I}{\mathrm{H}^+} + 2\,\overset{-I}{e^-}$	2	−0.5466	5
	Stannic dioxide	SnO_2	150.69	c	−519.65	$\overset{(IV)}{\mathrm{Sn}}\,\overset{-II}{\mathrm{O}_2}$	0	–	5
Te	Tellurium	Te	127.60	c	0.00	$\overset{0}{\mathrm{Te}} + 3\,\overset{I}{\mathrm{H}_2}\overset{-II}{\mathrm{O}} = \overset{I}{\mathrm{H}_2}\overset{(IV)}{\mathrm{Te}}\,\overset{-II}{\mathrm{O}_3} + 4\,\overset{I}{\mathrm{H}^+} + 4\,\overset{-I}{e^-}$	4	0.6035	5
	Tellurium oxide	TeO	143.60	c	38.49	$\overset{(II)}{\mathrm{Te}}\,\overset{-II}{\mathrm{O}} + 2\,\overset{I}{\mathrm{H}_2}\overset{-II}{\mathrm{O}} = \overset{I}{\mathrm{H}_2}\overset{(IV)}{\mathrm{Te}}\,\overset{-II}{\mathrm{O}_3} + 2\,\overset{I}{\mathrm{H}^+} + 2\,\overset{-I}{e^-}$	2	0.1925	5
	Tellurate	$H_2TeO_3/HTeO_3^-/TeO_3^{=}$	177.61	aq	−318.82	$\overset{I}{\mathrm{H}_2}\overset{(IV)}{\mathrm{Te}}\,\overset{-II}{\mathrm{O}_3}$	0	–	5
W	Tungsten	W	183.85	c	0.00	$\overset{0}{\mathrm{W}} + 3\,\overset{I}{\mathrm{H}_2}\overset{-II}{\mathrm{O}} = \overset{(VI)}{\mathrm{W}}\,\overset{-II}{\mathrm{O}_3} + 6\,\overset{I}{\mathrm{H}^+} + 6\,\overset{-I}{e^-}$	6	−0.5048	5
	Tungsten dioxide	WO_2	215.85	c	533.92	$\overset{(IV)}{\mathrm{W}}\,\overset{-II}{\mathrm{O}_2} + \overset{I}{\mathrm{H}_2}\overset{-II}{\mathrm{O}} = \overset{(VI)}{\mathrm{W}}\,\overset{-II}{\mathrm{O}_3} + 2\,\overset{I}{\mathrm{H}^+} + 2\,\overset{-I}{e^-}$	2	−0.3777	5
	Tungsten trioxide	WO_3	231.85	c	−764.08	$\overset{(VI)}{\mathrm{W}}\,\overset{-II}{\mathrm{O}_3}$	0	–	5

† ¶ †† ‡‡ Derived in accordance with Table 4–1b.

§ Derived in accordance with Table 4–1a. Parentheses identify derived oxidation numbers. For simplicity, parentheses have been omitted for derived (Table 4–2a) oxidation numbers for e(−I) in e^-, H(I) in H^+, and O(−II) in H_2O.

‡ §§ As for Table 4–2a.

\# Concentrations are standard (gas, 1 atm; aqueous, 1 mole/liter) except for pH 7 conditions (H^+, [10^{-7} M].

Table 4–2c. Redox and energy properties of biogeochemicals: Selected complex organic (C, H and/or O, and another element) monomers, polymers and biomass.

Group	Name (In general, decreasing reductance degree within groups)	pH 7 dominant form † Formula weight FW (g/mole)	State ‡	Redox and available electron energy properties: Empirical formula identifying oxidation numbers of elements (above symbol) §	Available electrons: Reduct. deg. C ¶ (e^-/C)	Content # Av e^- (e^-/mole)	Energy †† $Eh^{07}c$ (e^-V)	Ref. ‡‡
Monomers	Methylamine	32.07	aq	$\overset{(-II)}{C}\overset{I}{H}_3\overset{-III}{N}\overset{I}{H}_3^{+}$	6	6	−0.3441	2
	Methylsulfide	48.11	g	$\overset{(-II)}{C}\overset{I}{H}_3\overset{-II}{S}\overset{I}{H}$	6	6	−0.3080	5
	Methylchloride	50.49	aq	$\overset{(-II)}{C}\overset{I}{H}_3\overset{-I}{Cl}$	6	6	−0.4828	5
	Dichloromethane	84.93	aq	$\overset{(0)}{C}\overset{I}{H}_2\overset{-I}{Cl}_2$	4	4	−0.9233	5
	Chloroform	119.38	aq	$\overset{(II)}{C}\overset{I}{H}\overset{-I}{Cl}_3$	2	2	−2.2965	5
	Atrazine	215.69	aq	$\overset{(0.25)}{C}_8\overset{I}{H}_{14}\overset{-III}{N}_5\overset{-I}{Cl}$	3.750	30	(−0.3)	6
Polymers §§	Phospholipid	691.17	aq	$\overset{(-1.622)}{C}_{37}\overset{I}{H}_{74}\overset{-II}{O}_8\overset{-III}{N}\overset{V}{P}$	5.622	208	(−0.31)	6
	Polyhydroxybutyrate	86.09	aq	$\overset{(-0.50)}{C}_4\overset{I}{H}_6\overset{-II}{O}_2$	4.500	18	(−0.31)	6
	Protein	82.39	aq	$\overset{(-0.167)}{C}_{3.6}\overset{I}{H}_{5.9}\overset{-II}{O}_{1.1}\overset{-III}{N}\overset{-II}{S}_{0.05}$	4.167	15	(−0.33)	6
	Carbohydrate	162.14	aq	$\overset{(0)}{C}_6\overset{I}{H}_{10}\overset{-II}{O}_5$	4.000	24	(−0.43)	6
	Deoxyribonucleic acid	82.53	aq	$\overset{(0.596)}{C}_{2.6}\overset{I}{H}_{3.3}\overset{-II}{O}_{1.6}\overset{-III}{N}\overset{V}{P}_{0.27}$	3.404	8.85	(−0.42)	6

– continued –

Table 4–2c. Continued

	Name (In general, decreasing reductance degree within groups)	Formula weight FW (g/mole)	State ‡	Empirical formula identifying oxidation numbers of elements (above symbol) §	Reduct. deg. C ¶ (e^-/C)	Content # Av e^- (e^-/mole)	Energy †† $Eh^{07}c$ (e^-V)	Ref. ‡‡
				pH 7 dominant form † — Redox and available electron energy properties	Available electrons			
	Ribonucleic acid	85.92	aq	$\overset{(0.940)}{C}_{2.5}\overset{I}{H}_{3.1}\overset{-II}{O}_{1.9}\overset{-III}{N}\overset{V}{P}_{0.27}$	3.060	7.65	(−0.45)	6
Biomass ¶¶	Fungal cells	220.21	s	$\overset{(-0.380)}{C}_{9}\overset{I}{H}_{15}\overset{-II}{O}_{4.5}\overset{-III}{N}\overset{V}{P}_{0.1}\overset{-II}{S}_{0.08}$ ...(97%)	4.380	39.42	(−0.36)	6
	Bacterial cells	103.45	s	$\overset{(-0.250)}{C}_{4}\overset{I}{H}_{7}\overset{-II}{O}_{1.5}\overset{-III}{N}$... (90%)	4.250	17	(−0.36)	6
		103.45	s	$\overset{(-0.335)}{C}_{4}\overset{I}{H}_{7}\overset{-II}{O}_{1.5}\overset{-III}{N}\overset{V}{P}_{0.08}\overset{-II}{S}_{0.03}$ ...(93.2%)	4.335	17.34	(−0.36)	6

† Derived in accordance with Table 4–1b.

‡ As for Table 4–2a.

§ Derived in accordance with Table 4–1a.

¶, #, †† Derived in accordance with Table 4–1b for available e^- half reactions in which C is the only redox active element (only element to show a change in oxidation state in the reactant as compared to the product side of the half reaction). CO_2(IV) is the oxidized e^- donor product, and organic N(−III), P(V), and S(−II) are released as NH_4^+(−III), $H_2PO_4^-$(V) and H_2S (−II), respectively, where the numbers in parentheses are the oxidation numbers of C, N, P, and S. From Table 4–1b, Reductance degree of C = 4 − oxidation number of C; Av e^- = Reductance degree of C * coefficient for C in material.

‡‡ ΔGf^0 as for Table 4–2a. Reference 6: $Eh^{07}c$ properties in parentheses were approximated (a) for atrazine from ΔH^0c = 5,300 cals/g (Weiss, 1986) --> ΔHf^0 --> $\Delta Hox = \Delta G^{07}ox$ (approx, Harris, 1982) --> $Eh^{07}c$, and (b) for polymers and biomass from ΔGf^{07} components (Harris and Adams, 1979; Harris, 1982; Harris, unpublished data), see below.

§§ Polymer monomers. Phospholipid: phosphatidyl (palmitic acid) ethanolamine. Protein: typical bacterial. Carbohydrate: glucose. RNA and DNA: standard statistical pentose sugar/purine and pyrimidine bases/phosphate.

¶¶ Fungal cells: 41% carbohydrate, 30% protein, 20% phospholipid, 5.8% RNA, 0.2% DNA, 3% other (5% ash).
Bacterial cells: 8% carbohydrate, 58% protein, 8% phospholipid, 16% RNA, 4% DNA, 6% other (10% ash).

content (ave e^-) and potential energy ($Eh^{07}c$) properties of biogeochemicals (Table 4-1b) are presented in stand alone form for easy reference. Application of these rules to characterize simple monomers of H and O (Table 4-2a), other biogeochemically redox active elements (Table 4-2b), and more complex organic compounds, biopolymers, and microbial biomass (Table 4-2c) is presented in a way clearly identifying the role of oxidation numbers in the derivation of the reductance degree, and available electron content and potential energy. Table 4-3 presents an oxidation state interpretation of methyl transfer reactions involving toxic elements. Rules for the use of key redox properties and chemical composition in deriving growth yields with respect to the electron donor are documented in Table 4-4, expanding particularly on redox aspects of assimilatory reactions and the energetics of dissimilatory electron transfer reactions introduced earlier in Fig. 4-2. Table 4-4 also includes consideration of the concentration dependence of electron transfer reactions; this, together with a Latimer diagram pathway approach to presenting redox and energy properties of biogeochemicals, is amplified on in Fig. 4-3. Lastly, drawing on concepts and principles established in the preceding figures and tables, rules for deriving mass balanced dissimilatory and growth reactions for predicting and tracking the nature and kinetics of nutrient and contaminant metabolism in soil are presented in Table 4-5a and 5b, respectively.

NUTRITIONAL NEEDS OF SOIL ORGANISMS

Nutrients and contaminants are potential candidates for consumption by the chemotrophic soil organism bioreactor (Fig. 4-1) for use as body building raw materials (assimilatory substrate) and/or as energy (dissimilatory substrate) sources (Fig. 4-2).

The object of assimilatory reactions is to produce the carbon, nitrogen, phosphorus, sulfur, and other nutrient element components of the cell biomass. The relative amounts of different elements immobilized into biomass, conveniently summarized in the form of empirical molar formulae (Fig. 4-2; Table 4-2c), varies between different organisms depending largely on the relative amounts of polymers they contain (Table 4-2c). The monomer and empirical molar composition of biopolymers is relatively constant across all life forms. Similarly, as identified in Tables 4-1b and 4-2b,c, the oxidation states (numbers) of biomass elements in major biopolymers such as carbohydrates, proteins, phospholipids and nucleic acids generally show little variation. Organic carbon in biomass and soil organic matter is close to an average oxidation state of about zero and biomass nitrogen behaves as though it is in the fully reduced -III oxidation state (Harris & Adams, 1979); phosphate is uniformly in the fully oxidized V oxidation state, and biomass sulfur is generally considered to be in the fully reduced -II oxidation state (Tables 4-1b and 4-2c; Brock & Madigan, 1991). Assimilatory nutrients vary in molar elemental composition and range in oxidation state from fully reduced to fully oxidized for C (-IV to IV), N (-III to V), and S (-II to VI) (Tables 4-2b,c; Brock & Madigan, 1991; Harris, 1982; Fenchel & Blackburn, 1979). Assimilatory substrate selection by chemotrophs recognizes that the sole purpose of the assimilatory reaction is substrate conversion to biomass, and tends to be dominated by the principle of maximized growth with minimum effort and conservation of electron donors (Harris, 1982). The object of dissimilatory reactions is to produce energy to drive energy consuming

Table 4–3. Redox aspects of methyl transfer reactions†.

Methyl transfer donors, carriers, and acceptors	Methyl group	Mass and charge balanced methyl transfer equations for pH 7 dominant forms‡, identifying the oxidation numbers of the elements (above symbol)§
Methyl donors/carriers		
N–methyl tetrahydrofolate	Carbonium$^+$	$\overset{II}{(R)}\,\overset{-III}{N}\,\overset{(-II)}{C}\,\overset{I}{H_3} + \overset{I}{H^+} = \overset{(II)}{(R)}\,\overset{-III}{N}\,\overset{I}{H} + \overset{(-II)}{C}\,\overset{I}{H_3^+}$
S–adenosylmethionine	Carbonium$^+$	$\overset{II}{(R)}\,\overset{-II}{S^+}\,\overset{(-II)}{C}\,\overset{I}{H_3} = \overset{II}{(R)}\,\overset{(-II)}{S} + \overset{(-II)}{C}\,\overset{I}{H_3^+}$
Methyl cobalamin	Carbanion$^-$	$\overset{-I}{(R)}\,\overset{III}{Co^+}\,\overset{(-IV)}{C}\,\overset{I}{H_3} + \overset{I}{H_2}\overset{-II}{O} = \overset{-I}{(R)}\,\overset{(III)}{Co^+}\,\overset{-II}{O}\,\overset{I}{H} + \overset{I}{H^+} + \overset{(-IV)}{C}\,\overset{I}{H_3^-}$
	Free radical	$\overset{-I}{(R)}\,\overset{III}{Co^+}\,\overset{(-IV)}{C}\,\overset{I}{H_3} + \overset{I}{H_2}\overset{-II}{O} = \overset{-I}{(R)}\,\overset{(II)}{Co^\bullet}\,\overset{-II}{O}\,\overset{I}{H} + \overset{I}{H^+} + \overset{(-III)}{C}\,\overset{I}{H_3^\bullet}$
	Carbonium$^+$	$\overset{-I}{(R)}\,\overset{III}{Co^+}\,\overset{(-IV)}{C}\,\overset{I}{H_3} + \overset{I}{H^+} = \overset{-I}{(R)}\,\overset{(I)}{Co^+}\,\overset{I}{H} + \overset{(-II)}{C}\,\overset{I}{H_3^+}$
Methyl acceptors		
Lead	Carbanion$^-$	$4\overset{(-IV)}{C}\,\overset{I}{H_3^-} + 4\overset{I}{H^+} + \overset{(IV)}{Pb}\,\overset{-I}{X_4} = \overset{(IV)}{Pb}\,(\overset{-IV}{C}\,\overset{I}{H_3})_4 + 4\overset{I}{H}\,\overset{-I}{X}$
Mercury	Carbanion$^-$	$\overset{(-IV)}{C}\,\overset{I}{H_3^-} + \overset{(II)}{Hg^{2+}} = \overset{(II)}{Hg^+}\,\overset{-IV}{C}\,\overset{I}{H_3}$
		$\overset{(-IV)}{C}\,\overset{I}{H_3^-} + \overset{(II)}{Hg^+}\,\overset{-IV}{C}\,\overset{I}{H_3} = \overset{(II)}{Hg}\,(\overset{-IV}{C}\,\overset{I}{H_3})_2$
Tin	Free radical	$4\overset{(-III)}{C}\,\overset{I}{H_3^\bullet} + 4\overset{-I}{e^-} + \overset{(IV)}{Sn}\,\overset{-I}{X_4} = \overset{(IV)}{Sn}\,(\overset{-IV}{C}\,\overset{I}{H_3})_4 + 4\overset{-I}{X^-}$
Selenium	Carbanion$^-$	$\overset{(-IV)}{C}\,\overset{I}{H_3^-} + 4\overset{-I}{e^-} + 7\overset{I}{H^+} + \overset{(IV)}{Se}\,\overset{-II}{O_3^=} = \overset{I}{H}\,\overset{(-II)}{Se}\,\overset{-II}{C}\,\overset{I}{H_3} + 3\overset{I}{H_2}\overset{-II}{O}$
	Free radical	$\overset{(-III)}{C}\,\overset{I}{H_3^\bullet} + 5\overset{-I}{e^-} + 7\overset{I}{H^+} + \overset{(IV)}{Se}\,\overset{-II}{O_3^=} = \overset{I}{H}\,\overset{(-II)}{Se}\,\overset{-II}{C}\,\overset{I}{H_3} + 3\overset{I}{H_2}\overset{-II}{O}$
	Carbonium$^+$	$\overset{(-II)}{C}\,\overset{I}{H_3^+} + 6\overset{-I}{e^-} + 7\overset{I}{H^+} + \overset{(IV)}{Se}\,\overset{-II}{O_3^=} = \overset{I}{H}\,\overset{(-II)}{Se}\,\overset{-II}{C}\,\overset{I}{H_3} + 3\overset{I}{H_2}\overset{-II}{O}$
		$\overset{(-II)}{C}\,\overset{I}{H_3^+} + \overset{I}{H}\,\overset{(-II)}{Se}\,\overset{-II}{C}\,\overset{I}{H_3} = \overset{(-II)}{Se}\,(\overset{-IV}{C}\,\overset{I}{H_3})_2 + \overset{I}{H^+}$

– continued –

Table 4–3. Continued.

Methyl transfer donors, carriers, and acceptors	Methyl group	Mass and charge balanced methyl transfer equations for pH 7 dominant forms‡, identifying the oxidation numbers of the elements (above symbol)§
Arsenic	Carbanion$^-$	$\overset{(-IV)}{C}\overset{I}{H}_3^- + 4\overset{-I}{e^-} + 5\overset{I}{H^+} + \overset{I}{H}_3\overset{(III)}{As}\overset{-II}{O}_3 = \overset{I}{H}_2\overset{(-I)}{As}\overset{-IV}{C}\overset{I}{H}_3 + 3\overset{I}{H}_2\overset{-II}{O}$
		$\overset{(-IV)}{C}\overset{I}{H}_3^- + 2\overset{-I}{e^-} + 3\overset{I}{H^+} + \overset{I}{H}\overset{(V)}{As}\overset{-II}{O}_2(\overset{-IV}{C}\overset{I}{H}_3)_2 = \overset{(III)}{As}(\overset{-IV}{C}\overset{I}{H}_3)_3 + 2\overset{I}{H}_2\overset{-II}{O}$
	Free radical	$2\overset{(-III)}{C}\overset{I}{H}_3^{\bullet} + \overset{I}{H}_3\overset{(III)}{As}\overset{-II}{O}_3 = \overset{I}{H}\overset{(V)}{As}\overset{-II}{O}_2(\overset{-IV}{C}\overset{I}{H}_3)_2 + \overset{I}{H}_2\overset{-II}{O}$
		$\overset{(-III)}{C}\overset{I}{H}_3^{\bullet} + 3\overset{-I}{e^-} + 3\overset{I}{H^+} + \overset{I}{H}\overset{(V)}{As}\overset{-II}{O}_2(\overset{-IV}{C}\overset{I}{H}_3)_2 = \overset{(III)}{As}(\overset{-IV}{C}\overset{I}{H}_3)_3 + 2\overset{I}{H}_2\overset{-II}{O}$
	Carbonium$^+$	$2\overset{(-II)}{C}\overset{I}{H}_3^+ + 2\overset{-I}{e^-} + \overset{I}{H}_3\overset{(III)}{As}\overset{-II}{O}_3 = \overset{I}{H}\overset{(V)}{As}\overset{-II}{O}_2(\overset{-IV}{C}\overset{I}{H}_3)_2 + \overset{I}{H}_2\overset{-II}{O}$
		$\overset{(-II)}{C}\overset{I}{H}_3^+ + 4\overset{-I}{e^-} + 3\overset{I}{H^+} + \overset{I}{H}\overset{(V)}{As}\overset{-II}{O}_2(\overset{-IV}{C}\overset{I}{H}_3)_2 = \overset{(III)}{As}(\overset{-IV}{C}\overset{I}{H}_3)_3 + 2\overset{I}{H}_2\overset{-II}{O}$
		$2\overset{(-II)}{C}\overset{I}{H}_3^+ + \overset{I}{H}_2\overset{(-I)}{As}\overset{-IV}{C}\overset{I}{H}_3 = \overset{(III)}{As}(\overset{-IV}{C}\overset{I}{H}_3)_2 + 2\overset{I}{H^+}$

†From information presented in Ridley et al. (1977)
‡Established in accordance with Table 4–1b.
§Derived in accordance with Tables 4–1a and 4–2b.

Table 4–4. Rules and equations for deriving biomass growth yield.

Consistent with Fig. 4–2, the growth yield of an organism with respect to the e^- donor is given by,

Growth yield = 1/growth requirement
= 1/(assimilatory growth requirement + dissimilatory growth requirement)

A. Assimilatory growth requirement = Cell need for e^- from e^- donor/av e^- value of e^- donor

1. Cell need for e^- from e^- donor = C–bound av e^- content of biomass + sum (Δ weighted N–, P–, and S–bound av e^- of biomass versus N, P, and S substrates, respectively)

   where the weighted qualifier recognized the need to include the stoichiometric coefficients for N, P, and S in the calculation.

   a. The av e^- (C–bound e^-) content of biomass is derived as described in Table 4–1b. Conversion to a dry weight basis involves division by the molar formula weight of the biomass

   b. Recognizing the oxidation states of biomass N (–III), P (V), and S (–II) (Tables 4–1a, 4–2c),
   Δ N–bound av e^- for NO_3^-(V), ½ N_2(0), NH_4^+(–III) are 8, 5, and 0 e^-/mole N, respectively
   Δ P–bound av e^- for $H_2PO_4^-$(V) is 0 e^-/mole P
   Δ S–bound av e^- for $SO_4^=$(VI), S(0), and H_2S(–II) are 8, 6, and 0 e^-/mole S, respectively

2. The av e^- value of the e^- donor is derived as described in Table 4–1b. Conversion to a dry weight basis involves division by the molar formula weight of the e^- donor

B. Dissimilatory growth requirement = Cell need for energy from e^- donor/energy value of e^- donor

1. Cell need for energy from e^- donor is derived by theoretical considerations (see text)

   For an aerobic microbial heterotroph an energy need of 12.5 kJ/g cells is consistent with a typical growth yield of about 0.5 g cells/g organic substrate

2. Energy value of e^- donor is a function of the relative electron transfer potentials of the e^- donor/oxidized e^- donor and the e^- acceptor/reduced e^- acceptor couples. Conversion of mole based energy to a dry weight basis involves division by the formula weight of the e^- donor.

   a. Consistent with Fig. 4–2 and the concept of thermodynamic force (Harris, 1982), the equation for the energy value of the e^- donor is given by,

   Energy value of e^- donor = thermodynamic force of dissimilatory e^- transfer reaction
   = – Gibbs free energy of the dissimilatory e^- transfer reaction (ΔG^7)
   = nF(Eh^7 of e^- acceptor/reduced e^- acceptor couple – Eh^7 of e^- donor/oxidized e^- donor couple)

   where n is the number of e^- transferred and F is the Faraday.

   For dissimilatory reactions involving the most oxidized forms of the oxidized e^- donor and e^- acceptor, the standard energy value of the e^- donor becomes,

   *Energy value of e^- donor =(av e^- of e^- donor – av e^- of reduced e^- acceptor) * F * ($Eh^{07}c$ of reduced e^- acceptor – $Eh^{07}c$ of e^- donor)*

   b. The Eh^7 of a redox couple is a function of Eh^{o7} and the activity (concentration) of the components of the ½ reaction ,

   Eh^7 of couple = Eh^{07} of couple + (0.0592/n)log([ox]/[red])

   where 0.0592 is 2.303RT/F, and [Ox] and [R] represent the weighted concentrations of the components on the oxidized and reduced sides of the redox couple half reaction, respectively.

   c. The general equation for the Eh^{07} of a redox couple L(more reduced)/R is given by,

   *Eh^{07} of L/R couple = {(av e^- of L * $Eh^{07}c$ of L) – (av e^- of R * $Eh^{07}c$ of R)}/{av e^- of L – av e^- of R}*

   Av e^- content and $Eh^{07}c$ properties are listed in Table 4–2 and Fig. 4–5.

   If the R component of the redox couple represents the most oxidized form of the element, then av e^- for R = 0, and the above equation becomes,

   Eh^{07} of L/R couple = $Eh^{07}c$ of L

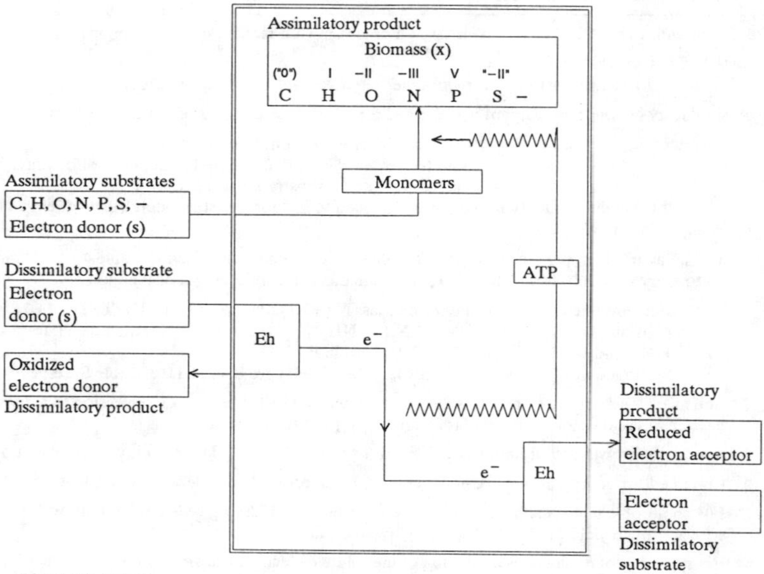

$$\text{Growth Yield (x/s)} = \frac{1}{\dfrac{\text{Assim need cells } (e^- x)}{\text{Assim value sub } (e^-/s)} + \dfrac{\text{Energy need cells (energy/x)}}{\text{Energy value sub (energy/s)}}}$$

Substrate energy = #e^- * Δ Eh e^- donor and acceptor couples * Faraday

Growth yield of bacterial biomass ($\overset{(-0.335)}{C_4}\ \overset{I}{H_7}\ \overset{-II}{O_{1.5}}\ \overset{-III}{N}\ \overset{V}{P_{0.08}}\ \overset{-II}{S_{0.03}}$ −−, 103.5 g FW, 12.5 kJ/g hetero−trophic energy need) on formaldehyde (CH_2O(0), 30.0 g FW) as C/e^- donor, NH_4^+(−III) as N, $H_2PO_4^-$(V) as P, and $SO_4^=$(VI) as S source.

Electron Donor			Electron Acceptor			Growth Yield
Couple	Eh^{07} (V)	Av e^-	Couple	Eh^{07} (V)	ΔE (V)	Yx/s (g/g)
CH_2O/CO_2	−0.5	4	$H_2PO_4^-/PH_3$	−0.7	−0.2	−
			$H^+/½H_2$, WO_3/WO_2	−0.4	0.1	0.09
			CO_2/CH_4, $SO_4^=/H_2S$	−0.2	0.3	0.22
			ReO_4^-/Re, H_3AsO_3/As	−0.1	0.4	0.27
			$H_2AsO_4^-/H_3AsO_3$, Unsat/Sat Org	0.0	0.5	0.31
			$SeO_4^=/HSe^-$, $Fe^{3+}/Fe^{2+}[10^{-12}:1]$	0.1	0.4	0.35
			MnO_2/Mn^{2+}, Org Cl/(Org, Cl^-)	0.4	0.9	0.42
			$SeO_4^=/SeO_3^=$, H_2TeO_3/Te	0.5	1.0	0.45
			$NO_3^-/½N_2$	0.7	1.2	0.48
			$O_2/2H_20$, Hg^{2+}/Hg	0.8	1.3	0.49

Figure 4−2. Assimilatory and dissimilatory nutrition and growth yields of chemotrophs.

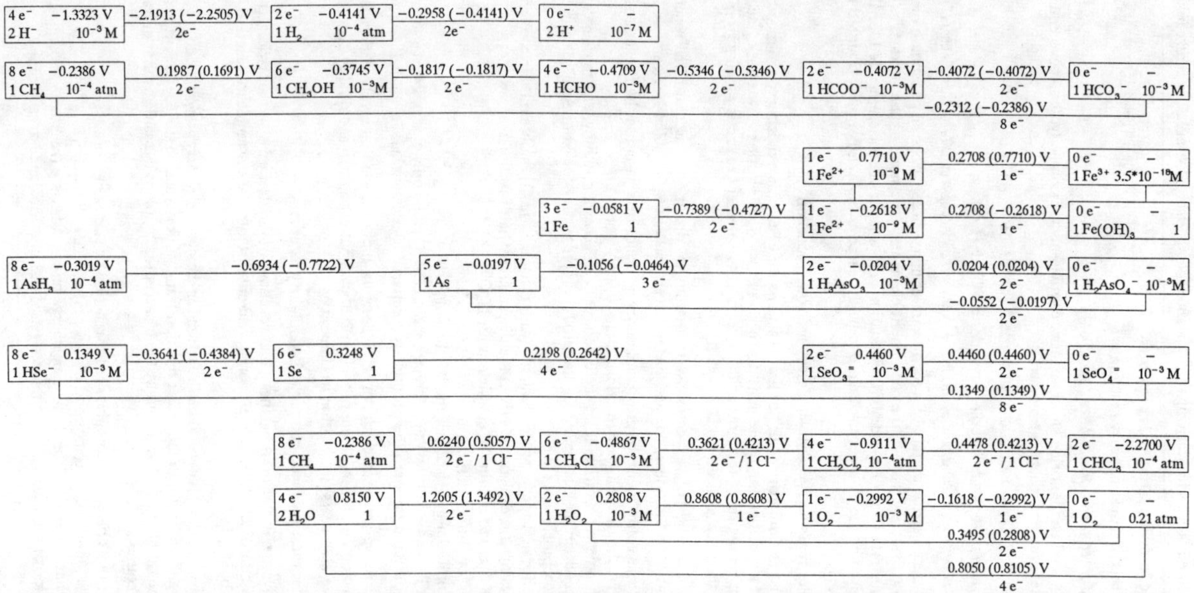

Figure 4−3. Latimer diagrams of pH 7 electron transfer and energy potential of representative redox forms of elements: H, C, Fe, As, Se and O. Within the boxes, the number of moles of compound is identified in the bottom left corner, the total available e^- content (av e^- * # of moles) in the top left, the av e^- energy potential ($Eh^{07}c$) in the top right, and the activity/conc. (atm = atmospheres, M = moles/L) in the bottom right. The av e^- and $Eh^{07}c$ data are from Tables 4−2a−c; slight differences in Eh^{07} for C compounds in the figure versus Tables 4−2b,c are due to the use of HCO_3^- rather than CO_2 as the most oxidized form of C. Between the boxes, the horizontal lines join the two components of a redox couple (L/R), with the more reduced component on the left (L) and the more oxidized component on the right (R). Above the line is identified the e^- transfer potential of the concentrations specified in the boxes (Eh^7), and, in parentheses, the standard e^- transfer potential of the couple (Eh^{07}). The Eh^7 and Eh^{07} data are derived as described in Table 4−3.

Table 4–5a. Rules and equations for deriving mass balance metabolism: Dissimilatory redox reactions.

A. Basic components of a dissimilatory redox reaction

The basic components of a dissimilatory reaction are,

Dissimilatory substrates = dissimilatory products

Consistent with Fig. 4–2 and ignoring stoichiometric coefficients for simplicity, the above equation expands into the dissimilatory reaction framework,

$$e^- \text{ donor} \pm H_2O \;=\; \text{oxidized } e^- \text{ donor} \pm NR \pm H^+ + e^-$$

$$w\,(e^- \pm H^+ + e^- \text{ acceptor} \;=\; \text{reduced } e^- \text{ acceptor} \pm NR \pm H_2O)$$

$$E^- \text{ donor} + e^- \text{ acceptor} \;=\; \text{oxidized } e^- \text{ donor} + \text{reduced } e^- \text{ acceptor} \pm NR \pm H_2O \pm H^+$$

where NR represents non redox metabolites such as ammonium (heterotrophs), phosphate or chloride in which the oxidation state of the component elements is the same on the product as compared to the substrate (reactant) side of the equation, and w is the coefficient needed for the number of e^- accepted by the e^- acceptor couple to be the same as the number of e^- donated by by the e^- donor couple.

B. Derivation of a molar mass balanced dissimilatory redox half reaction

1. Tables 4–2a,b list dissimilatory e^- donor half reactions for e^- donors oxidizing to the most oxidized form of a specified target element. Reversal of these half reactions gives the e^- acceptor half reactions for reduction of the most oxidized form of the element to forms ranging from an intermediate to the fully reduced oxidation state.
2. For general case derivation of an element M–containing e^- donor or e^- acceptor half reaction involving element M as the only redox active element (consistent with Tables 4–1a and 4–2a,b),
   a. Identify the basic components of the dissimilatory e^- donor or e^- acceptor half reaction, and enter them into the dissimilatory reaction framework (A, above).
   b. Set the stoichiometric coefficient for the e^- donor or e^- acceptor at unity, and mass balance for element M to obtain the coefficient for the product representing the oxidized e^- donor or reduced e^- acceptor, respectively
   c. Mass balance for non redox active elements (other than H and O) to obtain their coefficient(s) (such components are designated NR in the dissimilatory half reaction framework)
   d. Mass balance for O to obtain the coefficient for H_2O
   e. Mass balance for H to obtain the coefficient for H^+
   f. Balance the ionic charges to obtain the coefficient for e^-
   g. Cross check (Table 4–1a rules) that the target element M is the only element that shows a change in oxidation number in the product as compared to reactant side of the half reaction.
3. For general case derivation of an e^- donor or e^- acceptor half reaction that involves more than one redox active element, follow the approach in 2. above, but expand 2b and qualify 2g to include other redox active element(s) in addition to element M.

C. Derivation of a molar mass balanced complete dissimilatory reaction

1. Identify and derive the molar mass balanced half reaction for the desired e^- donor/oxidized e^- donor couple, standardize it to 1 mole e^- donor, and enter it into the dissimilatory reaction framework (A, above).
2. Identify and derive the mass balanced half reaction for the desired e^- acceptor/reduced e^- acceptor couple and enter it into the redox reaction framework. Establish the coefficient w for the couple so that the number of e^- accepted by the e^- acceptor couple is the same as that donated by the e^- donor couple.
3. Derive the molar mass balanced complete redox reaction by standard procedures,
   a. Combine the e^- matched e^- donor and e^- acceptor couple half reactions
   b. Cross check that the elements and the ionic charge on the substrate (reactant) as compared to the product side of the reaction are balanced (start with the H balance to save time).

Table 4–5b. Rules and equations for deriving mass balance metabolism: Growth reactions.

A. Basic components of a growth reaction

Recognizing that certain substrates may fulfill more than one function, the basic components of a growth reaction are,

Dissimilatory substrates + assimilatory substrates
=
Dissimilatory products + biomass + other products

Consistent with Fig. 4–2 and ignoring stoichiometric coefficients for simplicity, the above equation expands into the growth reaction framework,

$$(E^- \text{ donor} + e^- \text{ acceptor}) + (C\text{ source} + N\text{ source} + P\text{ source} + S\text{ source} + \text{other assim sources})$$
$$=$$
$$(\text{Oxidized } e^- \text{ donor} + \text{reduced } e^- \text{ acceptor}) + (\text{cell } C, H, O, N, P, S, ...) \pm NR \pm H_2O \pm H^+$$

where NR represents non redox metabolites such as ammonium (heterotrophs), phosphate or chloride in which the oxidation state of the component elements is the same on the product as compared to the substrate (reactant) side of the equation.

B. Derivation of a molar mass balance of growth reaction

1. Identify the basic components of the growth reaction and enter them into the growth reaction framework (A, above).

The molar stoichiometric coefficients for the substrates, and biomass (cells) and other products of the growth reaction are obtained and entered sequentially into the growth reaction framework.

2. Standardize the coefficient for the e^- donor at 1.
3. Derive the dry weight based growth yield (Table 4–4), and convert it to a molar form to provide the coefficient for the cells,

   Cell coefficient = Molar growth yield
   *= Dry weight growth yield * (formula weight of e^- donor/formula weight of cells)*

4. Derive the coefficients for the oxidized e^- donor, the e^- acceptor and reduced e^- acceptor as follows (rules 5–8 take precedent for CO_2, H_2O and H^+)
   a. Derive the mass balanced dissimilatory reaction (Table 4–5a)
   b. Derive the coefficients for the oxidized e^- donor, the e^- acceptor and reduced e^- acceptor as,

      Oxidized e^- donor coefficient for growth reaction =
      *Fraction of e^- donor dissimilated * oxidized e^- donor coefficient for dissimiatory reaction*

      E^- acceptor coefficient for growth reaction =
      *Fraction of e^- donor dissimilated * e^- acceptor coefficient for dissimilatory reaction*

      Reduced e^- acceptor coefficient for growth reaction =
      *Fraction of e^- donor dissimilated * reduced e^- acceptor coefficient for dissimilatory reaction*

   c. The fraction of the e^- donor dissimilated compared to that assimilated is derived as,

      Fraction of e^- donor dissimilated =
      *1 – {cell coefficient * (cell av e^- + ΔN–bound e^- + ΔS–bound e^-)}/e^- donor av e^-*

      where cell av e^- and ΔN– and ΔS–bound e^- are as defined in Table 4–3.

5. Mass balance for non redox elements to obtain the coefficient(s) for NR
6. Mass balance for C to obtain the coefficient for CO_2
7. Mass balance for O to obtain the coefficient for H_2O
8. Balance the ionic charges to obtain the coefficient for H^+
9. Mass balance for H as a cross check that the reaction is totally balanced

assimilatory and maintenance reactions. Dissimilatory reactions are almost exclusively redox reactions, involving the transfer of electrons from an electron donor/oxidized electron donor couple of relatively high electron transfer potential to an electron acceptor/reduced electron acceptor couple of lower electron transfer potential: The energy released is directly proportional to the number of electrons transferred and the difference in electron transfer potential of the electron donor and electron acceptor couples involved (Fig. 4-2). Since soil environments tend to be electron donor limited, electron donor oxidation of fully or partially reduced forms of C, H, S, Fe, Mn, N, tends to proceed concurrently in soil within the limitations of electron acceptor thermodynamics, whereas practical experience confirms that sequential reduction of oxidized forms of nutrient elements occurs in line with their thermodynamic ranking for maximized exploitation of limited electron donor resources (Harris, 1982; Fig. 4-2).

The growth of an organism is a function of assimilatory and dissimilatory reactions, as quantified in the growth yield equation (Fig. 4-2; Table 4-4). The growth yield equation integrates the partitioning of an e^- donor substrate into an assimilatory and dissimilatory function. The assimilatory component of the growth yield equation recognizes the assimilatory need of the cell versus the assimilatory value of the substrate, and the dissimilatory component recognizes the energy need of the cell versus the energy value of the substrate. The illustrative example in Fig. 4-2 holds constant in the growth yield equation the two assimilatory components (the assimilatory electron cell need for the electron donor substrate, and the assimilatory electron value of the electron donor substrate), one (the energy need of the cells) of the two energy components, and two (the electron transfer content and energy potential of the electron donor substrate) of the three subcomponents of the remaining energy component (energy value of the electron donor substrate). The remaining subcomponent (electron transfer potential of the electron acceptor couple) of the growth yield equation is the only variable, and thus the example in Fig. 4-2 illustrates the effect of electron transfer potential of the electron acceptor pair on growth yield (calculation details given in the growth yield section). In accordance with the maximized growth yield rule, electron acceptor use in soil would be predicted to follow the sequence, reduction of oxygen(0) to water(-II), denitrification of nitrate(V) to dinitrogen(0), reduction of selenate(VI) to selenite(IV) and reduction of tellurate (VI) to Te(0), reduction of manganic(IV) to manganous (II) and reduction of chlorinated organics(-II) to saturated organics(-IV), reduction of selenate(VI) to selenide(-II) and ferric(III) to ferrous(II), reduction of arsenate(V) to arsenite(III) and fermentation of high energy organics (such as sugars and amino acids) to lower energy organics (such as alcohols and saturated organic acids) and carbon dioxide, reduction of rhenate(VII) to rhenium(0) and arsenite(III) to arsenic(0), proton(I) reduction to hydrogen(-II), sulfate(VI) reduction to sulfide(-II), and carbon dioxide(IV) reduction to methane (-IV). Tungsten trioxide(VI) reduction to tungsten dioxide(IV) and phosphate(V) reduction to phosphine(-III) would not be predicted to occur at biological pH because of competitive reduction of protons(I) to hydrogen(-I). Predictions on relative reduction sequence for nutrients and contaminants must take into account, especially for species maintained at very low concentrations in soil by low solubility (e.g., Fe and Hg) or metabolic scavenging (e.g., H_2) mechanisms, strong effects of concentration on electron transfer potential (Table 4-4; Fig. 4-3).

The practical importance of redox and energy in microbial ecology is illustrated by the classical trophic classification system (Fenchel & Blackburn, 1979; Brock & Madigan, 1991) based on light versus chemical energy (photo versus chemotrophs), oxidation state of assimilatory carbon source (autotrophs using C(IV) versus heterotrophs using C(<IV)), nature of electron donor (lithotrophs using inorganic versus organotrophs using organic electron donors), and nature of electron acceptor (aerobes using oxygen versus anaerobes using alternate electron acceptors). Contaminant degraders are located largely in the chemo-hetero-organo trophic groups, particularly the obligately aerobic fungal, actinomycete and bacterial groups, but also facultative and obligately anaerobic bacterial group. Dechlorinating contaminant degraders are also represented within the methylotroph and methanogen (aerobic and anaerobic, respectively, chemo-auto-litho trophs) and fermenter (facultatively anaerobic, chemo-hetero-organo-trophs) bacterial groups. Recent work identifies a potential role for the photo-auto/hetero-litho/organo trophic, facultatively aerobic *Rhodospirillaceae* in reductive bioremediation of heavy metal oxyanion and oxide contamination (Moore & Kaplan, 1994).

Successful nutritional use of substrates for dissimilatory and particularly assimilatory purposes is favored by retention of pathway intermediates. In contrast, in addition to use of toxic substrates for assimilatory and/or dissimilatory functions, successful detoxification of toxic substrates may also include (1) cometabolism and release of non toxic, degradation resistant products potentially toxic to other organisms, (2) cometabolism and release of toxic products readily metabolized by other organisms, and (3) cometabolism and release of toxic products susceptible to rapid chemical immobilization.

REDOX AND ENERGY PROPERTIES OF BIOGEOCHEMICALS

Oxidation State (Number), Reductance Degree, and Available Electron Content

With appropriate theoretical qualifications, the oxidation state (number) concept and related available electron content and energy potential of biogeochemicals (Tables 4-1a,b and 4-2a,b,c) provide a powerful framework for interpreting the literature and understanding and tracking electron transfer reactions dominating microbially mediated transformations of biogeochemicals. Metabolic reactions frequently support the concept that covalently bonded elements "possess" shared electrons in accordance with assigned oxidation states based on relative electronegativity (Table 4-1a). For example, catabolic pathways of organic heterocyclic N and amino compounds exclusively involve release of nitrogen in the form of ammonium as an N degradation product, as does dichromate oxidation of such organic N compounds (Harris & Adams, 1979). Similarly, biochemical pathway considerations identify a -II oxidation state for methyl carbon bonded to sulfur and nitrogen in the methyl transfer S-adenosylmethionine and N^5-methyltetrahydrofolate coenzymes, respectively, resulting in methyl transfer as the carbonium$^+$ (CH_3^+, C(-II)) group (Ridley et al., 1977; Table 4-3). The methyl carbon bonded to cobalt in the methylcobalamin coenzyme exists in the -IV oxidation state, but can be transferred as the carbanion$^-$ (CH_3^-, C(-IV)), methyl radical ($CH_3^\cdot$, C(-III)), or carbonium$^+$ (CH_3^+, C (-II)) group, with diverse redox implications for the oxidation state of cobalt in demethylated cobalamin, and the oxidation state of the methyl carbon and methyl-bonded element in the final methylated product. Practical implications of such phenomena are prediction of the methyl accepting chemical forms of elements such as tin, sulfur, arsenic, selenium, and mercury as a function of metal oxidation state and electron accepting potential (Ridley et al., 1977; Table 4-3).

Table 4-1a presents rules governing assignment of oxidation numbers to specific elements associated with a target element, so that the oxidation number of the target element may be calculated by difference. Simple monomers of H and O (Table 4-2a) and other biologically redox active elements (Table 4-2b), and more complex organics (Table 4-2c), are listed in accordance with the boundary most reduced at the top, through intermediate oxidation state, to the boundary most oxidized forms of each of the elements. The concepts and rules for deriving available electron half reactions, and related available electron content, reductance degree (Erickson & Lee, 1989; Heijnen et al., 1992), and element-bound electrons (Harris, 1982) are summarized in Table 4-1b, and applied in Tables 4-2a,b,c.

Energy of Formation, Available Electron Potential Energy, Electron Transfer Energy, and Latimer Diagrams

Sources for standard Gibbs energy of formation (DGf^0) of biogeochemicals are footnoted in Table 4-2a. The DGf^0 for toxicants such as halogenated aromatic compounds may be estimated from chemical structure considerations (Dolfing and Harrison, 1992). Standard enthalpy of formation can be used to derive heat of biological oxidation, and because of common similarity between enthalpy and Gibbs energy of biological combustion of organics (Harris, 1982), enthalpy of formation can be used to obtain a gross approximation of Gibbs energy properties of organics.

As identified in Table 4-1b, standard Gibbs energy of formation properties allow determination of the solubility and proton ionization state of the aqueous forms of compounds predominating at a target pH (e.g., pH 7).

The potential energy of the available electrons of a material ($Eh^{07}c$) is derived from the standard Gibbs energy of formation of the components of the completely balanced available electron half reaction, with the proton concentration standardized to pH 7 (Table 4-1b). Derivation of numbers and energy of available electrons for a redox couple (L/R, where L is more reduced than R) is accomplished using available electron content and energy ($Eh^{07}c$) of each component of the couple as described in Table 4-4 and Fig. 4-3. $Eh^{07}c$ properties of materials are listed in Tables 4-2a,b,c and located within and between (in parentheses) the Latimer boxes in Fig. 4-3.

Latimer diagrams (Fig. 4-3; Harris, 1982) provide predictive and interpretive information on oxidation and reduction pathways, including the presence of energy generating and energy requiring steps. Such steps have implications, for example, with respect to pathway intermediates most unlikely and most likely to be released as dissimilatory products. Latimer diagrams also allow presentation of redox and energy properties of redox related materials in a form readily conducive to derivation of electron transfer content and energy properties of diverse combinations of redox couples.

The Latimer diagrams in Fig. 4-3 are expanded versions of those described earlier (Harris, 1982). As identified in the footnote in Fig. 4-3, and by comparison of data in Tables 4-2a,b,c and Fig. 4-3, each Latimer diagram box contains the content and standard pH 7 energy potential of the available electrons of the material, and also identifies a designated concentration. The boxes are arranged horizontally to proceed from the most reduced oxidation state on the left to the most oxidized state on the right. The number below a horizontal line between two boxes identifies the number of electrons transferred if the material in the box on the left (box L) acts as an electron donor to produce the material in the box on the right as the oxidized electron donor (box R) (or conversely, the number identifies the number of electrons transferred if the material in box R acts as an electron acceptor to produce the material in box L as the reduced electron acceptor). The number in parentheses above the line identifies the standard energy potential of the electrons involved in the electron transfer (Eh^{07}) and is derived, as described in Table 4-4 and Fig. 4-3., from the available electron content (av e^-) and energy potential ($Eh^{07}c$) data in the boxes. The number not in parentheses above the horizontal line identifies the electron transfer potential (Eh^7) corresponding to the concentrations identified in the boxes (derived as described in Table 4-3).

The concentrations used for illustration in the Latimer diagram boxes were standardized somewhat arbitrarily to represent an aerobic environment as follows. Organic compounds and other compounds with high activation energies for redox reactions, and whose concentrations do not tend to be controlled by formation of highly insoluble oxides or hydroxides, were set at 10^{-3} mole/L (aq) and 10^{-4} atm (g). The concentrations of inorganic ions controlled by oxide/hydroxide solubility were set according to solubility criteria for the most oxidized forms and at a low level of 10^{-9} mole/L for reduced forms to reflect aerobic conditions. The oxide/hydroxide was used for highly insoluble forms showing < 10^{-12} moles ion /L solubility at pH 7 (e.g., $Fe(OH)_3$ rather than the Fe^{3+} ion for Fe(III), Fig. 4-3).

The redox composition of biomass and the electron acceptor ranking in relation to the relatively high organic C ($Eh^{07}c$ = -0.5 V) electron donor illustrated in Fig. 4-2 has practical implications beyond the effect on growth yield. For example, the unfavorably high negative Eh^7 for reduction of tungsten oxides would predict that tolerance of microorganisms to tungsten toxicity is more likely to follow the reduced uptake or increased efflux rather than the oxide reduction mechanism (Moore and Kaplan, 1994). Proton reduction to hydrogen gas (Fig. 4-2, 3) becomes increasingly more attractive as the hydrogen gas concentration is lowered, as occurs under highly anaerobic microbial consortia conditions. The redox inactivity of P biogeochemicals (constant oxidation state of V) is explained by the lack of need for reduced P as an assimilatory product, and the endergonic nature of all inorganic P reduction steps from phosphate to phosphine (Fig. 4-2; Table 4-2b). Similarly, inorganic tin and cadmium would be predicted to be biologically redox inactive as electron acceptors. However, methylation of tin is known to occur by a pathway consistent with the representation in Table 4-3 (Ridley et al., 1977), and thus it is important to recognize that particularly for transition metals such as iron, manganese, tin and cobalt, the redox energetics of metallo-organic forms may deviate considerably from the redox energetics of strictly inorganic forms.

The energetically unfavorable first steps in carbon dioxide (bicarbonate) reduction to methane by common electron carriers such as NADH/NAD^+ (Eh^{07}), and a similar endergonic step in the reduction of sulfate to sulfide (calculated from $Eh^{07}c$ properties as described in Table 4-3 and footnoted in Fig. 4-3), are overcome by the obligately anaerobic methanogenic and sulfate reducing bacteria because of the energetically favorable subsequent steps in the pathways. The oxyanion and oxide of rhenium, arsenate and arsenite reduction to the elemental zero oxidation state, and molybdate reduction to molybdenum dioxide show reduction potentials in the same range ($Eh^7 = 0.0 \pm 0.2$) as typical fermentation reductions such as aldehyde reduction to alcohol (e.g., glucose/ethanol; formaldehyde/ methanol). Confirming microbial reduction of arsenate and arsenite is well established (Frankenberger & Losi, 1994). The energetics of arsenic reduction to arsine suggest that arsine production would not be expected even under strongly anaerobic conditions. Methylation of arsenite rather than arsenate by methyl transfer via carbonium (CH_3^+) would be predicted (Ridley et al., 1977; Table 4-4); subsequent reduction of methylated arsenic to arsine might be more energetically favorable than for the strictly inorganic route. Based on electronegativity criteria, the oxidation state of methylated arsenic would tend to be relatively positive (Table 4-4). Use of reduced arsenic compounds as electron donors by aerobic organisms occurs as would be predicted (Frankenberger & Losi, 1994).

In redox energy terms, selenium is closer to arsenic than to sulfur, particularly in terms of the energetically attractive electron acceptor steps in the reduction pathway (Fig. 4-3). Oxidized copper, iron, tellurium, mercury, chromate, antimony, and manganese are good electron acceptors, in the same positive Eh^7 range (0.4 ± 0.2 V) as reductive dechlorination (Dolfing & Harrison, 1992; Brown et al., 1987). Further biological reduction to metallic Fe, Cr, Sb, and Mn would not be predicted. Denitrification offers more electron acceptor potential than nitrate reduction to ammonium in terms of maximizing electron donor resources, and shows an Eh^7 sufficiently close to the oxygen/water couple to merit being called anaerobic respiration (Table 4-2b). In accordance with energy efficiency maximization for limiting electron donors, oxygen is used exclusively as the terminal dissimilatory electron acceptor in aerobic systems.

MICROBIAL GROWTH YIELDS

Growth yield efficiency is a primary determinant driving nutrient and contaminant transformation choices by microorganisms. Growth yields are also the starting point for determining mass balance substrate needs and product release relationships needed to monitor and potentially control microbial transformations. The word form of the growth yield equation used in Fig. 4-2 and derived in Table 4-4 may be expressed in symbol form as,

$$Y_{x/s} = \frac{1}{A_{s/x}} = \frac{1}{A_{as/x} + A_{ds/x}} = \frac{1}{\dfrac{N_{ae/x}}{V_{ae/s}} + \dfrac{N_{kJ/x}}{V_{kJ/s}}}$$

where $Y_{x/s}$ is the specific growth yield of biomass(x) on electron donor substrate (s), $A_{s/x}$ is the corresponding specific growth requirement which subdivides into an assimilatory ($A_{as/x}$) and a dissimilatory ($A_{ds/x}$) requirement. The assimilatory requirement expands to become the available electron need of the cells for the electron donor ($N_{ae/x}$) divided by the available electron value of the electron donor ($V_{ae/s}$). Similarly, the dissimilatory requirement expands to become the energy need of the cells ($N_{kJ/x}$) divided by the energy value of the electron donor substrate ($V_{kJ/s}$). As identified in Table 4-4 and recognizing the data base in Tables 4-2a,b,c: $N_{aw/x}$ is a function of the C-bound available electron content (av e^-) of the electron donor and the difference in N-bound and S-bound available electron content of the biomass versus that of the N and S substrates; $V_{aw/s}$ is the available electron content of the electron donor. The energy need of the cells, $N_{kJ/x}$, is the most difficult component to estimate. The value of 12.5 kJ/g cells was obtained by back calculation to reflect a typical aerobic heterotrophic growth yield of about 0.5 g cells/g electron donor substrate, but is also consistent with generally accepted ATP relations (Harris, 1981, 1982). For heterotrophs, the C-bound available electron energy potential of the biomass is at a level similar to that of a typical assimilatory organic electron donor. The cell energy need is undoubtedly a function of the available electron energy potential of the biomass C and the available energy potential of the assimilatory electron donor substrate but the relationship is mechanistically complex (Heijnen et al., 1992). The energy value of the electron donor, $V_{kJ/s}$, is a function of the concentration-dependent electron transfer potential of the electron donor couple and that of the electron acceptor couple, and the number of electrons transferred in the dissimilatory reaction (Table 4-4). As identified in Table 4-4, conversion between mole and dry weight based growth yields involves use of substrate and biomass empirical formula weights.

Illustrative application of the growth yield equation to the example in Fig. 4-2 with oxygen/water as electron acceptor couple involves recognition of the redox composition, energy and formula weight properties listed in Fig. 4-2, together with derivation of the available electron content of the biomass (17.34 C-bound electrons/mole cells) from first principles (Table 4-1b), or by obtaining it from Table 4-2c. For this example, the basic components of the growth yield equation become: (1) $N_{ae/x}$, the assimilatory cell need for available electrons from the e^- donor with ammonium and sulfate as assimilatory sources of N and S, = (17.34 + (0.03 x 8)) /103.5 = 0.1699 av e^-/g cells; (2) $V_{ae/s}$, the assimilatory available electron value of the e^- donor substrate, = 4/30.0 = 0.1332 av e^- /g substrate; (3) $N_{kJ/x}$, the energy need of the cells = 12.5 kJ/g cells; and (4) $V_{kJ/s}$, the energy value of the electron donor, = (4 x 96.49 x (0.8-(-0.5))/30.0 = 16.72 kJ/g cells. Continuation gives: (5) $A_{as/x}$, the assimilatory requirement, = 0.1699/0.1332 = 1.2766 g substrate/g cells; (6) $A_{ds/x}$, the dissimilatory requirement = 12.5/16.72 = 0.7476 g substrate/g cells; (7) $A_{s/x}$, the growth requirement, = 1.2766 + 0.7476 = 2.0242 g substrate/g cells; and (8) $Y_{x/s}$, the growth yield = 1/2.2042 = 0.494 g cells/g substrate. The growth yield for the other electron acceptor couples in Fig. 4-2 were derived by the same approach, with the only variable being the electron transfer potential of the electron acceptor couple.

MASS BALANCE METABOLISM

Mass balanced dissimilatory and growth reactions provide a mechanistic basis for predicting and tracking the nature and kinetics of nutrient and contaminant metabolism by microorganism in soil and related systems (Heijnen et al., 1992; Cook, 1987). Comparison of experimental and predicted relative stoichiometric coefficients can validate or cause reexamination of assumed pathways. This includes tracking of complete versus incomplete metabolism of contaminants as a function of direct analysis for a target compound or indirect analysis for a more readily measurable reaction component of known stoichiometric relationship to the target compound. Mass balancing growth reactions also provides quantitative information on boundary amendment levels for assimilatory nutrients such as nitrogen, phosphorus and sulfur to support accelerated contaminant degradation by organic carbon and energy manipulation.

Rules for deriving mass balanced dissimilatory and growth reactions are given in Tables 4-5a and 4-5b, respectively. Application of these rules to a bacterial example with similar properties to those described in Fig. 4-2, in which atrazine ($C_8H_{14}N_5Cl$, $Eh^{07}c$ » -0.3 V, FW = 215.69 g/mole) replaces formaldehyde as electron/C source and ammonium as N source, gives a growth yield of 0.486 g cells/g atrazine, and the following molar mass balance dissimilatory electron donor half reaction [EDR], dissimilatory electron acceptor half reaction [EAR], complete dissimilatory reaction [DR], and growth reaction [GR],

(0.25) I -III -I I -II (IV) -II (-III) I (-I) I -I

$$C_8H_{14}N_5Cl + 8\,H_2O = 8\,CO_2 + 5\,NH_4^+ + Cl^- + 26\,H^+ + 30\,e^- \qquad \text{[EDR]}$$

-I I 0 I -II

$$(30/4)\,(4\,e^- + 4\,H^+ + O_2 = 2\,H_2O) \qquad \text{[EAR]}$$

(0.25) I -III -I 0 I -II (IV -II (-III) I (-I)

$$C_8H_{14}N_5Cl + 7.5\,O_2 + H_2O = 8\,CO_2 + 5\,NH_4^+ + Cl^- \qquad \text{[DR]}$$

(0.25) -III -I 0 I (V) -II (V) -II I -II I

$$C_8H_{14}N_5Cl + 3.05\,O_2 + 0.081\,H_2PO_4^- + 0.030\,SO_4^= + 2.88\,H_2O + 3.13\,H^+$$

$$=$$

(-0.335) I -I -III V -II (IV) -II (-III) I (-I)

$$1.013\,C_4H_7O_{1.5}NP_{0.08}S_{0.03} + 3.95\,CO_2 + 3.99\,NH_4^+ + Cl^- \qquad \text{[GR]}$$

Consistent with previous work (Andrews & Harris, 1986; Harris, 1981, 1982), the stoichiometric coefficients on the above growth equation are termed $Y_{w/s}$, where w is a substrate (s), cells (x), or a metabolic byproduct (p).

Specific rates of metabolism, q_w (units: w/x×t), are given by:

$$q_w = q_s \text{ x } Y_{w/s}$$

where q_s, the specific rate of substrate uptake and metabolism, is a function of the maximum specific growth rate ($q_x = m_{max}$) under conditions of saturating C_s (substrate concentration), and a function of C_s under C_s limiting conditions. Since substrate affinity coefficients for biological uptake systems are commonly in the 10^{-7} to 10^{-3} M range (Button, 1985), it should be recognized that substrate uptake rates for concentrations one or more orders of magnitude lower than this range will inevitably be relatively low.

Rates of metabolism, Q_w (units: w/system×t), are given by:

$$Q_w = q_w \text{ x } X = q_w \text{ x } (V \text{ x } C_x)$$

where X is the amount of cells in the system, V is volume of the system, and C_x is the cell concentration.

Mass balance yield and rate equations may be estimated by the above approach for a diversity of biogeochemical cycles and interfacing contaminant transformations, using the data base in Table 4-2a,b,c and Fig. 4-3 and other sources of redox properties of nutrients and contaminants (Thauer et al., 1977; Harris, 1982; Erickson & Lee, 1989; Heijnen et al., 1992). Such a reference base, together with an understanding of the forces driving microbial ecology and biochemical transformations in soil, should provide a useful framework for development and monitoring of bioremediation techniques for accelerating the degradation of target contaminants in soil.

REFERENCES

Andrews, J.A., and R.F. Harris. 1986. r- and K-selection and microbial ecology. Adv. Microbiol. Ecol. 9:99-147.

Brock T.D., and M.T. Madigan. 1991. Biology of microorganisms. Prentice Hall, Englewood Cliffs, New Jersey.

Brown, J.F., R.E. Wagner, H. Feng, D.L. Bedard, M.J. Brennan, J.C. Carnahan, and R. J. May. 1987. Environmental dechlorination of PCB's. Environ. Tox. Chem. 6:579-593.

Button, D.K. 1985. Kinetics of nutrient-limited transport and microbial growth. Microbiological Reviews 49:270-297.

Chang, R. 1984. Chemistry. Random House, New York.

Cook, A.M. 1987. Biodegradation of s-triazine xenobiotics. FEMS Microbiol. Rev. 32: 93-116.

Cotton, F.A., and G. Wilkinson. 1972. Advanced inorganic chemistry. Interscience Publishers, New York.

Dolfing, J., and B.K. Harrison. Gibbs free energy of formation of halogenated aromatic compounds and their potential role as electron acceptors in anaerobic environments. Environ. Sci. Technol. 1992. 26:2213-2218.

Erickson, L.E., and K.H. Lee. 1989. Degradation of atrazine and related s-triazines. Critical. Rev. Env. Control. 19:1-14.

Fenchel, T., and T.H. Blackburn. 1979. Bacteria and mineral cycling. Academic Press, Inc., New York.

Frankenberger, W.T., and M.E. Losi. 1995. Applications of bioremediation in the clean-up of heavy elements. p.170-207. *In* H.D. Skipper and R.F. Turco (ed.) Bioremediation: Science and applications. SSSA Spec. Publ. 43. ASA,CSSA and SSSA, Madison WI.

Harris, R.F. 1981. Effects of water potential on microbial growth and activity. p. 23-95. *In* J.F. Parr and W.R. Gardner (ed.) Water potential relations in soil microbiology. ASA Spec. Pub. no. 9, Madison, WI.

Harris, R.F. 1982. Energetics of nitrogen transformations. *In* F.J. Stevenson, J.M. Bremner, R.D. Hauck and D.R. Keeney (eds.) Nitrogen in agricultural soils. Agronomy Monograph No. 22. Am. Soc. Agron., Madison, WI. p. 833-890.

Harris, R.F., and S.S. Adams. 1979. Determination of the carbon-bound electron composition of microbial cells and metabolites by dichromate oxidation. Appl. Environ. Microbiol. 37:237-243.

Heijnen, J.J., M.C.M. van Loosdrecht, and L. Tijhuis. 1992. A black box mathematical model to calculate auto- and heterotrophic biomass yields based on gibbs energy dissipation. Biotechnol. Bioeng. 40:1139-1154.

Latimer, W.L. 1952. Oxidation Potentials. Prentice-Hall, Inc., Englewood Cliffs, New Jersey.

Moore, M.D., and S. Kaplan. 1994. Members of the family *Rhodospirillaceae* reduce heavy-metal oxyanions to maintain redox poise during photosynthetic growth. Amer. Soc. Microbiol. News 60:17-23.

Ridley, W.P., L.J. Dizikes, and J.M. Wood. 1977. Biomethylation of toxic elements in the environment. Science 197:329-332.

Sadiq, M., and W.L. Lindsay. 1979. Selected standard free energies of formation for use in soil science. p. 385-422. *In* W.L. Lindsay. Chemical equilibria in soils. Wiley-Interscience, New York.

Thauer, R.K., K. Jungerman, and K. Decker. 1977. Energy conservation in chemotrophic anaerobic bateria. Bacteriol. Rev. 41:100-180.

Wagman, D.D., W.H. Evans, V.B. Parker, I. Halow, S.M. Bailey, and R.H. Schum. 1968. Selected values of thermodynamic properties. Technical Note 270-3. U.S. Dep. of Commerce, National Bureau of Standards, U.S. Government Printing Office, Washington D.C.

Weiss, G. 1986. Hazardous chemicals data book. Noyes Data Corporation, Park Ridge, New Jersey.

Zumdahl, S.S. 1989. Chemistry. D.C. Heath and Co. Massachusetts.

5 The Microflora of Bioremediation

R.F. Turco
Laboratory for Soil Microbiology
Environmental Soil Science
Purdue University
West Lafayette, Indiana

M. Sadowsky
Soil Science Department
University of Minnesota
St. Paul, Minnesota

The inappropriate disposal or discharge of toxic organic materials into the environment has resulted in the pollution of soils and water supplies. The responsibility for removal and remediation of these contaminated ecosystems rests squarely on current and future generations. The processes used to cleanup or remediate these contaminants can follow many paths. By definition and function, biological remediation, often abbreviated as bioremediation, makes use of microbial processes to reduce contaminants in soil or water. Madsen (1991) provides a very useful definition of bioremediation: "a managed or spontaneous process in which biological, especially microbial, catalysis acts on pollutant compounds, thereby remedying or eliminating environmental contamination." While bioremediation is generally viewed as equivalent to only biological mineralization, others have suggested that microbially-based enzymatic reactions that bind the pollutant into the soil matrix may be a good route to immobilize pollutants and prevent future environmental deterioration (Bollag, 1992). Microbial-mediated mineralization of pollutants, or a combination of bioremediation with chemical/physical treatment, resulting in mineralization, is one the most cost effective and exemplary approaches to the resolution of the problem (Madsen, 1991). However, as indicated by Stroo (1992), despite the conjecture and research on the potential for biotechnology to enhance the treatment of hazardous materials in soil, little progress has been made to date in developing commercially available products and processes. It is clear that the use and manipulation of indigenous microorganisms remains at the core of most in situ bioremediation programs (Brown et al., 1993).

Successful bioremediation of soils is a function of three independent but interrelated factors: the contaminant, microorganisms, and the environment (Madsen, 1991; Stroo 1992; Tiedje, 1993). That is, the contaminant's structure and function, concentration and distribution, and the physiological and genetic makeup

 Bioremediation: Science and Applications. SSSA Special Publication 43.

of the resident or introduced microbial populations, all of which are situated in an environmental framework, control the success of bioremediation. Tiedje (1993) points out that historically most considerations of the bioremediation process have been made from an engineering approach that considers soil as a type of tank reactor. He indicates that this perspective may have adversely affected our views, and to some degree the successes that have been achieved with bioremediation. He argues that developing a full understanding of the ecological principles at work within the system, will result in a far more satisfying approach than simply relying on new engineering concepts. It should be apparent, though often overlooked, that soil bioremediation is only needed when a soil system is not functioning at an optimal level. If one or more of the critical components, contaminant chemistry, microbial function or environment is suboptimal in terms of the conversion of organic constituents, then degradation will be slow or not take place at all.

Soil is an irregular mixture of three phases: i) inorganic and organic solids, ii) soil solution containing dissolved organics and inorganics, and iii) gases. Soils are generally considered to be nutrient poor as they tend to be depleted of microbially available carbon and nitrogen. Impacting this environment and resident microbial populations are radical and often abrupt changes in temperature and moisture. Clearly, microorganisms must contend with an ever changing environment that offers little in terms of nutrients or short-term environmental stability. In essence, the goal of a successful bioremediation program is to overcome the natural heterogeneity within the soil systems, optimize the conditions by removing the rate limiting factor, and promote the microbial population to remove organic pollutants. A discussion of how soil microbial populations, either indigenous or introduced, respond to pollutants is the purpose of this paper.

DISTRIBUTION OF SOIL MICROORGANISMS

Microorganisms can be viewed as the biochemical machinery that drive the bioremediation process. This idea is a logical extension of many studies that show that soil organic matter turnover and nutrient cycling are microbially driven processes (Coleman et al., 1983). On a gram dry weight basis, microbial numbers in soil often exceed 10^8 for bacteria, 10^6 for actinomycetes, and 10^5 for fungi. Bacterial species diversity has been shown to exceed 10,000 different species per gram soil (Torsvik, et al., 1990a,b). Microorganisms can exceed 500 mg biomass C kg^{-1} soil (Jenkinson & Ladd, 1981). Microorganisms, however, generally comprise less than 3% of the soil organic carbon, and take up only 0.001% of the soil's volume (Sparling, 1985). The microorganisms are situated throughout the soil profile which can be delineated into a series of linked ecosystems. These systems can be defined as: surface-root zone soils, the vadose zone, and the aquifer. The dispersion of microbial biomass within the profile is related to distance from the soil surface and the placement and availability of organic materials (Hissett & Gray, 1976). Microbial biomass declines as you move away from the surface. Microbial density and activity in soils and subsurface materials have been correlated with the type of organic matter present and the net water recharging of a particular site (Konopka & Turco, 1991).

Richaume et al. (1993) showed the distribution of bacteria in soil is non-uniform and the distribution reflects soil structure and available nutrients. Rao and

Wagenet (1985) suggested that the variability in a soil is the sum of intrinsic and extrinsic factors. Intrinsic variability reflects soil developmental factors, while extrinsic factors reflect outside forces, such as land use. Parkin and Shelton (1992) indicated that few studies have addressed the temporal and spatial variability associated with the field scale distribution of microorganism or their abilities to degrade recalcitrant organic compounds. The distribution of microorganisms in soil can also be thought of from the applied ecological perspective of "patch dynamics," where patch formation is a reflection of intrinsic and extrinsic forces (Rao et al., 1986). The distribution of microorganisms in patches seems to be at work on a number of scales within the soil system. Rao et al. (1986) showed spatial variability in the degradation of pesticides applied to soil. Parkin and Shelton (1992) suggested the formation of carbofuran-enhanced-degrading microbial populations may reflect planting row placement, tillage practices, and time of year (probably due to temperature and moisture relationships). Roberston et al. (1988) evaluated microbial-mediated N transformations across a landscape. They found that there was a significant patterning of N fluxes within the landscape indicating localized differences in microbial ability. In one of the few studies to evaluate spatial variability in a soil microbial population, Wollum and Cassel (1984) found that major differences in *Rhizobium* populations across a field were possible at sampling distances of less than 0.2 m. They found that the extrinsic factor of planted row direction had a major influence on the location of the population. In all of the above studies, an outside force (row crop planting) had allowed the development of a patch where the population size had been altered. When microorganisms are able to use applied organics as a significant nutrient source, a shift in the population may take place. This appears to be the case for enhanced degradation of pesticides such as carbofuran (Turco & Konopka, 1990a,b; Parkin et al., 1991).

Our thinking about the residency of the microorganisms in soil is tempered by the way in which we typically recover microorganisms from soil. Most often soil is homogenized using mechanical and chemical means to disrupt and remove the cells from the surface of soil particles and the number of microorganisms are reported on a per gram basis (Wollum, 1982). As a consequence, we have a poor conception of microorganisms in their natural resident state. This, coupled with their relatively high numbers, leads us into thinking microorganisms are everywhere in soil. Hattori and Hattori (1976) indicated that pore size influences the distribution of microbial populations in soil. Killham et al. (1993) demonstrated that the location of substrates within the soil matrix will control its subsequent turnover. They found that when a simple nutrient, such as glucose, is forced predominately into pore spaces with neck diameters smaller than 6 μm, degradation was greater than when the nutrient was maintained in pores with neck diameters greater than this size. They reasoned that the active population was resident in the former, and that the microbial population size in the larger pores had been suppressed by predation from protozoa. Heijnen and van Veen (1991) indicated that protozoal predation of bacteria in pore spaces with neck diameters larger than 6 μm, had a major influence on the survival of introduced strains. They rationalized that smaller pore spaces offered a protective microhabitat for soil bacteria and that bacteria outside of these habitats were scavenged due to predation. Consequently, before substantial degradation can occur in soils, it may be necessary for pollutants to diffuse into the internal regions of aggregates.

The amount of moisture on a soil surface is directly related to the pore spaces. The pore size distribution controls water retention; larger pores will retain less water, as they drain more quickly. These larger pores are a less hospitable environment. The inside of the aggregate is wetter, reflecting the ability of the smaller pores within the aggregate to retain water (Papendick & Campbell, 1981) Given the small size of most nonrhizosphere soil bacteria, less than 0.3 μm in diameter, (Bae et al., 1972), small pores may be an ideal environment. Hissett and Gray (1976) estimated that microorganisms are resident on less than 0.17% of the surface of soil organic matter and less than 0.02% of the soil mineral surfaces. It appears that gram-positive bacteria tend to occupy the outside of soil micro-aggregates while gram negative bacteria are found within the aggregate structure (Kilbertus et al., 1980). This patch distribution is thought to reflect the gram-negative organism's preference for the more hospitable microenvironment within the aggregate as compared to the harsh and less environmentally stable external surfaces. Microorganisms on the ped surface are exposed to repeated desiccation and may die or shift to a dormant phase. The predominant microbial activity in soil may be occurring in the internal portions of aggregates (Martin & Foster, 1985; Foster, 1988; Killham et al., 1993).

Movement of bacteria in soil is a function of water content and water potential relationships. Movement of bacteria in soil is tempered by their sorption onto a soil particle or transport into small soil pores (Smith et al., 1985). Movement of bacteria in soil can occur via passive transport with water or by active transport through the expenditure of energy (Huysman & Verstraete, 1993; Issa et al. 1993a,b). In general, passive movement will carry bacteria greater distances, over several centimeters, than active movement (Madsen & Alexander, 1982). Tan et al. (1992) found that passive movement was retarded by the adsorption of bacteria onto surfaces. Microorganisms generally adhere to soil surfaces by electrostatic interactions, London-van der Waals forces, and hydrophobic interactions (van Loosdrecht et al., 1987; van Loosdrecht et al., 1990). The ionic strength of the suspending solutions and the soil surface properties were found to be the predominant factors affecting bacterial transport. Trevors et al. (1990) found that the movement of a genetically engineered strain of *Pseudomonas* through a soil column was a function of water flow rate and the number of times the column was flushed. Without the addition of water, only a limited number of the introduced bacterium were detected below the surface layer Huysman and Verstraete (1993) found that cell retention in soil was a function of the type of bacterium as well as the soil surfaces. They reported that the migration of a strain with hydrophobic surface properties was 2 to 3 times slower compared to that with hydrophilic properties. They were able to correlate this with an increased adhesion of the hydrophobic strains to soil particles. Issa et al. (1993a,b) suggested that movement of motile bacteria, ones with high numbers of flagella, may be impaired as the large number of flagella may increase the cells overall size preventing its movement through small pore necks. Other work has shown that the contribution of the flagella to movement declines as the soil dries and the water film thickness decreases. It is suggested that presence of flagella may enhance cell sorption on to soil surfaces (Issa et al., 1993b)

SOIL MICROBIAL NUTRITION

When considering bioremediation processes in soil, attention to the relationship between the active population and available nutrient source is critical. Soil is rapidly depleted of available carbon and nitrogen due to the activity of resident soil microorganisms (Atlas & Bartha, 1993). On one hand nutrient limitation can be seen as a factor that suppresses the overall microbial activity in the soil, repressing the population's ability to function at optimal levels. Alternatively, the finding that soil contains some 10 000 different species and has 10^9 cells per gram, implies that natural adaptive forces are capable of responding to the varied conditions present in the soil matrix. However, the recent introduction of halogenated organic chemicals, which contain minimal energy for microbial growth, has outpaced the ability of the adaptive process to keep up. Given the tendency of the microorganisms is to be resident inside of soil aggregates and have limited movement, water dissolved organic carbon (DOC) is a key component in meeting the nutritional needs of the microbial biomass.

In general microorganisms make use of dissolved C and N that is derived from plant decomposition, root exudates, microbial decomposition or the mineralization of soil organic matter (Coleman et al., 1983). Fulvic acid, which is derived from soil organic mater, can serve as a significant nutrient source (Turco et al., 1991). Of particular note is the finding that manures, which are high in DOC, can significantly increase the size of soil microbial biomass (Collins et al., 1992). In general, bacteria are found in pores with neck diameters of less than 6 μm. This may offer the population protection from protozoa which require a pore diameter of greater than 6 μm in order to invade the space (Kilbertus et al., 1980; Martin & Foster, 1985; Foster, 1988; Killham et al., 1993). However, as indicated by Killham et al. (1993), little is actually known about the location of the *in situ* transformations of soil carbon and soil nitrogen or how utilization is related to the positioning of the active microbial biomass. Borchers and Perry (1992) showed that in a silt loam soil there are large pools of physically protected N, which acts as a slow-release sorce of N. Powlson (1980) demonstrated that grinding soils will also increase the available fraction of soil C.

It should be noted, however, that Sawhney et al. (1988) and Steinberg et al. (1987) showed that 1,2-dibromoethane (EDB), a water soluble, biodegradable, and weakly-sorbed pesticide, persisted in soil for a long period of time. They reasoned that the chemical is entrapped in soil micropores where it is protected from microbial degradation. Taken together with the work of Heijnen and van Veen (1991), this finding implies microorganisms are physically resident in only a certain size fraction of the pore space in the soil structure. It seems that the pore neck size is most likely between 0.25 and 6 μm in diameter. The smaller size limits the entry of bacteria while sizes in excess of 6 μm allow for the entry of protozoa. There residency controls, to some degree, their access to certain nutrients. It also suggests that one needs to consider soil structure and the place of the resident microflora as a factor affecting the transformation of organic compounds. Efforts to describe the interaction of soil structure and degradation have been attempted with simulation modeling. Dhawan et al. (1993) described a macropore flow model that accounts for bioremediation in the interstitial spaces among soil aggregates. They found, using numerical models, that the total time for remediation depended on the rate at which contaminants are consumed within the large aggregates. This rate depended on the biodegradation kinetics (chemical degradability) and the rate of diffusion of the substrate and oxygen into the aggregates and internal pore space. The simulation results indicated that for soils with aggregates whose radius is 1 cm or larger, that the remediation time was significantly longer when compared to soils with smaller aggregates.

From a bacterial-point-of-view, the balance of C to N should approach 30:1 (Paul & Clark, 1989) if optimal growth conditions are to be met. During active growth, microorganisms need 30 parts of carbon for each part of N they utilize. Others indicated that for bioremediation the ratio should be in the range of 120:10:1, C:N:P, respectively (Sims, 1993). However, as Pope and Matthews (1993) pointed out, nutrient requirements for biodegradation in the field have not been thoroughly studied and most of our working knowledge is derived directly from the agronomics associated with crop growth and residue decomposition.

The effort to remediate the Exxon Valdez oil spill in Prince William Sound, Alaska, included fertilizer applications to accelerate natural microbial degradation of the oil (see Chapter 12 by Hinton). The applied fertilizer materials included a slow release material Customblend (28% N and 8% P) and Inipol EAP22 (a microemulsion of saturated urea). In this case fertilizer addition gave higher hexadecane and phenanthrene mineralization potential than did untreated soils (Lindstrom et al., 1991). It is clear that the addition of water soluble minerals will supplement the needs of microbial populations. However, a consideration of the importance of minor elements is lacking. Lewis and Crawford, (1993) indicated that presence of dissolved iron(II), iron(III), and cobalt(II), as well as an insoluble iron(III) compound, ferric oxyhydroxide, inhibited carbon tetrachloride transformation under anaerobic conditions. Others have shown that mineral salt additions did not significantly affect the degradation of Aroclor 1221 by *Pseudomonas cepacia* strain JHR22 (Havel & Reineke, 1993).

Morgan et al. (1993) found that the successful soil inoculation with *Phanerochaete chrysosporium* and *Chrysosporium lignorum* CL1 for the mineralization of 3,4-dichloroaniline and benzo(a)pyrene required supplementary carbon sources, but additions of inorganic nutrients had no stimulatory effect. Additions of glucose, hay, pine bark, wood chips, and peat all promoted growth and the degradation process by possibly enhancing the co-oxidation of the contaminants.

ELECTRON ACCEPTORS

Regardless of the availability of C, microorganisms must utilize an electron acceptor if degradation is to occur. In fermentation reactions the substrate is both the electron source and acceptor since the substrate is both oxidized and reduced in the degradation process (Stanier et al., 1976). Oxygen can be used as an electron acceptor with a general stoichiometry for the reaction can be expressed by:

$$\text{Substrate} + O_2 \rightarrow \text{Biomass} + CO_2 + H_2O + \text{Metabolites} + \text{Energy}$$

A general reaction to illustrate this process is the oxidation of glucose:

$$C_6H_{12}O_6 + 6\,O_2 \rightarrow 6CO_2 + 6H_2O + \text{energy}.$$

This translates to a need of 2 moles of oxygen per mole of carbon utilized. A more specific type of reaction is given for the oxidation of toluene:

$$C_6H_5\text{-}CH_3 + 9O_2 \rightarrow 7CO_2 + 4H_2O + \text{energy}.$$

This translates to 2.5 moles oxygen per mole of carbon utilized. If we consider a typical contaminate substrate such as ethylene glycol we find the following reaction:

$$2(CH_2OH)_2 + 5O_2 \rightarrow 4CO_2 + 6H_2O$$

Along with the needs for N and P the oxidation reaction demands 2.5 moles or 80 g of O_2 per mole of the ethylene glycol. This translates to 1.29 g of oxygen per gram of ethylene glycol. A 10 000 kg plume contained in soil would require 1.29×10^4 kg of O_2 to completely mineralize the plume. Given that air contains about 21% O_2 by weight, we would need about 6.14×10^4 kg of air. This translates to roughly 5.11×10^7 m^3, assuming air has a specific weight of 1.20 kg m^{-3}.

Given that the solubility of O_2 in water is only 0.028 ml O_2 mL^{-1} H_2O atm^{-1} (Greenwood, 1975) the depletion of the available supplies will be rapid when demand is high. The situation is further complicated by the fact that the diffusion of O_2 in water is about 1×10^{-4} of the diffusion of O_2 in air, with a diffusion coefficient of 2.5×10^{-5} cm^2 sec^{-1} in water versus 0.189 cm^2 sec^{-1} for air (Papendick & Campbell, 1981). It stands to reason that the aerobic degradation rate in soil is a function of O_2 resupply to the regions undergoing degradation processes.

Campbell and Papendick (1981), extending the work of Currie (1965), showed that diffusion of gas in a non-reactive solid is defined by:

$$D = D_o S^m$$

where D is the apparent diffusion coefficient, D_o the diffusion coefficient of oxygen in air, S is the air filled porosity ($S = \theta_s - \theta$ where θ_s is water content at saturation) and m is fitting factor which ranges from 1 to 3 (and is set at 2 for most soils). Their calculations indicate that, in general, the diffusion of air into soil is proportional to the square of the air-filled porosity. However, the active microbial populations appear to be constrained to locations with pore neck diameters of 6 µm or less. Using the 6 µm as a limiting pore neck size, we find that the total exchange area for gases to a microbial colony is limited by the pore neck, to an area of 28 µm^2 or less. At many remediation sites this situation is further complicated by the presence of a second material such as oil. Dowing and Truesdale (1955) showed that oil situated on water, will act as a diffusion barrier for oxygen moving into water. Schwarzenbach et al. (1993) reported a diffusivity of about 2×10^{-3} cm^2 s^{-1} for oxygen moving in oil. The film thickness will decreases the transfer rate, with films greater than 100 µm diminishing the overall transfer velocity by as much as one-half. Soils containing large amounts of oil will have a limited ability to transfer oxygen.

The generally poor oxygen status of most soils is thought to be a reason why the rapid introduction of O_2 into the soil profile following plowing stimulates microbial activity and decrease residual soil-C levels. Follett and Schimel (1989) showed that when a site which was in native sod was plowed and an alternate wheat-fallow rotation was established, that available N, an indicator of stability, in the top 10 cm of soil had decreased to 50% of native sod plow treatments six years after the start of cultivation. Staley et al. (1988) showed that biomass-C and organic-C levels in the soil surface layers, under no-till conditions, were up to 83% greater than those found using conventional tillage practices. Blum et al. (1991) found that total phenolic acid contents in the surface soils under wheat-no till conditions were significantly higher than those from tilled systems, reflecting the introduction of O_2 through soil mixing.

Modeling efforts for subsurface systems support the finding that O_2 is a limitation in the degradation of xenobiotic compounds (Dhawan et al., 1993). Kaiser and Bollag (1992) reported that under aerobic conditions, pyridine and hydroxylated pyridines were more rapidly transformed than under anaerobic conditions. They tested surface and subsurface soils which had been exposed to these compounds for several decades. Presumably this allowed for the establishment of an adapted microbial population. All the pyridine derivatives tested were biotransformed within two weeks in the presence of oxygen. Under anaerobic conditions, however, longer time periods were required. Song and Bartha (1991) reported that jet fuel degradation rates were reduced in subsurface environments by a lack of available O_2.

Under oxygen-limited conditions strict and facultative anaerobic bacteria are able to metabolize and grow if supplied with one of a number of alternate electron acceptors. The use of anaerobic organisms offers many new possibilities in bioremediation applications (Stroo, 1992). The full extent of the possibilities for the use of anaerobic organism to degrade xenobiotics is only now coming to light (J. Suflita, personal communication).

In contrast to aerobic respiration, anaerobic organisms can make use of number of alternate electron acceptors. A general reaction scheme for various electron acceptors typically used by microorganism is given by:

$$Substrate + \{NO_3, Mn^{4+}, Fe^{3+}, SO_4^{2-}, CO_2\} \rightarrow Biomass + CO_2 + \{N_2, Mn^{2+}, Fe_2+, S_2, CH_4\}$$

The favored reductant is often a reflection of the availability of the material, the biochemical abilities of populations and the amount of carbon substrate. The type of electron acceptor favored will reflect the redox status of the system. Therefore, the sequence of reduction, NO_3, Mn^{4+} , Fe^{3+} , SO_4^{2-}, CO_2 reflects the redox potential and oxidizing capacity of the chemical half-reaction (Zehnder & Strum, 1988). The reduction of NO_3 can occur at a standard reduction potential of +0.74 volts. Other materials can be reduced after the NO_3 is depleted. In general reduction of Mn^{4+}, occurs at +0.52 volts, reduction of Fe^{3+} at -0.05 volts, reduction of SO_4^{2-} at -0.22 volts and reduction of CO_2 to form CH_4 at -0.24 volts. These values are for aqueous solution with a pH of 7 (Stumm & Morgan, 1981). Smith and Arah (1986) and others (Gottschal, 1986) demonstrated the presence of anaerobic microsites or conditions, in otherwise aerobic soils. However, Parkin and Meisinger (1989) demonstrated that C levels below the root zone in typical agricultural soils are too low to support a widespread denitrification mode of growth. In contrast, oxygen levels in xenobiotic contaminated soils and aquifers are low. Wilson et al. (1990) showed that at the center of a subsurface hydrocarbon plume, the conditions are highly reduced. At the edges of the plume, the conditions were aerobic and better oxygenated as the contaminated water mixes with pure water near the plume edge. Others, (Payne & Durgin, 1988) have used the subsurface formation of CH_4, an indicator of extremely reduced conditions (E^o=-0.22 volts), to map plume migration in soils and aquifers. As the plume enters uncontaminated materials, oxygen and subsequent electron acceptors are depleted until CO_2 can be reduced to CH_4.

Kuhn et al. (1988) have shown that it is possible to transform alkylated benzenes under denitrifying conditions. Denitrification can commence when oxygen concentrations drop below 10 µmol L^{-1} (Roberston & Kuenen, 1984). Nozawa and Maruyama (1988) demonstrated that the anaerobic metabolism of phthalate and other aromatic compounds can occur by the denitrifying soil bacterium *Pseudomonas* sp. strain P136. In addition, dinoseb which is not degraded in contaminated soils under aerobic conditions can be degraded when anaerobic conditions are established (Stevens et al., 1991; Kaake et al., 1992). Anaerobiosis was developed by treating the soil with starchy potato-processing waste materials and allowing the aerobic population to deplete the available O_2. The anaerobic microbial consortium degraded dinoseb completely, without the formation of polymerization products produced under aerobic conditions. This approach has been applied to the removal of munition-contaminated soils. Funk et al. (1993) reported that a wide spectrum of explosives materials, including 2,4,6-trinitrotoluene, could be rapidly reduced and then degraded in soil systems when anaerobic conditions were established. Anaerobic processes are particularly effective at removing halogens from haloorganic compounds, however, under these conditions, biodegradation rates are relatively slow. The anaerobic transformation of polychlorinated biphenyls, DDT and percholoroethylene have also been reported (Suflita et al., 1982). Moreover, the nematicide 1,2-dibromo-3-chloropropane (DBCP), which persists in groundwater and soil, has been shown to be converted to organic products and CO_2 when soil suspensions are placed under anaerobic conditions (Bloom & Alexander, 1990).

USE OF ISOLATED MICROORGANISMS FOR BIOREMEDIATION

The bioremediation of many hazardous wastes frequently involves the application of microorganisms which have the ability to degrade recalcitrant pollutants. The applied microorganisms can be indigenous or exogenous with respect to the contamination site. While many of the degrading microorganisms used are naturally occurring and members of stable microbial communities (Lindstrom et al., 1991; Stevens et al., 1991), a major emphasis of recent bioremediation research has been the development and use of genetic engineering and selection techniques for the construction or isolation of microorganisms with enhanced degradation capabilities. However, establishing significant numbers of these naturally occurring or genetically modified strains in field sites and bioreactors has been, and remains, a critical problem. Faced with competition from stable members of indigenous populations, even highly efficient and desirable pollutant-degrading strains are destined to have a limited impact on the cleanup of hazardous waste sites.

At present, little information is known on what gives a microbial population the capability to become "indigenous." Moreover, there is much confusion and a lack of clear-cut definition on what allows the organism to become indigenous. What is known, however, is that strains of microorganisms which are stable members of a microbial community appear to have a selective, competitive advantage in occupying available niches in that local environment (Atlas & Bartha, 1993). When and how an organism becomes indigenous or established is presently not known, however some data indicate that removing an organism from a given soil environment, for a short time, and replacing it into the same environment does not appreciably change it's competitiveness and ability to reestablish itself as an indigenous member of the microbial community (T. McLoughlin, personnel communication). Primary or early preemptive colonization, followed by prolonged periods of stable maintenance in a soil population leads to the establishment of the "indigenous" state. Several studies, however, suggest that a nonindigenous microbe can become a member of the indigenous population by prolonged and repeated applications of soil inoculants (Dunigan et al., 1984; Yeung et al., 1989; McLoughlin et al., 1990a, b). Some studies suggest that while microbial numbers per se do not determine competitive success in soils and the establishment of indigenousness (Moawad et al., 1984), there is clearly a threshold level of indigenous microorganisms which precludes inoculation success with a similar or genetically related strain (McLoughlin et al., 1990a,b; Thies et al., 1991). This number has been estimated to be about 10 bacteria/gm soil of a particular organism. The ability of native strains to prevent successful colonization and persistence in a given soil ecosystem presents major problems for many *in situ* soil remediation technologies.

Perhaps one of the best models to examine competitiveness and the affects of indigenousness on the introduction of desired strains into soils deals with the *Rhizobium/Bradyrhizobium* legume-symbioses. It is recognized that the major problems in utilizing symbiotic-fixed N to enhance crop yield is the establishment of an introduced strain of rhizobia into the nodules of legumes grown in soils which contain indigenous populations of root-nodule bacteria (Weaver & Frederick, 1974; Ham 1980; Mead et al., 1985; Triplett & Sadowsky 1992; Dowling & Broughton, 1986). A number of studies with legumes have demonstrated that even at high inoculation rates, the proportion of nodules formed by the inoculum strain is generally less than 20% (Ham et al., 1971; Caldwell & Vest, 1970). Weaver and Frederick (1974) indicated that in order to establish an inoculum strain in 50% of the nodules of a soybean plant, the rate of inoculum application would have to be at least 1000 times greater than the indigenous population. Several studies have also indicated that in order for an inoculum strain to become established in a field soil, that population levels of indigenous organisms need to be low, less than 10 cells gm^{-1} soil, (Thies et al., 1991) or soil inoculations need to be done at high rates over successive growing seasons (Dunigan et al., 1984; McLoughlin et al., 1990a,b). These studies suggest that with the rhizobial strains and inoculation techniques presently available, the displacement of indigenous strains by those carried in inocula will be difficult and economically unfeasible. While many studies have addressed the issue of the competition for nodulation problem, and several mutants have been isolated with increased (or reduced) competitive capabilities (Triplett & Sadowsky, 1992) no information is available about the genetic basis for competitive success. It should be noted, however, that competition for nodulation of legumes is complex and is due to the fitness of the organism in the soil and the ability of the microbe to initiate a symbiotic interaction with a leguminous plant. It is very likely that many genes are acting in concert to control competitiveness of soil-borne bacteria.

Several factors are been thought to influence the competitiveness of indigenous and applied microorganisms in soils. These can be divided into abiotic and biotic factors which include: nutrient limitations and specificity, moisture requirements, pH, temperature relations, soil texture and porosity, organic matter content, bacteriocin and antibiotic production, solute types and concentrations, numbers and types of indigenous microorganisms, selective predation by protozoa, residence time in soils, mobility in soil, and time in laboratory culture prior to reintroduction into soils. All of these factors, either singly or in combination, play major roles in the introduction of organisms into natural environments for bioremediation purposes. The problems encountered, however, are largely determined by the type of bioremediation technology utilized. While repeated, high dosage applications of a particular biodegrading organism might be possible in *ex situ* and bioreactor applications, such practices are difficult and expensive for *in situ* processes. In addition to the requirement for controlled abiotic parameters for optimal microbial growth, one also must contend with the presence of indigenous microorganisms. For *in situ* applications, however, much promise has been shown using bioaugmentation procedures, where the growth of indigenous, biodegrading microorganisms are stimulated by the addition of fertilizers (Atlas & Bartha, 1993).

REFERENCES

Atlas, R.M., and R. Bartha. 1993. Microbial ecology: fundamentals and applications. The Benjamin/Cummings Publishing Company, Redwood City, California.

Bae, H.C., E.H. Cota-Robles, and L.E. Casida. 1972. Microflora of soil as viewed by transmission electron microscopy. Appl. Microbiol. 23:637-648.

Bloom, R.A., and M. Alexander. 1990. Microbial transformation of 1,2-dibromo-3-chloropropane (DBCP). J. Environ. Qual. 19:722-726.

Blum, U., T.R. Wentworth, K. Klein, A.D. Worsham, L.D. King, T.M Gerig, and S.W. Lyu. 1991. Phenolic acid content of soils from wheat-no till, wheat-conventional till, and fallow-conventional till soybean cropping systems. J. Chem. Ecol. 17:1045-1068.

Bollag, J.-M., 1992. Decontaminating soil with enzymes. Environ. Sci. Technol. 26:1876-1881.

Borchers, J.G. and D.A. Perry. 1992. The influence of soil texture and aggregation on carbon and nitrogen dynamics in southwest Oregon forests and clearcuts. Can. J. For. Res. J. Can. Rech. For. 22:298-305.

Brown, R.A., W. Mahaffey and R.D. Norris. 1993. *In situ* bioremediation: the state of the practice. p121-135. *In In situ* bioremdiation: When does it work. National Research Council. National Academy Press, Washington DC.

Caldwell, B.E. and G. Vest. 1970. Effect of *Rhizobium japonicum* strains on soybean yields. Crop Sci. 10:19-21.

Coleman D.C., C.P.P. Reid, and C. Cole. 1983. Biological strategies of nutrient cycling in soil systems. Adv. Ecol. Res. 13:1-55.

Collins, H.P., P.E. Rasmussen, C.L. Douglas Jr. 1992. Crop rotation and residue management effects on soil carbon and microbial dynamics. Soil. Sci. Soc. Am. J.56: 783-788.

Currie J.A. 1965. Diffusion within soil microstructure: a structural parameter for soils. J. Soil Sci. 16:279-289.

Dhawan, S., L.E. Erickson, and L.T. Fan. 1993. Model development and simulation of bioremediation in soil beds with aggregates. Groundwater 31:271-284.

Dowling, D. N., and W. J. Broughton. 1986. Competition for noidulation of legumes. Ann. Rev. Microbiol. 40:131-157.

Downing, A. L., and G. A. Truesdale. 1955. Some factors affecting the rate of solution of oxygen in water. J. Appl. Chem. 5:570-581.

Dunigan, E.P., P.K. Bollich, R.L. Hutchinson, P.M. Hicks, F.C. Zaunbrecher, S.G. Scott, and R.P. Mowers. 1984. Introduction and survival of an inoculant strain of *Rhizobium japonicum* in soil. Agron. J. 76:463-466.

Follett R.F., and D.S. Schimel. 1989. Effect of tillage practices on microbial biomass dynamics. Soil Sci. Soc. Amer. J. 53:1091-1096.

Foster, R.C. 1988. Microenvironments of soil microorganims. Biol. Fertil. Soils. 6:189-203.

Funk, S.B., D.J. Roberts, D.L. Crawford, and R.L. Crawford. 1993. Inital-phase of optimization for bioremediation of munition compound-contaminated soils. Appl. Environ. Micrbiol. 59:2171-2177.

Gottschal, J.C. 1986. Occurrence and functioning of anaerobic bacteria in oxidized environments. F.E.M.S. Symp. Fed. Eur. Microbiol. Soc. 33:263-273

Greenwood, D.J. 1975. Measurment of soil aeration in soil physical conditions and crop production. Min. Agric. Fish Food Tech. Bull. 29:261-390.

Ham, G.E. 1980. Inoculation of legumes with *Rhizobium* in competition with naturalized strains. p. 131-138. *In* W.E. Newton and W.H. Orme-Johnson (eds.) Nitrogen Fixation, Vol. II. University Park Press, Baltimore.

Ham, G.E., V.B. Cardwell, and H.W. Johnson. 1971. Evaluation of *Rhizobium japonicum* inoculants in soils containing naturalized populations of rhizobia. Agron. J. 63:301-303.

Hattori, T. and R. Hattori. 1976. The physical environment in soil microbiology: An attempt to extend the principles of microbiology to soil microbiology. Crit. Rev. Microbiol. 4:423-461.

Havel, J. and W. Reineke. 1993. Degradation of Aroclor 1221 in soil by a hybrid pseudomonad. F.E.M.S. Microbiol. Lett. 108: 211-218.

Heijnen, C.E. and J.A. van Veen. 1991. A determination of protective microhabitats for bacteria introduced into soil. FEMS Micro. Ecol. 85:73-80.

Hissett R. and T.R.G. Gray. 1976. Microsites and time changes in soil microbial ecology. p.23-39. *In* J. M. Anderson and A. MacFadyen (ed.) The role of terrestrial and aquatic organisms in decomposition processes. Blackwell, Oxford Great Britan.

Hinton, S.M. 1995. Bioremediation of the Valdez and Prall's Island oil spills. p. 208-216. *In* H.D. Skipper and R.F. Turco (ed.) Bioremediation: Science and applications. SSSA Spec. Publ. 43. ASA,CSSA and SSSA, Madison WI.

Huysman, F. and W. Verstraete. 1993. Water-facilitated transport of bacteria in unsaturated soil columns: influence of cell surface hydrophobicity and soil properties. Soil Biol. Biochem. 25:83-90.

Issa, S., L.P. Simmonds, and M. Woods. 1993a. Passive movement of chickpea and bean rhizobium through soils. Soil Biol. Biochem. 25:959-965.

Issa, S., L.P. Simmonds, and M. Woods. 1993b. Active movement of chickpea and bean Rhizobium in dry soils. Soil Biol. Biochem. 25:951-958.

Jenkinson, D.S., and J.M. Ladd. 1981. Microbial biomass in soil: Movement and turnover. p. 415-471. *In*: E. A. Paul and J. M. Ladd (ed.). Soil Biochemistry. Marcel Dekker. New York.

Kaake, R.H., D.J. Roberts, T.O. Stevens, R.L. Crawford, and D.L. Crawford. 1992. Bioremediation of soils contaminated with the herbicide 2-sec-butyl-4,6-dinitrophenol (dinoseb). Appl. Environ. Microbiol 58:1683-1689.

Kaiser, J.P., and J.M. Bollag. 1992. Influence of soil inoculum and redox potential on the degradation of several pyridine derivatives. Soil Biol. Biochem. 24:351-357.

Kilbertus, G., J. Proth, and B. Vervier. 1980. Effects de la dessiccation sur les bacteria gram-negatives d'un sol. Soil Biol. Biochem. 11:109-114.

Killham, K., M. Amato, and J.N. Ladd. 1993. Effect of substrate location in soil and soil pore-water regime on carbon turnover. Soil Biol. Biochem. 25:57-62.

Konopka A.E., and R.F Turco. 1991. Biodegradation of organic compounds in vadose zone and aquifer sediments. Appl. Environ. Micro. 57:2260-2268.

Kuhn, E.P., J. Zeyer, P. Eicher and R.P. Schwarzenbach. 1988. Anaerobic degradation of alkylated benezenes in denitrifying laboratory aquifer columns. Appl. Environ. Microbiol. 54:490-496.

Lewis, T.A. and Crawford, R.L. 1993. Physiological factors affecting carbon tetrachloride dehalogenation by the denitrifying bacterium *Pseudomonas sp.* strain KC. Appl. Environ. Microbiol. 59:1635-1641.

Lindstrom, J. E, R.C. Prince, J. C. Clark, M. J. Grossman, T. R. Yeager, J. F. Braddock, and E. J. Brown. 1991. Microbial populations and hydrocarbon biodegradation potentials in fertilized shoreline sediments affected by the T/V Exxon Valdez oil spill. Appl. Environ. Micro. 57:2514-2522.

Madsen E.L. 1991. Determining *in situ* biodegradation facts and challenges. Environ. Sci. Technol. 25:1663-1673.

Madsen E.L. and M. Alexander. 1982. Transport of *Rhizobium* and *Pseudomonas* through soil. Soil Sci. Soc. Am. J. 46:557-560.

Martin, J. K. and R. C. Foster. 1985. A model system for studying the biochemistry and biology of the root-soil interface. Soil Biol. Biochem. 17:261-269.

McLoughlin, T.J., S. Hearn, and S.G. Alt. 1990a. Competition for nodule occupancy of introduced *Bradyrhizobium japonicum* strains in a Wisconsin soil with low numbers of indigenous bradyrhizobia population. Can. J. Microbiol. 36:839-845.

McLoughlin, T.J., S.G. Alt, and P.A. Merlo. 1990b. Persistence of introduced *Bradyrhizobium japonicum* strains in forming nodules in subsequent years after inoculation in Wisconsin soils. Can. J. Microbiol. 36:974-800.

Mead, J., P. Higgins, and F.O. Gara. 1985. Studies on the inoculation and competitiveness of a *Rhizobium leguminosarum* strain in soils containing indigenous rhizobia. Appl. Environ. Microbiol. 49:899-903.

Moawad, H.A., W.R. Ellis, and E.L. Schmidt. 1984. Rhizosphere response as a factor in competition among three serogroups of indigenous *Rhizobium japonicum* for nodulation of field grown soybeans. Appl. Environ. Microbiol. 47:607-612.

Morgan, P. S.A. Lee,. S.T. Lewis, S.T., A.N. Sheppard, and R.J.Watkinson. 1993. Growth and biodegradation by white-rot fungi inoculated into soil. Soil-Biol-Biochem. 25:279-287.

Nozawa, T., Y. Maruyama. 1988. Anaerobic metabolism of phthalate and other aromatic compounds by a denitrifying bacterium. J. Bacteriol. 170:5778-5784.

Papendick, R. I. and G. S. Campbell 1981. p.1-22. Theory and measurment of water potential. *In* Elliott et al. (ed.) Water potential relations in soil microbiology. Soil Science Society of America. Madison WI.

Paul E. and F. E. Clark. 1989. Soil microbiology and biochemistry. Academic Press.

Parkin T.B., and D.R. Shelton, 1992. Spatial and temporal variability of carbofuran degradation in soil. J. Eniv. Qual. 21:672-678.

Parkin, T.B., J.A. Robinson, and D. R. Shelton. 1991. Evaluation of methods for characterizing carbofuran hydrolysis in soil. J. Eniv. Qual. 20:763-769

Parkin,T.B., and J.J. Meisinger. 1989. Denitrification below the crop rooting zone as influenced by surface tillage. J. Environ. Qual. 18:12-16.

Payne, T. B., and P. B. Dugin. 1988. Hydrocarbon vapor concentrations adjacent to tight underground gasoline storage tanks. p. 1173-1188. *In:* Proceedings of 2nd Outdoor Action Conf. on Aquifer Restoration, Ground Water Monitoring and Geophysical Methods. National Well Water Assoc. Dublin Ohio.

Pope, D.F., and J.E. Matthews 1993. Soil Treatment: Land Treatment. *In* Bioremediation of hazardous waste sites: practical approaches to implementation. U.S. EPA. EPA/600/K-93/002. p.7.1-7.5.

Powlson, D. S. 1980. The effects of grinding on microbial and non-microbial organic matter in soil. J. Soil Sci. 31:77-85.

Rao, P. S. C., K. S. V. Edvardsson, L. T. Ou, R. E. Jessup, P. Nkedi-Kizza, and A. G. Hornsby. 1986. Spatial variability of pesticide sorption and degradation parameters. p. 100-115. *In* W. Y. Garner, R. C. Honeycutt and H. N. Nigg (ed.) Evaluation of pesticide in groundwater. American Chem. Soc., Washington, D.C.

Rao, P.S.C., and R.J. Wagenet. 1985. Spatial variability of pesticides in field soils: methods of data analysis and consequences. Weed Sci. 33 (suppl 2):18-24.

Richaume, A., C. Steinberg, and L. Jocteru-Monrozier. 1993. Differences between direct and indirect enumeration of soil bacteria: the influence of soil structure and cell location. Soil Biol. Biochem. 25:641-643.

Roberston L.A., and J.G. Kuenen. 1984. Aerobic denitrification: a controversy revived. Aech. Microbiol. 139:351-354.

Roberston, G.P., M.A. Huston, F.C. Evans, and J.M. Tiedje. 1988. Spatial variability in a successional plant community: patterns of nitrogen avalability. Ecol. 69:1517-1524.

Sawhney, B.L., J.J. Pignatello, and S.M. Steinderg. 1988. Determination of 1,2-Dibromoethant (EDB) in field soils: Implications for volatile organic compounds. J. Environ. Qual. 17:149-152.

Schwarzenbach, R. P., P. M. Gschwend, and D. M. Imboden. 1993. Environmental Organic Chemistry. John Wiley and Sons, Inc. New York.

Sims, R.C. 1993. Site Characterization Requirements. *In* Bioremediation of hazardous waste sites: practical approaches to implementation. U.S. EPA. EPA/600/K-93/002. p.3.1-3.8.

Smith, K.A., and J.R.M. Arah. 1986. Anaerobic micro-environments in soil and the occurrence of anaerobic bacteria. F.E.M.S Symp. Fed. Eur. Microbiol. 33:247-261

Smith, M. S., G.W. Thomas, R.E. White, and N. Sitar. 1985. Transport of *Escherichia coli* through intact and disturbed soil columns. J. Environ. Qual. 14:87-91.

Song, H.G., and R. Bartha. 1990. Effects of jet fuel spills on the microbial community of soil. Appl. Environ. Microbiol. 56:646-651.

Sparling, G.P. 1985. The soil biomass. p 223-262. *In* D. Vaughan, R. E. Malcolm (ed.) Soil Organic matter and biological activity. Martinus Nijhoff Junk, Dordrecht.

Staley, T. E., W. M. Edwards, C.L Scott, and L.B. Owens. 1988. Soil microbial biomass and organic component alterations in a no-tillage chronosequence. Soil Sci. Soc. Am. J. 52:998-1005.

Stanier, R. Y, E. A. Adelberg, and J. Ingraham. 1976. The microbial world. Prentice-Hall Englewood Cliffs, New Jersey.

Steinberg, S.M. J.J. Pignatello, and B.L. Sawhney. 1987. Persistence of 1,2-dibromoethane in soils: entrapment in intraparticle micropores. Environ. Sci. Tech. 21:1201-1208.

Stevens,T.O., R.L.Crawford, and D.L. Crawford. 1991. Selection and isolation of bacteria capable of degrading dinoseb (2-sec-butyl-4,6-dinitrophenol). Biodegrad. 2:1-13.

Stroo, H.F. 1992. Biotechnology and hazardous waste treatment. J. Environ. Qual. 21:167-175.

Stumm, W., and J. J. Morgan. 1981. Aquatic chemistry. 2nd Ed. John Wiley and Sons. Inc. New York, New York.

Suflita, J.M., A. Horowitz, D.R. Shelton, and J.M. Tiedje. 1982. Dehalogenation: A novel pathway for the anaerobic biodegradation of haloaromatic compounds. Science. 218:1115-1117.

Tan,Y., W.J. Bond, and D.M. Griffin. 1992. Transport of bacteria during unsteady unsaturated soil water flow. Soil Sci. Soc. Am. J. 56:1331-1340.

Thies, J.E., P.W. Singleton, and B.B. Bohlool. 1991. Influence of size of indigenous rhizobial populations on establishment and symbiotic performance of introduced rhizobia on field-grown legumes. Appl. Environ. Microbiol. 57:19-28.

Tiedje, J.M. 1993. Bioremediation from an ecological perspective. p110-120. *In In situ* bioremdiaiton: When does it work. National Research Council. National Academy Press, Washington DC.

Torsvik, V., J. Goksoy, and F. L. Daae. 1990a. High diverstiy in DNA of soil bacteria. Appl. Envir. Micro. 56:782-787.

Torsvik, V., K. Salte, R. Sorheim, and J. Goksoyr. 1990b. Comparison of phenotypic diversity and DNA heterogeneity in a population of soil bacteria. Appl. Envir. Micro. 56:776-781.

Trevors, J.T., J.D. Elsas, L.S. Overbeek, and M.E. Starodub. 1990. Transport of a genetically engineered *Pseudomonas fluorescens* strain through a soil microcosm. Appl. Environ. Microbiol. Washington, 56:401-408.

Triplett, E.W., and M.J. Sadowksy. 1992. Genetics of competition for nodulation. Ann. Rev. Microbiol. 46:399-428.

Turco, R. F. and A. E. Konopka. 1990a. Response of microbial populations to carbofuran in soils enhanced for its degradation. p. 152-166. *In* R. D. Racke and J. R. Coats Enhanced Biodegradation of Pesticides in the Environment. ACS Symposium Series 426. Amer. Chem. Soc, Washington D.C.

Turco, R. F. and A. Konopka. 1990b. Biodegradation of carbofuran in enhanced and non-enhanced soils. Soil Biol. Biochem 22:195-201.

Turco, R.F., A. Konopka, and C. B. Roth. 1991. Changes in structure and adsorptive behavior of dissolved humic and fulvic acid mediated by groundwater microorganisms. Indiana Water Resource Research Center Report. West Lafayette, IN.

van Loosdrecht, M.C., J. Lyklema, W. Norde, G. Schraa and A. J. B. Zehnder. 1987. The role of the bacterial cell wall hydrophobicity in adhesion. Appl. Environ. Microbiol. 53:1898-1901.

van Loosdrecht, M.C.M., J. Lyklema, W. Norde, and A.J.B. Zehnder. 1990. Influence of interfaces on microbial activity. Microbiol. Rev. 54:75-87.

Weaver, R.W., and L.R. Frederick. 1974. Effect of inoculum rate on competitive nodulation of *Glycine max (L.)* Merrill. II. Field studies. Agron. J. 66:233-236.

Wilson, B. H., J. T. Wilson, D. H. Kampbell, B. E. Bledsoe, and J. M. Armstrong. 1990. Biotransformation of aromatic and chlorinated hydrocarbons at an aviation gasoline spill site. Geomicrobiology J. 8:225-240.

Wollum A.G., and D.K. Cassel. 1984. Spatial variability of *Rhizobium japonicum* in two North Carolina soils. Soil. Sci. Soc. Am. J. 48:1082-1087.

Wollum, A.G. 1982. Cultural methods for soil microorganims. p 781-801. *In* A. L. Page, R. Miller and D. Keeney (ed.) Methods of soil analysis, part 2. Soil Scien. Soc., Madison WI.

Yeung, K.-H., A.,M. Schell, and P.G. Hartel. 1989. Growth of genetically engineered *Pseudomonas aeruginosa* and *Pseudomonas putida* in soil and rhizosphere. Appl. Environ. Microbiol. 55:3243- 3246.

Zehnder A.J.B., and W. Strumm. 1988. Geochemistry and biogeochemisrty of anaerobic habitats. *In* Zender A.J.B. (ed.) Biology of Anaerobic Microorganism. John Wiley & Sons, New York.

6 Innovative Molecular Tools for Understanding Bioremediation -- 16s RNA Sequences and Panmers

S.M. Hinton, V. Minak-Bernero, and L.G. Keim
Exxon Corporate Research Center
Annandale, New Jesery

There is a debate over bioremediation strategies for petroleum pollution in open systems. Alternative strategies involve either, the addition of nutrients, or the "seeding" of microbes. The debate and advances in bioremediation technology are impeded by our limited understanding of natural microbial ecosystems. Historically, the study of microbial ecology has been hampered by the limitations in the classical techniques of microbiology, which were developed to reduce the complexity of natural consortium to simplify the study of microbes. Enrichment and pure culture techniques have been the foundation of microbiology and are essential for the phenotypic characterization of a microorganism. The power of the enrichment culture technique lies in the selection of individual organisms from a diverse community; its limitation is that the organisms must be grown in the laboratory.

Once a microorganism is in pure culture, the microbiologist must address the fundamental problem of taxonomy based upon arbitrary phenotypes (Woese, 1987) which has provided inadequate definitions of microbes. One classical strategy for the study of microbial ecosystems involves reassemblage of cultured members into a more complex community. This strategy assumes that the essential membership of the ecosystem can be cultured and the ecosystem is equal to the sum of the parts. These constraints yield an incomplete, and therefore, biased view of the bacteria in an environment. As a result of these technical constraints, the "picture" of the microbial community in developing a bioremediation strategy is usually of little consequence and is treated as a "black box". The ecosystem is not defined; therefore, it can not be adequately considered for the application of bioremediation. Monitoring bioremediation in the field typically consists of measuring inputs and outputs from the ecosystem. The input is the pollutant and the output is the transformation of the component(s) of the contamination. Therefore, the efficacy of bioremediation is usually defined by (limited) chemical analysis. The lack of "tools" to study the biological component of BIOremediation and the cost of monitoring the biodegradation process in the field have limited our understanding of the role of the microbial community in the biodegradative process. As a result, the microbiologisthas limitations to help design bioremediation strategies for open systems, such as oil spills.

In the past decade, advances in biotechnology have greatly expanded the "tools" available to the microbial ecologist. For example, molecular genetic technology has revolutionized the comparative studies of biopolymers, i.e., the amino acid sequence of homologous proteins and the nucleotide sequence of homologous nucleic acids. Comparison of macromolecular sequences has been used to infer natural or phylogenetic relationships and has spawned the field of molecular evolution. Central to molecular evolution studies is the concept that a biopolymer is a historical record (Zuckerkandl & Pauling, 1965) and differences in that record or sequence of homologous biopolymers is related to evolutionary divergence. The identification of specific biopolymers from microbial communities, rather than the microbes themselves, alleviates the need for cultivation, provides a definition of the genetic diversity, and can provide an understanding of the natural relationship among microorganisms. Woese (1987) championed the use of ribosomal RNAs (rRNA) as the biopolymer for identifying organisms and inferring natural relationships. The rRNAs can be used to characterize uncultivated microbes (Seewaldt & Stackebrandt, 1982), and to infer their phylogeny because these molecules are universally distributed, constant in function, slow to change, easily manipulated, and contain ample sequence information (Woese, 1987).

THE STUDY OF MICROBIAL DIVERSITY

Pace and colleagues pioneered rRNA-based analysis of natural microbial communities (Pace et al., 1985). Figure 6-1 is a diagram of the strategy employed for phylogenetic analysis of microbial populations (Lane et al., 1985). The essential steps in this strategy are the extraction and purification of nucleic acids, the amplification of the desired rRNA sequences from bulk genomic DNA, the isolation and purification of the individual rRNA sequences by cloning, and rapid methods for DNA sequence analysis. This approach has been applied to a few selected environments and the tentative conclusion is that cultured microbes represent only a small proportion of the microbial diversity (Giovannoni et al., 1990; Ward et al., 1990).

If it is true, that the vast majority of microbes and the biochemical diversity they represent are unknown, we are faced with the question how to study or probe this heretofore "black box"? The full extent of microbial diversity is unknown but could be appreciated with an understanding of their evolutionary divergence. Therefore, the conceptual framework for our studies has its basis in the principles of molecular evolution. The technical challenge is to devise techniques to identify biomolecules from natural populations and relate them to more characterized microbes or enzymes which would assist in the prediction of traits, attributes, or enzymatic activities. The objectives of our work are to determine the microbial diversity and the relative abundance of each microbe in a consortium, and determine the potential that a microbial consortium has for a biodegradation activity.

PURIFICATION OF ENVIRONMENTAL NUCLEIC ACIDS

There are a number of key technical hurdles in this rRNA strategy (see Fig. 6-1) and in other molecular studies that have to be overcome for them to be used for routine analysis of environmental samples. The first hurdle for any molecular analysis is the extraction and purification of nucleic acids from environmental samples. Humic substances found in aquatic, soil, and sediment environments are the most notorious contaminants co-extracted with nucleic acids that inhibit molecular reactions, i.e., the polymerase chain reaction (Tsai & Olson, 1992). The

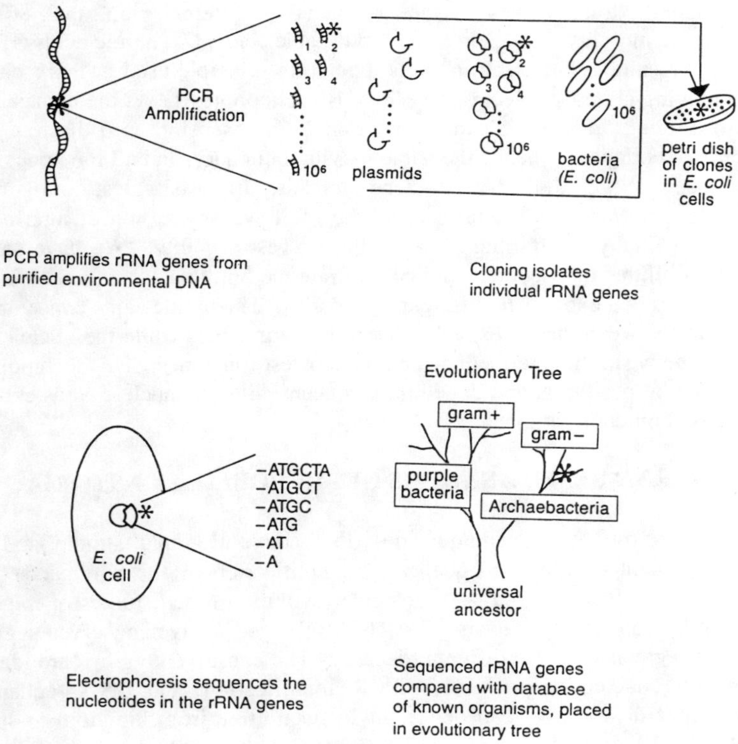

Fig. 6-1. Outline of the 16S rRNA strategy for analysis of microbial populations.

interfering humic substances have phenolic groups that denature biomolecules by bonding to N-substituted amides or oxidize to form a quinone which covalently bonds to DNA or proteins. To address this problem, we have developed two electrophoretic techniques that purify nucleic acids from humic acids. One technique used to purify environmental nucleic acids is a simple modification of the standard submarine agarose gel procedure. Humic acids comigrate with nucleic acids under standard electrophoretic conditions for anionic migration of ions, but the electrophoretic mobility of humic compounds can be altered by adding a substance that binds to them. Polyvinylpyrrolidone (PVP) (Loomis & Battaile, 1966; Maliyakal, 1992) is such a compound as it forms hydrogen bonds with phenolic compounds and has been successfully used to remove them from plant extracts. Purification of DNA is achieved by the addition of (0.2% W/V) PVP to the gel buffer which retards the electrophoretic mobility of phenolic compounds (i.e., the tracking dye bromophenol blue) by H-bonding to them, thereby preventing their comigration with the nucleic acids (Young et al., 1993).

The second electrophoretic technique used to purify nucleic acids from environmental extracts takes advantage of the difference in the effective electrophoretic mobility of nucleic acids and organic acids, i.e., humic acids (Hinton, unpublished results). The electrophoretic operating principle used in this technique is called moving boundary electrophoresis or isotachophoresis. As the names imply, ions with different effective electrophoretic mobilities separate into discrete zones which creates boundaries, hence the name moving boundary, and all ion zones move at equal speeds, hence the name iso-tacho-phoresis. In the simplest configuration, there is a leading ion with the fastest mobility, followed by an ion of intermediate mobility, and finally the trailing ion with the slowest mobility. We have selected ions with mobilities that separate and concentrate the humic acids and nucleic acids from environmental extracts in different boundaries. The humic acids concentrate in the boundary between the fastest and intermediate ion zones while the nucleic acids concentrate between the intermediate and the slowest ion zones. This technique has the advantage of purifying, and in addition, concentrating the nucleic acids extracted from an environmental sample.

16S rRNA-BASED ANALYSIS OF MICROBIAL CONSTORIA

The electrophoretic techniques described above are rapid, simple, and yield DNA of sufficient quantity and quality for amplification by the polymerase chain reaction (PCR). PCR is central to molecular evolution based studies of microbial consortium by 16S rRNA analysis. The 16S rRNA genes have conserved segments while other segments vary to different degrees. The conserved regions are ideal for designing oligonucleotide primers for PCR amplification (and DNA sequencing) which are specific for any level of the phylogenetic tree from kingdom to species (Lane et al., 1985; Giovannoni et al., 1988). The variable segments are the informative regions used to identify the microbes and to determine their relationship to other life on the planet. The closer the relationship that an unknown microbe has to characterized organisms the more confidence one has in correlating basic attributes of the uncultured microorganism. One of the limitations of this technique would be the size of the rRNA database of known microorganisms and the breadth of natural diversity it represents. The ribosomal RNA database project (Olsen et al.,

1991) manages the collection of prokaryotic rRNA sequences (1379 entries) which has tripled in size in the last few years. As the reference set of microorganisms in the rRNA database grows with time so does the power of this approach for studying microbial diversity.

The 16S rRNA-based strategy to study natural microbial populations (see Fig. 6-1) through DNA sequencing of individual rRNA clones reveals a picture of the membership of the microbial community. In a sense, the resolution or number of pixels, composing the "picture" of the community structure is dependent upon the number of clones sequenced. The inherent problems with this approach is that randomly selected rRNA clones probably represent the most abundant members. It is difficult to determine what would be the effort required to have confidence one has resolved the entire community structure. This approach is designed to study microbes in a natural consortium, but for routine analysis the effort required to study the community structure would be prohibitive.

DENATURING GRADIENT GEL ELECTROPHORESIS

We are attempting to modify the rRNA-based approach for community structure analysis. The challenge is to devise a method that resolves the collection of 16S rRNA PCR fragments so the community structure can be visualized, possibly as a pattern or map. The points in the pattern or map representing the community of microorganisms would provide basic population information and allow the investigator to choose any member for further scrutiny by DNA sequence analysis. With this concept in mind, the cloning step for the isolation of 16S rRNA fragments must be eliminated because the community structure is hidden in the clones. In place of cloning, we are developing a method to isolate the mixture of 16S rRNA fragments with the sequence-dependent separation technique called denaturing gradient gel electrophoresis (DGGE; Fischer & Lerman, 1983). The advantage of this technique is that PCR amplification from specific conserved sites in the 16S rRNA sequence generates a mixture of DNA fragments that are unresolvable in length, but differ significantly in nucleotide sequence. The operating principle of DGGE is that electrophoretic mobility of DNA in polyacrylamide gels is sensitive to the secondary structure of the molecule. Partially melted molecules of DNA migrating more slowly than double helices. The experimental system involves a fragment of DNA migrating into an increasing concentration of solvent until the denaturing conditions are greater than the energy of strand association which is dependent upon the nucleotide sequence. At the point where the conditions melt the DNA fragment, there is an abrupt decline in the effective electrophoretic mobility with a sharpening of the DNA band. One important condition for DGGE separation is that the difference in the DNA sequences be in the region that partially melts, in other words, the region with the lowest energy of strand association. Lerman and colleagues (Lerman & Silverstein, 1987) have developed a model for the simulation of DNA melting which can predict the region in a given sequence that has the least thermostability. In addition, they have simulations of the thermostability of the DNA helix correlated with the expected change in electrophoretic mobility. This information is the first step toward optimizing DGGE for the separation of related 16S rRNA sequences. Representatives from each phylogenetic group were analyzed to locate the lowest binding energy domain in their 16S rRNA sequence. The analysis revealed that between positions 100 and 300 (based upon *Escherichia coli* numbering) the 16S rRNA sequences have the least thermostability (Hinton, unpublished). For the majority of microorganisms this region contains enough phylogenetic information to identify an organism (Woese, private communication) and ample sequence variation to allow for sequence-dependent separation of rRNA molecules. Currently, we are in the process of determining standardized conditions for DGGE to allow the position of the band to be correlated with the identity of the organism and the relative size of the band to be correlated to the relative abundance of the individual. This approach would allow one to follow the dynamic changes in the microbial population during the biodegradation process.

BIODEGRADATIVE POTENTIAL

Another important measure is the biodegradative potential and the extent that potential is expressed. The biodegradative potential (and not activity) would be reflected in the DNA encoding the hydrocarbon-oxidizing enzymes, and the expression of that genetic potential would be reflected in the level of mRNA for these enzymes. Nucleic acid probe technology has provided invaluable tools for rapid and sensitive methods of detection. Typically, a DNA probe is specific for a genetic sequence, therefore, species-specific and function-specific probes can be designed to identify and quantitate a microbe or a degradative gene, respectively. An inherent problem with this approach for the study of microbial consortium is that the specificity limits the ability to probe unknown natural diversity. Conceptually, this problem is analogous to the limitations of conventional culture-based studies of microbial ecology. As an example, Westlake and colleagues probed aromatic-degrading microbes with gene-specific probes, and concluded that specific-gene probes are limited and would underestimate the potential for aromatic-degradation (Foght & Westlake, 1991). They suggested that a suite of gene probes would be required to study a mixed population. Reconsider for a moment the purpose of probing a microorganism versus a mixed microbial population. For an individual organism, probes are designed to determine if a specific enzyme is present or not. On the other hand, for microbial populations, the probe should be designed to determine if an activity is present, and it should be promiscuous enough to detect the extent of the diversity representing that activity. Therefore, we have adopted an approach to study natural diversity by designing activity probes specific for the structure which the function is dependent upon.

STRUCTURALLY CONSERVED ELEMENTS

The underlying premise for designing structure-function probes is explained by Kimura (1968). The neutral mutation theory of molecular evolution argues that the differences in the rate of evolution between molecules and different parts of the genome are accounted for by conservation of biologically significant sequences. The neutral theory predicts that divergent evolution gave rise to homologous family of proteins which are identified as related sequences by shared structurally conserved elements (SCE). The evolutionary process has conserved the crucial components or structure-function domains into "patches" or discontinuous patterns of identity (Hubbard & Blundell, 1987) even when the overall primary sequence similarity is low (Greer, 1990). These regions of identity are ideal candidates for designing primers for PCR-based detection of the homologous family of biomolecules whether all members have been characterized or not.

The first step in identifying "patches" of SCE is to align the amino acid sequences of homologous proteins. Ideally, the protein's primary sequences are aligned so there is a one-to-one correspondence between amino acids that are derived from a common ancestral residue. The difficulty in aligning primary sequences is the shear size of the sequence data, and the divergence of homologous sequences through internal mutations, i.e., amino acid substitution, deletion and insertion. In general, sequence analysis is best handled by the computer. The computer algorithm of Smith and Smith (1992) for pattern-induced multi sequence alignment (PIMA) is a method to construct an alignment of homologous sequences based on their common patterns of conserved sequence elements. As an example, a PIMA alignment of amino acid sequences for the benzene, toluene, naphthalene, and biphenyl dioxygenase family is shown in Fig. 6-2. The PIMA algorithm constructs a pattern of conserved (primary) sequence elements common to all members of the dioxygenase proteins by optimizing the alignment at each position based upon a hierarchy of amino acid identity or equivalence. These SCEs for the most part are regions of the protein sequence alignment with the minimal amount of ambiguous codons represented by the aligned residues. The reason for SCE having the least degeneracy for nucleotide probe design is twofold; one, identical amino acids at a position reduces the multiplicity, and two, at an alignment position with amino acids considered to be equivalent substitutions (of physical properties) frequently have similar codon usage.

The PIMA program has an additional feature that assists in the design of PCR primers for homologs. Initially, the set of related sequences is pair wise locally aligned which generates a similarity scores for clustering (maximal linkage rule; Sneath & Sokal, 1973) the sequences into subfamilies. This can be used as a measure of the diversity of the homolog sequences, and assists in determining if a single set of PCR primers can be designed for the entire family. If the family diversity is too great, PCR primers can be made to each subfamily of sequences.

PANMERS

There are a number of strategies to design oligonucleotide primers for a consensus of amino acid sequences. A consensus primer would be an oligonucleotide or mixture of oligonucleotides that represent all possible codons for the consensus protein sequence and which does not reflect a bias of codon usage for any organism. The consequence of this approach is a primer design with the maximum degeneracy. The problem with highly degenerate mixtures of primers is an increase in nonspecific priming (Compton, 1990). In addition, the theoretical abundance of the absolute homologous primer may be quite low, which predicts that if the primer multiplicity is great enough the theoretical maximum product yield would be quite small. However, the theoretical prediction is not empirically observed indicating that less than perfect matches between template and primer will lead to product synthesis (Gould et al., 1989). Studies have suggested that highly degenerate primers support PCR amplification because of the choice of less stringent annealing conditions. Primer-template mismatches are tolerated in the 5'-end of the oligonucleotide (Sommer & Tautz, 1989), and the use of longer oligomers increase the specificity of the reaction, which compensates for the ambiguity of high degeneracy. One other important factor effecting product yield is the efficiency of mismatched template-primer extension is critical only in the early cycles. Once a mismatch primer-template is extended, the products are exact compliments with the primers for subsequent cycles.

An alternate approach to designing primers from corresponding amino acid sequence involves incorporating degenerate nucleotides, i.e., deoxyinosine, at positions with ambiguous codons to reduce the multiplicity of primers (Knoth et al., 1988; Lin & Brown, 1992). Deoxyinosine can base pair with low discrimination with all other bases. This nonspecific binding results in the loss of specific sequence information resulting in an increase in nonspecific priming. The synthesis of oligonucleotides containing these degenerate nucleotides is more costly and reduces the efficiency of synthesis (approximately 5% per base). The limitation of mixed oligonucleotide primers and primers containing degenerate nucleotides prompted our attempts to develop a strategy for designing "pan" primers, or panmers.

Recently, there was a report on the effects of primer-template mismatches on the PCR product yield (Kwok et al., 1990). The effect of mismatches at the 3'-end of the primer-template are known to affect PCR more dramatically than mismatches elsewhere (Petruska et al., 1988), therefore, this would be the most sensitive test of mismatches on PCR efficiency. As one would predict, the effect of mismatches on PCR amplification was dependent upon both the position and nature of the mismatches. What was not anticipated was that certain 3'-terminal mismatches, either on the primer or template, amplified as efficiently as fully complimentary primer-template pairs. In addition, mismatches 6 to 7 bases from the 3'-terminal end had little effect, if any, on the efficiency of PCR amplification (Sommer & Tautz, 1989). These results are incongruous with the central dogma of discrete one-to-one pairing of purines and pyrimidines (the GC and TA pairing "rule"). Table 6-1 shows the degeneracy of each deoxynucleotide which was effective in primer-template pairing. Deoxythymidine, like deoxyinosine, has four fold degeneracy, in other words, it can effectively base pair with any of the four nucleotides in DNA. Deoxyguanine and deoxycytosine both have three fold degeneracy, but are effective in pairing with different subsets of nucleotides (G pairs with either T, C or G, and C pairs with either T, G, or A). And deoxyadenosine is the least promiscuous with two fold degeneracy (effectively pairing with T or C). Could this type of degeneracy be the reason for the observed mismatch priming with highly degenerate primers? If so, this would allow the rational design of a "panmer" utilizing the degeneracy of each dioxynucleotide to accommodate codon ambiguity. Potentially, the panmer, a single oligonucleotide, could be constructed to recognize a homologous family of sequences. The conventional nucleotides, in contrast to deoxyinosine, are not all four-fold degenerate, therefore, retain sequence specificity.

To test the validity and possible utility of this strategy, we constructed single oligonucleotides to recognize SCEs of the dioxygenase family (see Fig. 6-2) using the observed degeneracy of each deoxynucleotide as rules (see Table 6-1) for panmer design. In order to test the scope and limitations of this strategy, two sets of panmers were constructed. One panmer set was based upon the consensus of the nucleotide sequences which has less degeneracy than the second set of panmers

Table 6-1. Effective primer-template pairing for PCR amplification.

Primer	Template			
T	T	C	G	A
C	T	—	G	A
G	T	C	G	—
A	T	C	—	—

designed for the consensus of the amino acid sequences. Additional conditions for the panmer design parameters were the oligomer length between 28 to 32 nucleotides for specificity, the calculated Tm for each panmer pair were adjusted to within 2°C of each other in an attempt to balance the annealing temperatures, and an effort was made to conserve as much genuine sequence at the 3'-terminal end of the primer as possible. Total nucleic acids extracted from *Pseudomonas putida* containing the nahA gene was used as the model template for PCR.

Fig. 6-3 shows the nucleotide sequence of the dioxygenase panmers and the corresponding nucleotide sequence of *nahA* for comparison to indicate conventional base-pairing mismatches and their locations. The results are that both the nucleotide and amino acid consensus primer sets yielded the *nahA* PCR product. In general, we have observed that up to one-third of the primer-template pairing can be conventional mismatches and function effectively with the substitutions of the appropriate deoxynucleotides (see Table 6-1). Currently, we are performing experiments to determine the scope and limitations of deoxynucleotide degeneracy for PCR in order to establish panmer design parameters. The focus of this effort is to design molecular probes that will encompass genetic diversity and detect "key" enzymatic activities to monitor biodegradation.

Research to probe microbial diversity is driven by the realization that our knowledge of microbial ecosystems is rudimentary, and new or novel approaches are demanded to reverse the perspective of the classical reductionism of conventional biology. Similarly, research into the genomes of various organisms will expand the study of genetics of single genes to the population of genes called the chromosome. This parallel may seem disconnected because they are different biological systems or programs, but the technology driving both studies will be synergistic. A better understanding of the ecology of biodegradation will lead to the development of bioremediation strategies that are rationally designed on the biology of the process.

```
    PSBPHA-1+   0MSSAIKEVQGAPVKWVTNWTPEAIRGLVDQEKGLLDPRIYADQSLYELEL
    PSBPHA-2+   0MSSSIKEVQGAPVKWVTNWTPEAIRGLVDQEKGLLDPRIYADQSLYELEL
    PSEPUBDO+   0    MNQTDTSPIRLRRSWNTSEIEALFDEHAGRIDPRIYTDEDLYQLEL
  PSEPUTODC1+   0    MNQTDTSPIRLRRSWNTSEIEALFDEHAGRIDPRIYTDEDLYQLEL
  PSEPUNAHAc+   0                   MNYKNKILVSESGLTQKHLIHGDEELFQHEL
   PSEPUNDOA+   0                   MNYNNKILVSESGLSQKHLIHGDEELFQHEL
                 |---------|---------|---------|---------|---------
                 1         11        21        31        41

    PSBPHA-1+  50ERVFGRSWLLLGHESHVPETGDFLATYMGEDPVVMVRQKDKSIKVFLNQC
    PSBPHA-2+  50ERVFGRSWLLLGHESHVPETGDFLATYMGEDPVVMVRQKDKSIKVFLNQC
    PSEPUBDO+  46ERVFARSWLLLGHETQIRKPGDYITTYMGEDPVVVVRQKDASIAVFLNQC
  PSEPUTODC1+  46ERVFARSWLLLGHETQIRKPGDYITTYMGEDPVVVVRQKDASIAVFLNQC
  PSEPUNAHAc+  31RTIFARNWLFLTHDSLIPSPGDYVTAKMGIDEVIVSRQSDGSIRAFLNVC
   PSEPUNDOA+  31KTIFARNWLFLTHDSLIPAPGDYVTAKMGIDEVIVSRQNDGSIRAFLNVC
                 |---------|---------|---------|---------|---------
                 51        61        71        81        91

    PSBPHA-1+ 100RHRGMRICRSDAGNAKAFTCSYHGWAYDIAGKLVNVPFEKEAFCDKKEGD
    PSBPHA-2+ 100RHRGMRICRSDAGNAKAFTCSYHGWAYDIAGKLVNVPFEKEAFCDKKEGD
    PSEPUBDO+  96RHRGMRICRADAGNAKAFTCSYHGWAYDTAGNLVNVPYEAESF--------
  PSEPUTODC1+  96RHRGMRICRADAGNAKAFTCSYHGWAYDTAGNLVNVPYEAESF--------
  PSEPUNAHAc+  81RHRGKTLVNAEAGNAKGFVCSYHGWGFGSNGELQSVPFEKELY------GE
   PSEPUNDOA+  81RHRGKTLVSVEAGNAKGFVCSYHGWGFGSNGELQSVPFEKDLY------GE
                 |---------|---------|---------|---------|---------
                 101       111       121       131       141

    PSBPHA-1+ 150CGFDKAEWGPLQARVATYKGLVFANWDVQAPDLETYLGDARPYMDVMLDR
    PSBPHA-2+ 150CGFDKAEWGPLQARVATYKGLVFANWDVQAPDLETYLGDARPYMDVMLDR
    PSEPUBDO+ 139ACLNKKEWSPLKARVETYKGLIFANWDENAVDLDTYLGEAKFYMDHMLDR
  PSEPUTODC1+ 139ACLNKKEWSPLKARVETYKGLIFANWDENAVDLDTYLGEAKFYMDHMLDR
  PSEPUNAHAc+ 126SLNKKCLGLKEVARVESFHGFIYGCFDQEAPPLMDYLGDAAWYLEPIFKH
   PSEPUNDOA+ 126SLNKKCLGLKEVARVESFHGFIYGCFDQEAPPLMDYLGDAAWYLEPMFKH
                 |---------|---------|---------|---------|---------
                 151       161       171       181       191

    PSBPHA-1+ 200TPAGTVAIGGMQKWVIPCNWKFAAEQFCSDMYHAGTTTHLSGILAGIPPE
    PSBPHA-2+ 200TPAGTVAIGGMQKWVIPCNWKFAAEQFCSDMYHAGTMSHLSGILAGMPPE
    PSEPUBDO+ 189TEAGTEAIPGVQKWVIPCNWKFAAEQFCSDMYHAGTTSHLSGILAGLPED
  PSEPUTODC1+ 189TEAGTEAIPGVQKWVIPCNWKFAAEQFCSDMYHAGTTSHLSGILAGLPED
  PSEPUNAHAc+ 176S-GGLELVGPPGKVVIKANWKAPAENFVGDAYHVG-WTHASSLRSGESIF
   PSEPUNDOA+ 176S-GGLELVGPPGKVVIKANWKAPAENFVGDAYHVG-WTHASSLRSGESIF
                 |---------|---------|---------|---------|---------
                 201       211       221       231       241
```

Fig. 6-2. PIMA alignment of the amino acid sequences of biphenyl dioxygenases, PSBPHA-1 (Erickson & Mondello, 1992), and PSBPHA-2 (Taira et al., 1992); benzene dioxygenase, PSEPUBDO (Irie et al., 1987); toluene dioxygenase, PSEPUTODC1 (Zylstra & Gibson, 1989); and naphthalene dioxygenases, PSEPUNAHAc (Harcourt et al., 1992) and PSEPUNDOA (Kurkela et al., 1988). The SCE are outlined in boxes.

```
PSBPHA-1+     250MDLSQAQIPTKGNQFRAAWGGHGSGWYVDEPGSLLAVMGPKVTQYWTEGP
PSBPHA-2+     250MDLSHAQVPTKGNQFRAGWGGHGSGWFVDEPGMLMAVMGPKVTQYWTEGP
PSEPUBDO+     239LEMADLAPPTVGKQYRASWGGHGSGFYVGDPNLMLAIMGPKVTSYWTEGP
PSEPUTODC1+   239LEMADLAPPTVGKQYRASWGGHGSGFYVGDPNLMLAIMGPKVTSYWTEGP
PSEPUNAHAc+   224ASLAGNAVLPPEGAGLQMTSKYGSGMGVLWDGYSGVHSADLVPELMAFGG
PSEPUNDOA+    224SSLAGNAALPPEGAGLQMTSKYGSGMGVLWDGYSGVHSADLVPELMAFGG
                 |---------|---------|---------|---------|---------
                251       261       271       281       291

PSBPHA-1+     300A-AELAEQRLGHTGMPVRRMVGQHM--TIFPTCSFLPTFNNIRIWHPRGP
PSBPHA-2+     300A-ADLAEQRLGHT-MPVRRMFGQHM--SVFPTCSFLPAINTIRTWHPRGP
PSEPUBDO+     289A-SEKAAERLGSV-ERGSKLMVEHM--TVFPTCSFLPGINTVRTLASARA
PSEPUTODC1+   289A-SEKAAERLGSV-ERGSKLMVEHM--TVFPTCSFLPGINTVRTWHPRGP
PSEPUNAHAc+   274SKQERLNKEIGD---VRARIYRSHLNCTVFPNNSMLTCSGVFKVWNPIDA
PSEPUNDOA+    274AKQERLNKEIGD---VRARIYRSHLNCTVFPNNSMLTCSGVFKVWNPIDA
                 |---------|---------|---------|---------|---------
                301       311       321       331       341

PSBPHA-1+     347NEIEVWAFTLVDADAPAEIKEEYRRHNIRNFSAGGVFEQDDGENWVEIQK
PSBPHA-2+     346NEIEVWAFTLVDADAPAEIKEEYRRHNIRTFSAGGVFEQDDGENWVEIQK
PSEPUBDO+     335ERGEVWAFTVVDADAPDDIKEEFRAR-LRTFSPVACSSRTTGRTGSRSST
PSEPUTODC1+   335NEVEVWAFTVVDADAPDDIKEEFRRQTLRTFSAGGVFEQDDGENWVEIQH
PSEPUNAHAc+   321NTTEVWTYAIVEKDMPEDLKRRLADSVQRTFGPAGFWESDDNDNMETASQ
PSEPUNDOA+    321NTTEVWTYAIVEKDMPEDLKRRLADSVQRTFGPAGFWESDDNDNMETASQ
                 |---------|---------|---------|---------|---------
                351       361       371       381       391

PSBPHA-1+     397GLRGYKAKSQPLNAQMGLGRSQTGHPDFPGNVG-YVYAEEAARGMYHHWM
PSBPHA-2+     396GLRGYKAKSQPLNAQMGLGRSQTGHPDFPGNVG-YVYAEEAARGMYHHWM
PSEPUBDO+     384SCEA-TSRSRPFNAEMSMDQTVDNDPVYPGRISNNVYSEEAARGLYAHWL
PSEPUTODC1+   385ILRGHKARSRPFNAEMSMDQTVDNDPVYPGRISNNVYSEEAARGLYAHWL
PSEPUNAHAc+   371NGKKYQSRDSDLLSNLGFGKDVYGDAVYPGVVGKSAIGETSYRGFYRAYQ
PSEPUNDOA+    371NGKKYQSRDSDLLSNLGFGEDVYGDAVYPGVVGKSAIGETSYRGFYRAYQ
                 |---------|---------|---------|---------|---------
                401       411       421       431       441

PSBPHA-1+     446RMMSEPSWATLKP
PSBPHA-2+     445RMMSEPSWATLKP
PSEPUBDO+     433RMMTSPDWDALKATR
PSEPUTODC1+   435RMMTSPDWDALKATR
PSEPUNAHAc+   421AHVSSSNWAEFEDASSTWHTELTKTTDR
PSEPUNDOA+    421AHVSSSNWAEFEHASSTWHTELTKTTDR
                 |---------|---------|---------|---------|---------
                451       461       471       481       491
```

Fig. 6-2. Continued.

```
Primer 1 (nuc) 5' gcgttcctgaacgcttgccggcaccgcggc 3'
nahA           5' gccttcctgaacgtttgtcggcaccgtggc 3'
                  110111111111101110111111110111

Primer 1 (aa)  5' gctttccttaacgcttgccgtcaccgtggt 3'
nahA           5' gccttcctgaacgtttgtcggcaccgtggc 3'
                  110111110111101110110111111110

Primer 2 (nuc) 5' atcgccgggaaagttctctgcgggtagcttcca 3'
nahA           5' atcccccacaaagttttccgcgggtgccttcca 3'
                  111011000111111011011111100111111

Primer 2 (aa)  5' atctgctggaaatttctctgctggtaccttcca 3'
nahA           5' atcccccacaaagttttccgcgggtgccttcca 3'
                  111001000111011011011011101111111
```

Fig. 6-3. Primers 1 and 2 are the panmers designed for the nucleotide (nuc) and amino acid (aa) consensus of the dioxygenase family sequence alignment (see Figure 2) located between positions 95 to 104, and 220 to 230, respectively. The numbers below each set of nucleotide sequences indicate primer-nahA template conventional mismatch (0) and match (1).

REFERENCES

Compton, T. 1990. in PCR protocols, A guide to Methods and Applications (Eds. Innis, M.A., D.H. Gelfand, J.J. Sninsky, and T.J. White), Academic Press Inc., New York, Chapter 5, 39-45.

Erickson, B.D., and F.J. Mondello. 1992. Nucleotide sequencing and transcriptional mapping of genes encoding biphenyl dioxygenase, a multicomponent PCB-degrading enzyme in Pseudomonas strain LB400. J. Bacteriol. 174: 2903-2912.

Fischer, S.G., and L.S. Lerman. 1983. DNA fragments differing by single base-pair substitutions are separated in denaturing gradient gels: Correspondence with melting theory. Proc. Natl. Acad. Sci. USA 80: 1579-1583.

Foght, J.M., and D.W.S. Westlake. 1991. Cross hybridization of plasmid and genomic DNA from aromatic and polycyclic aromatic hydrocarbon degrading bacteria. Can.J. Microbiol. 37: 924-932.

Giovannoni, S.J., T.B. Britschgi, C.L. Moyer, and K.G. Field. 1990. Genetic diversity in Sargasso sea bacterioplankton. Nature 345: 60-63.

Giovannoni, S.L., E.F. DeLong, G.J. Olsen, and N.R. Pace. 1988. Phylogenetic group-specific oligodeoxynucleotide probes for identification of single microbial cells. J. Bacteriol. 170:720-726.

Gould, S.J., S. Subramani, and I. E. Scheffler. 1989. Use of the DNA polymerase chain reaction for homology probing: Isoation of partial cDNA or genomic clones encoding the iron-sulfur protein of succinate dehydrogenase from several species. Proc. Natl. Acad. Sci. USA 86: 1934-1938.

Greer, J. 1990. Comparative modeling methods - Application to the family of the mammalian serine proteases. Proteins 7: 317-334.

Harcourt, A.A., W.-c, Suen, D.L. Cruden, D.T. Gibson and G.J. Zylstra. 1992. Nucleotide sequences encoding the genes for naphthalene dioxygenase in *Pseudomonas putida* strain G7 and Pseudomonas strain NCIB9816-4. Genbank submission

Hubbard, T.J.P., and T.L. Blundell. 1987. Comparison of solvent-inaccessible cores of homologous proteins - definitions useful for protein modeling. Protein Engr. 1: 159-171.

Irie, S., S. Doi, T. Yorifuji, M. Takagi, and K. Yano. 1987. Nucleotide sequencing and characterization of the genes encoding benzene oxidation enzymes of Pseudomonas putida. J. Bacteriol. 169: 5174-5179.

Kimura, M. 1968. Evolutionary rate at the molecular level. Nature 217: 624-626.

Knoth, K., S. Roberds, C. Potect, and M. Tamkun. 1988. Highly degenerate, inosine-containing primers specifically amplify rare cDNA using the polymerase chain reaction. Nucleic Acids Res. 16: 10932.

Kurkela, S., H. Lehvaslaiho, E.T. Palva, and T.H. Teeri. 1988. Cloning, nucloetide sequence and characterization of genes encoding naphthalene dioxygenase of Pseudomonas putida strain NCIB9816. Gene 73: 355-362.

Kwok, S., D.E. Kellogg, N. McKinney, D. Spasic, L. Goda, C. Levenson, and J.J. Sninsky. 1990. Effects of primer-template mismatches on the polymerase chain reaction: Human immunodeficiency virus type 1 model studies. Nucleic Acids Res. 18: 999-1005.

Lane, D.J., B. Pace, G.J. Olsen, D.A. Stahl, M.L. Sogin, and N.R. Pace. 1985. Rapid determination of 16S ribosomal RNA sequences for phylogenetic analyses. Proc. Natl. Acad. Sci. USA 82: 6955-6959.

Lerman, L.S., and K. Silverstein. 1987. Computational simulation of DNA melting and its application to denaturing gradient gel electrophoresis. Methods Enzymol. 155: 482-500.

Lin, P.K.T., and D.M. Brown. 1992. Synthesis of oligodeoxyribonucleotides containing degenerate bases and their use as primers in the polymerase chain reaction. Nucleic Acids Res. 20: 5149-5152.

Loomis, W.D., and J. Battaile. 1966. Plant phenolic compounds and the isolation of plant enzymes. Phytochemistry 5:423-428.

Maliyakal, J. D. 1992. An efficient method for isolation of RNA and DNA from plants containing polyphenolics. Nucleic Acids Res. 20:2381.

Olsen, G.J., N. Larsen and C.R. Woese. 1991. The ribosomal RNA database project. Nucleic Acids Res. 19: 2017-2018.

Pace, N.R., D.A. Stahl, D.J. Lane, and G.J. Olsen. 1985. The analysis of natural microbial populations by ribosomal RNA sequences. Adv. Microbiol. Ecol. 9:1-55.

Petruska, J., M.F. Goodman, M.S. Boosalis, L.S. Sowers, C. Cheong, and I. Tinoco. 1988. Comparison between DNA melting thermodynamics and DNA polymerase fidelity. Proc. Natl Acad. Sci. USA 85: 6252-6256.

Seewaldt, E., and E. Stackebrandt. 1982. Partial sequence of 16S ribosomal RNA and the phylogeny of Prochloron. Nature 295: 221-271.

Smith, R.F. and T.F. Smith. 1992. Pattern-induced multi-sequence alignment (PIMA) algorithm employing secondary structure-dependent gap penalties for comparative protein modelling. Protein Eng. 5:35-41.

Sneath, P.H., and R.R. Sokal. 1973. Numerical Taxonomy (Freeman, San Francisco)

Sommer, R., and D. Tautz. 1989. Minimal homology requirements for PCR primers. Nucleic Acids Res. 17: 6749.

Taira, K., J. Hirosa, S. Hayashida, and K.K. Furukawa. 1992. Analysis of bph operon from polychlorinated biphenyl-degrading strain of *Pseudomonas pseudoalcaligenes* KF707. J. Biol. Chem. 267: 4844-4853.

Tsai, Y.-L. and B.H. Olson. 1992. Rapid method for separation of bacterial DNA from humic substances in sediments for polymerase chain reaction. Appl. Environ. Microbiol. 58: 2292-2295

Ward, D.M., R. Weller and M.M. Bateson. 1990. 16S rRNA sequence reveal numerous uncultured microorganisms in a natural community. Nature 345: 63-65.

Woese, C.R., 1987. Bacterial Evolution. Microbiol. Rev. 51: 221-271.

Young, C.C., R. L. Burghoff, L.G. Keim, V. Minak-Bernero, J.R. Lute, and S. M. Hinton.1993. Polyvinylpyrrolidone-agarose gel electrophorsis purification of polymerase chain reaction-amplifiable DNA from soils. Appl. Environ. Microbiol. 59: 1972-1974.

Zuckerkandl, E., and L. Pauling. 1965 Molecules as documents of evolutionary history. J. Theor. Biol. 8: 357.

Zylstra, G.J., and D.T. Gibson. 1989. Toluene degradation by *Pseudomonas putida* F1: Nucleotide sequence of the todC1C2BADE genes and their expression in *E. coli.* J. Biol. Chem. 264:14940-14946.

7 Lignin-Degrading Fungi as Degraders of Pentachlorophenol and Creosote in Soil[1]

J.A. Glaser
U.S. Environmental Protection Agency
Cincinnati, Ohio

R.T. Lamar
U.S. Department of Agriculture
Madison, Wisconsin

Fungi are known for their diversity and remarkable ability to degrade complex and persistent natural materials such as lignin. In contrast to bacteria, fungi are able to extend the location of their biomass through hyphal growth. These features distinguish fungi as organisms having great potential for use as components of a technology capable of treating soil contaminated with organic pollutants. Fungi capable of degrading lignin have received considerable attention for bioremediation application potential. The enzymes involved in lignin breakdown are extracellular and have low substrate specificity. Chemical structural similarities and expected interaction between lignin and organic pollutants have fostered the consideration of these fungi as potential pollutant degraders. Recent field applications of a treatment technology based on the use of *Phanerochaete spp.* (lignin degrading fungi) to treat pentachlorophenol (PCP) in soil have been very encouraging. In two widely different field settings having soil pH of 3.8 and 9.6, PCP concentrations of 600 to 700 mg kg^{-1} were reduced by 90% in soil bed application of the fungal treatment. When the PCP concentration exceeded 1000 mg kg^{-1}, 70% removal was observed. This superior to the PCP treatment encountered with bacterial based bioremediation systems where treatment effectiveness has been reported for concentrations a factor of ten lower (Edgehill & Finn, 1982, 1983; Saber & Crawford, 1985). Limiting features of the fungal technology are currently identified as the preparation and cost of inoculum and the optimal configuration or design of the field operations.

The environment has a limited capacity to degrade waste organic materials as confirmed by their persistence and toxicity. Few instances of environmental contamination are composed of a single component that could simplify the requirements for biological decomposition of the waste materials. Indeed,

[1]The EPA SITE Program has provided valuable support in the form of site preparation and sample analysis for the field studies. The EPA Bioremediation Field Initiative has supported the field work by providing funds for operation and data evaluation.

contamination at Resource Conservation and Recovery Act (RCRA) and Comprehensive Environmental Response Compensation and Liability Act (CERCLA) waste sites is usually composed of a complex mixture of waste materials. Organic and toxic metal components of the waste often require segregation for proper treatment. The use of microorganisms to convert waste materials into innocuous products is being investigated with the goal of developing treatment technologies that promise to be environmentally benign, economically competitive with other treatment options, and acceptable to the general public.

Detoxification of hazardous wastes is important as a means to reduce the risk associated with such waste. The potential of biological processes to detoxify hazardous waste has been recognized. The ability to degrade/detoxify organic and inorganic waste constituents requires two integral features: competent microbes able to degrade target pollutants and contact between the microorganism(s) and the pollutants. The degradative competence of an organism is best understood in terms of the biochemicals (enzymes) that enable the organism to convert the contaminant chemical to a nontoxic end product. Due to the variety of chemicals forming pollutant mixtures, either a single organism of exceptional competence or a mixture of organisms termed a microbial consortium is required. Bioavailability or contact between the organism, degradative enzyme, and the pollutant is a function of mass transfer and the suitable conditions to maintain the activity of the selected degrading organisms. Selection and development of reactor configurations and associated operating conditions are necessary to sustain economic operation of the treatment.

Two examples of organisms having high competence for a wide variety of organic pollutants are the wood degrading fungi, *Phanerochaete chrysosporium* and *P. sordida*. These fungi possess great potential to degrade aromatic components of toxic and hazardous waste. This ability is based in part, on their widely recognized ability to degrade lignin, a persistent biogenic polymer, and the non-specific enzyme systems of the fungus that support this degradation (Kirk, 1984). This report details the process to select competent fungi for the treatment of wood preserving chemicals contaminating soils and describes a testing program to develop a fungal treatment technology.

WOOD TREATMENT INDUSTRY WASTE

Wastes generated in the wood treating industry are a major hazardous waste problem confronting the United States (U.S. Department of Agriculture, 1980). Some 700 existing wood treating sites and their eventual remediation emphasizes the need for a reliable remediation technology. Three wood preservation technologies have contributed to the accumulated waste. Historically, creosote processing was followed sequentially by pentachlorophenol and then copper chromated arsenite treatment. These wastes have unique requirements for cleanup due to the toxicity of the chemical mixtures used in the treatment. Creosote derived from coal tar production, usually contains a host of compounds ranging from low molecular weight, volatile aromatic compounds to polyaromatic species including smaller quantities of aromatic nitrogen bases and an array of phenolic compounds. Diesel or fuel oil fractions were used to dilute the creosote mixture in the wood treating processing steps. The lower vapor pressure components of this mixture contribute to the residuals found at wood treating sites. Pentachlorophenol, a potent fungicide and a major contaminant at wood treating sites, exhibits significant toxicity towards microflora (Choudry & Hutzinger, 1982). The analysis of wastes derived from this technology have identified other potential toxic components. The copper chromated arsenite waste, due to the toxicity of the inorganic components of the treatment chemicals, is not targeted as an application of the fungal technology.

THE SELECTION OF FUNGI OVER BACTERIA

Microorganisms are known to possess a variety of biochemical means to transform chemical substances (Griffin, 1985). Substrates such as aliphatic straight chain hydrocarbons and polysaccharides provide energy to the microorganism sufficient to maintain growth. Many pollutants do not provide sufficient energy to sustain the growth of most microorganisms. The biological degradation of such substrates via cometabolism, occurs as part of a biological activity which maintains the organism's growth through the metabolism of energy producing substrates and concurrent biotransformation of the non-growth supporting (pollutant) substrates. Single bacteria or fungi may accomplish simple transformations on organic substrates. Often these same microorganisms may fail to complete the conversion of toxicant substrate to carbon dioxide or may generate a toxic intermediate(s) that can impair the growth of microorganisms. In the case of pentachlorophenol, many bacterial strains are intolerant to even small concentrations of this compound (Kozak et al., 1979). The use of bacterial consortia combines the utilization of many species where the biotransformation activity of one species supplants the inadequacies of another. Protection from environmental effects that may adversely affect the collective health of these communities is crucial to the success of treatment technology.

Fungi have not been evaluated extensively for use as degraders of waste materials until recently in spite of their recognized capabilities (Cooke, 1962; Bollag, 1979). Wastewater treatment operations have avoided the use of filamentous fungi wherever possible due to processing problems and the possibility that such fungi may be pathogenic. Exceptions to these generalizations do exist. A wood rotting basidiomycete, *Trametes versicolor*, was studied 30 years ago in an attempt to quantify its ability to degrade chlorinated phenols (Lyr, 1963). The ability of fungi to degrade wood preservative chemicals has been know for sometime (Duncan & Deverall, 1946; Unligli, 1968).

SELECTION OF FUNGAL STRAINS

Fungi are eukaryotic i.e., they possess a nuclear membrane which distinguishes them from prokaryotes and as microorganisms are considered to be plantlike, without chlorophyll or photosynthetic abilities (Deacon, 1984). Wood degrading fungi, such as *Phanerochaete chrysosporium* and *P. sordida*, are characterized by fast growth and reproductive cycles. Under laboratory conditions these organisms have been shown to degrade an extensive list of hazardous waste constituents (Lamar et al., 1992; Bumpus & Aust, 1987; Higgins, 1991; Bumpus, 1993). This ability to degrade hazardous pollutants appears to, in general, correlate with the ability of fungi to degrade lignin, a complex natural polymer composed of phenylpropane units that are generally resistant to microbial decay. The lignin degrading ability has been attributable to a complex mixture of extracellular enzymes secreted by the fungus. The enzymes are laccase and/or peroxidases that utilize hydrogen peroxide from complementary enzyme systems to perform the initial oxidative conversion of some pollutant substrates (Lamar et al., 1992).

These fungi have been isolated from wood, wood products and soil. There is no published evidence that *P. chrysosporium* can act as a plant pathogen. Isolates of this fungus have been identified in Canada, Sweden, and Russia as well as the United States of America, leading to the perception that the fungus is widely distributed in the world. Furthermore, this fungus has been evaluated as a feedstuff for rats, pigs, and sheep, having no apparent toxic effects to higher animals (Tompke et al., 1980).

FUNGAL CARBON SUBSTRATE DEGRADING ABILITIES

White rot fungi are the primary wood degraders in nature (Rayner & Boddy, 1988). They excel in their ability to recycle carbon of wood of angiosperm origin when compared with other fungi. Naturally occurring polymers of cellulose and lignin are degraded by these organisms with cellulose forming the major sources of carbon for fungal growth. Lignin is more difficult to degrade than cellulose because of lignin's composition as a heteropolymer formed from the cross linking of three precursor cinnamyl alcohols. Lignin cannot serve as the sole carbon source for growth of *P. chrysosporium* (Kirk & Shimada, 1985). The fungus must be able to switch its ability to degrade these various polymers as the concentration of polymer varies with the composition of the wood. This switching ability for naturally occurring strains of *P. chrysosporium* is controlled by the absence of certain nutrients. Nitrogen deficiency is generally thought to induce this secondary metabolic cycle of lignin depolymerization (Kirk, 1984).

The enzyme systems responsible for the initial attack on lignin require unusual abilities due to the complexity and resistance of the lignin structure (Table 7-1). The molecular weight of lignin is too large to allow it to enter the cells of microorganisms by known transport systems. An enzyme system permitting the microorganism to overcome this limitation is likely to be extracellular and non-specific (due to the heterogeneity and large molecular weight of the substrate). These wood degrading fungi are endowed with such an extracellular enzyme system. The ability of *P. chrysosporium* to degrade lignin by these extracellular enzymes occurs in a secondary metabolic cycle. The fungus uses cellulose as its primary growth substrate but when large quantities of lignin are encountered or certain nutrients are not present, the secondary metabolic cycle is entered (Kirk & Farrell, 1987).

SUMMARY OF WASTE CONSTITUENT DEGRADATION STUDIES

An expanding list of hazardous waste organic chemicals have been shown to be degraded to carbon dioxide or biotransformed by lignin-degrading fungi (Bumpus & Aust, 1987; Higgins, 1991; Lamar et al., 1992; Bumpus, 1993). In the case of DDT, it appears that ligninolytic conditions may not be necessary for the fungus to

Table 7-1. Biotransfromation reactions catalyzed by fungal extracellular lignin peroxidase enzymes†.

Reaction	Structural Change
Aromatic Ring Oxidation	$ArH \rightarrow ArOH$
Aryl Ether Oxidation	$ArOR \rightarrow ArOH$
Diol Cleavage	$ArCH(OH)CH_2OH \rightarrow ArCHO + HCHO$
Benzylic Methylene Hydroxylation	$ArCH_2R \rightarrow ArCH_2(OH)R$
Stryl Olefin Hydroxylation	$ArCR{=}CR_2 \rightarrow ArCR(OH)CR_2OH$
Benzyl Alcohol Oxidation	$ArCH_2OH \rightarrow ArCHO$
Oxidative Coupling	$ArH + Ar'H \rightarrow Ar\text{-}Ar'$

†From Kirk, 1984.

degrade this substrate (Bumpus & Aust, 1987; Kohler et al., 1988). Decomposition of DDT under these conditions may be due to corrin or porphyrin reactions that serve to dehalogenate the DDT reductively (Miskus et al., 1965; Marks et al., 1989). There may be additional instances of organic substrates not requiring ligninase for breakdown. The search for other organic hazardous waste substrates not requiring ligninolytic conditions is under investigation.

Radiorespirometric studies of the degradation of [U-^{14}C] pentachlorophenol in aqueous media under ligninolytic conditions indicated the substrate was converted to carbon dioxide (Lamar et al., 1989). Enzyme studies showed pentachlorophenol is converted to 1,4-tetrachlorobenzoquinone (Hammel & Tardone, 1988). The quinone was difficult to quantify due to its propensity to form charge-transfer complexes with cellular materials. Further elucidation of the metabolic pathway is in progress. Several aromatic hydrocarbons, benzo[a]anthracene, anthracene, pyrene, benzo[a]pyrene, and pyrelene (constituents of creosote), were converted to carbon dioxide by the fungus in liquid culture (Hammel et al., 1986; Haemmerli et al., 1986). Recent research directed to investigate the metabolite formation in these oxidation processes has determined that in the case of anthracene, anthraquinone is the initial metabolite of fungal oxidation (Hammel & Glaser, 1990). The anthraquinone was shown to have a comparable rate of mineralization in culture. However, napthalene appears not to be a substrate for the lignin peroxidases and may be oxidized through other oxidation routes such as cytochrome P-450 catalyzed reactions. This finding serves to differentiate the fungus from bacterial species since few bacteria have the ability to utilize the higher molecular weight aromatic polycyclics.

SOIL DETOXIFICATION TECHNOLOGY DEVELOPMENT

Bench Studies

Attempts to inoculate environmental matrices with nonindigenous microorganisms have met with varying degrees of success (Zaidi et al., 1988). The elucidation of optimal practices leading to successful inoculation of contaminated environmental materials remains to be discovered (Goldstein et al., 1985). Three well characterized soils (two topsoils and one subsoil) were used in the bench work (Table 7-2). At the outset of this research, *P. chrysosporium* was not known to inhabit the soil. Laboratory work, indicated that *P. chrysosporium* did not grow well in non-sterile soils; this may be attributable in part due to ineffective competition with the indigenous microflora (Table 7-2). These results were anticipated since the soil is not the normal habitat of *P. chrysosporium.* However, it has been found that growth within the soil can be accomplished when the amount of applied inoculum is increased (Lamar et al., 1990).

Table 7-2. Biological, physical chemical characteristics of test soils† .

	Soil		
Soil Characteristic	Marsham	Zurich	Batavia
Bacteria (CFU‡)	1.15 X 10^6	1.39 X 10^8	4.71 X 10^6
Fungi (CFU)	6.26 X 10^3	2.76 X 10^6	4.29 X 10^3
Actinomycetes (CFU)	1.51 X 10^6	1.77 X 10^7	9.71 X 10^6
Texture	silty clay-loam	sandy loam	sandy loam
pH	5.4	6.8	7.1
Organic Matter (%)	0.5	12.0	39.0
Nitrogen (%)	0.05	0.46	0.18

†From Lamar et al. 1990
‡CFU=Colony forming units

The effect of soil type, pH, temperature, and water potential on the growth of *P. chrysosporium* in three sterile soils was evaluated in a factorial experiment. Soils were sterilized by fumigating with methyl bromide to avoid confounding effects from native microflora. Growth of the fungus was evaluated at five soil temperatures ranging from 25 to 39°C and four water potentials ranging from -0.03 to -1.5 MPa. The extent of growth was determined by measuring the amount of ergosterol that could be extracted from two sub-samples of the soil from each test at the end of a two week incubation period and reported as µg of ergosterol g^{-1} of soil (Seitz et.al., 1979).

Growth of the fungus was the greatest in the Marsham soil, intermediate in the Xurich soil, and least in the Batavia soil. The same trend was observed when a visual assessment system of growth was used. Soil water potential, a controllable soil factor (Sommers et.al., 1989), had a significant effect on the growth of the fungus in soil. As the soil water potential was increased (corresponds to decrease in soil water content), fungal growth decreased. Growth of the fungus between 25 and 35°C was unaffected but significantly decreased at 39°C. These results disagree with earlier work using the visual growth estimation technique. Differences may be attributable to an increase of sporulation of the fungus at the soil surface, with increased temperature leading, to a biased measurement of growth. Soil temperatures under field conditions can be controlled by selecting the normal warm months for operation or by cultivating the fungus in a more controlled treatment environment. Biomass accumulations as well as growth habit of *P. chrysosporium* were greatly influenced by soil type.

Growth Measurement

Assessment of growth in the early stages of these studies was done by means of visual estimation on a ranking basis. Due to the three dimensional growth patterns of the fungus in a solid substrate such as soil, it was necessary to find a more reliable means to determine the extent of fungal growth. The normal means of assessing growth in bacterial systems have no application to the present study. Based on work investigating the infestation of cereal grains by fungi, a mycosteroid, ergosterol (ergosta-5:6,7:8,22:23-trien-3-ol), has been employed as a quantitative means to determine growth of the fungus (Seitz et al., 1979). It has been shown to correlate well with the visual estimation technique.

The degradation of ^{14}C[UL]-pentachlorophenol was studied in the laboratory over an eight week period in the three soils (Lamar & Dietrich, 1990). Mineralization, volatile losses, extractable PCP in the soil, and soil residuals containing bound PCP and transformed products were measured to develop a tight material balance. A small percentage of the total ^{14}C was accounted for by mineralization and volatilization. Both mineralization and volatilization were significantly greater in inoculated than in non-inoculated cultures of the three soils. Extractable quantities of PCP were greatly reduced by inoculation with *P. chrysosporium.* The greatest rate of PCP removal due to fungal activity was found in one of the topsoils. Extractable PCP after 14 days was about 2 $\mu g\ g^{-1}$ of the original 50 $\mu g\ g^{-1}$ spiked amount and that was reduced to roughly 1 $\mu g\ g^{-1}$ after an additional 14 days of treatment. Decreases of PCP concentration in the test controls are attributable to several potential causes: abiotic avenues of degradation or loss, irreversible binding to the soil or degradation due to the regrowth of native organisms.

The degradative ability of the fungus in the soil has been evaluated through the measurement of evolved labeled carbon dioxide. Separation of the soil into solvent extractable, humic acid, fulvic acid, and humin fractions permitted material balance evaluations (Stevenson, 1982). The fate of ^{14}C-PCP in the soil was determined by analysis of the recoverable carbon label from: an organic extractable fraction, the soil organic matter (humic and fulvic fractions), and the non-extractable humin fraction. Combustion analysis was used to assay the amount of ^{14}C associated with humic and fulvic acid fractions, and the non-extractable humin fraction.

The percent of total ^{14}C recovery ranged from 55 to 84% over a 56 day period in one of the topsoil tests. Volatilization losses were less than 3% for the three soils. There was a significant amount of labeled carbon activity associated with nonextractable fractions, indicating there is possible incorporation of the pollutant substrates in the soil material during metabolism. Possible polymerization reactions between pentachlorophenol and soil components such as syringic acid have been investigated (Bollag & Liu, 1985). Polymeric forms of a series of pollutants were constructed by Haider and Martin, who showed that *P. chrysosporium* would degrade these higher molecular weight materials but at a slower rate than the parent pollutant substrate (Haider & Martin, 1988). A more recent study showed that *P.chrysosporium* can degrade humic acid materials (Blondeau, 1989).

A parallel study investigating the possible utility of other wood degrading fungi was undertaken (Lamar et al., 1990). *Phanerochaete spp.* in the collection of the United States Department Agriculture Forest Products Laboratory, Madison, WI, were scrutinized for the capability of degrading pentachlorophenol (Table 3). In addition to *P. chrysosporium*, a *P. sordida* species was selected for its ability to degrade pentachlorophenol (Lamar & Dietrich, 1990). The two fungal species were then incorporated as part of the field trials.

Table 7-3. Loss of pentachlorophenol as affect by selected fungi†.

Fungus strain	Mineralization	Volatilization	Total ^{14}C Evolved
		%	
P. chrysosporium	2.0	13.8	15.8
P. sordida -- 7	2.7	12.9	15.6
P. sordida -- 8	1.9	8.9	10.8
P. sordida -- 9	1.2	11.9	13.1
P. sordida --10	11.6	8.5	20.1

†From Lamar et al., 1990

FIELD STUDIES

The investigation of fungal treatment systems has been mainly confined to laboratory or bench scale studies. The systematic approach to fungal treatment application has permitted the involvement in a treatment effectiveness study under field conditions at a wood preservative spill site in Wisconsin and an abandoned wood treating site in Mississippi .

Oshkosh Site

A site in Oshkosh, WI was selected for field trial applications of the white rot fungi to contaminated soil. The area of application was a former "tank farm" where above-ground storage tanks contained a wood preservative formulation know as "Woodlife". The composition of this product was predominantly mineral spirits (high boiling point pentanes and hexanes) and 5% pentachlorophenol. Extensive screening for pentachlorophenol in the tank farm soils showed concentrations between 1 and 4435 mg kg^{-1} to depths of 30 cm. Within the confines of a

Table 7-4. Soil treatment study design.

Component	*P. chrysosporium*	*P. sordida*	Sterile Woodchips	Organic Matter	Mechanical Aeration
A1	+	—	—	+	+
A2	—	+	—	+	+
B	—	—	+	+	+
C	—	—	+	—	+
D	—	—	—	+	+
E	—	—	—	—	+

protective berm for the tank farm, the field trial was laid out according to the treatment design in Table 7-4. After thorough mixing of the soil, nine plot borders were installed in a three-by-three configuration. The plot borders were constructed of galvanized steel sheet and measured 1 m x 1 m by 35 cm deep. Plot borders were worked into the soil surface and filled to a depth of 25 cm with approximately 370 kg (dry weight equilivent) of soil from outside the border. Two different fungi (*P. chrysosporium* and *P. sordida*) were used as the treatment species. In each case, the fungi were added to the contaminated area using inoculated wood chips. The treatability trial ran from early August to the end of September 1989, a total of 46 days. The pentachlorophenol concentration was depleted by 82% or 85% by *P. chrysosporium* or *P. sordida,* respectively (Table 7-5).

Table 7-5. Decrease in percentage pentachlorophenol concentrations.

	Time (day)				
Component	8	15	22	29	46
			%		
A1	9.1	33.3	76.5	70.5	82.3
A2	9.7	42.2	50.6	75.9	85.8
B	4.9	13.7	19.4	20.9	27.5
C	0.5	-10.0	13.2	7.1	16.1
D	15.3	26.1	20.6	10.7	3.1
E	10.9	13.8	3.8	23.8	19.1

Brookhaven Site

The study site was located 60 miles south of Jackson in Brookhaven, MS. and had been identified as a removal action site by EPA Region 4 (Fig. 7-1). The wood treating operation was based on both creosote and pentachlorophenol technology. During the time of facility operation, soil had been excavated in response to a spill of treatment chemicals (Fig. 7-2). The excavated soil was mounded above the ground surface in a RCRA hazardous waste treatment unit. The soil pile was sampled for pentachlorophenol and creosote components prior to developing the test site. Analysis of the results permitted the selection of sections of the pile, having less than 700 ppm pentachlorophenol, for the treatment effectiveness study. Contaminated soil was sized through a 2.5 cm (1 inch) mesh screen using a Read Screen All shaker screen having a 10 yd^3 $hour^{-1}$ capacity. The soil was deposited in separate portions on a polyethylene tarp. Further homogenization was accomplished by the mixing of different portions of screened soil which was then applied to the treatment plots by a front end loader (Fig. 7-3).

The field study was undertaken as a USEPA SITE Program demonstration involving two phases. The first phase was designed to evaluate the ability of three different fungal species to deplete pentachlorophenol in soil. Inoculum was developed by L.F. Lambert Spawn Co. of Coatesville, PA. The prepared inoculum and inoculum carrier were shipped to the site by refrigerated transportation. Three inoculum loadings were used with the required controls. The experimental design consisted of a randomized complete block without replication (RCB) and a balanced

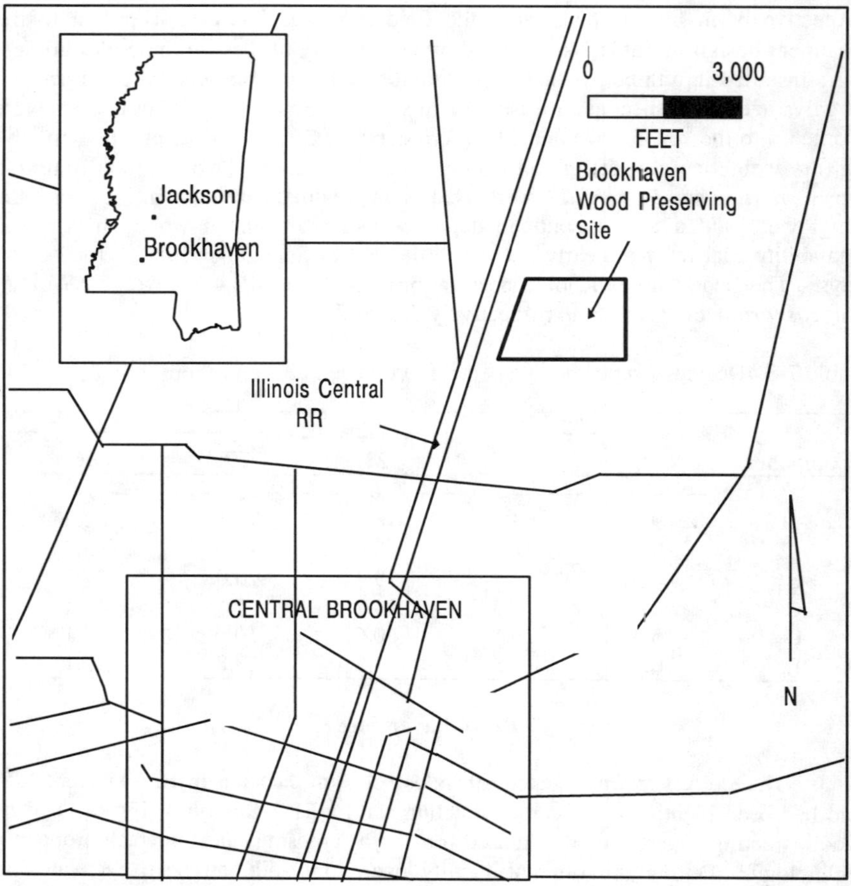

Fig. 7-1. Brookhaven MS site location.

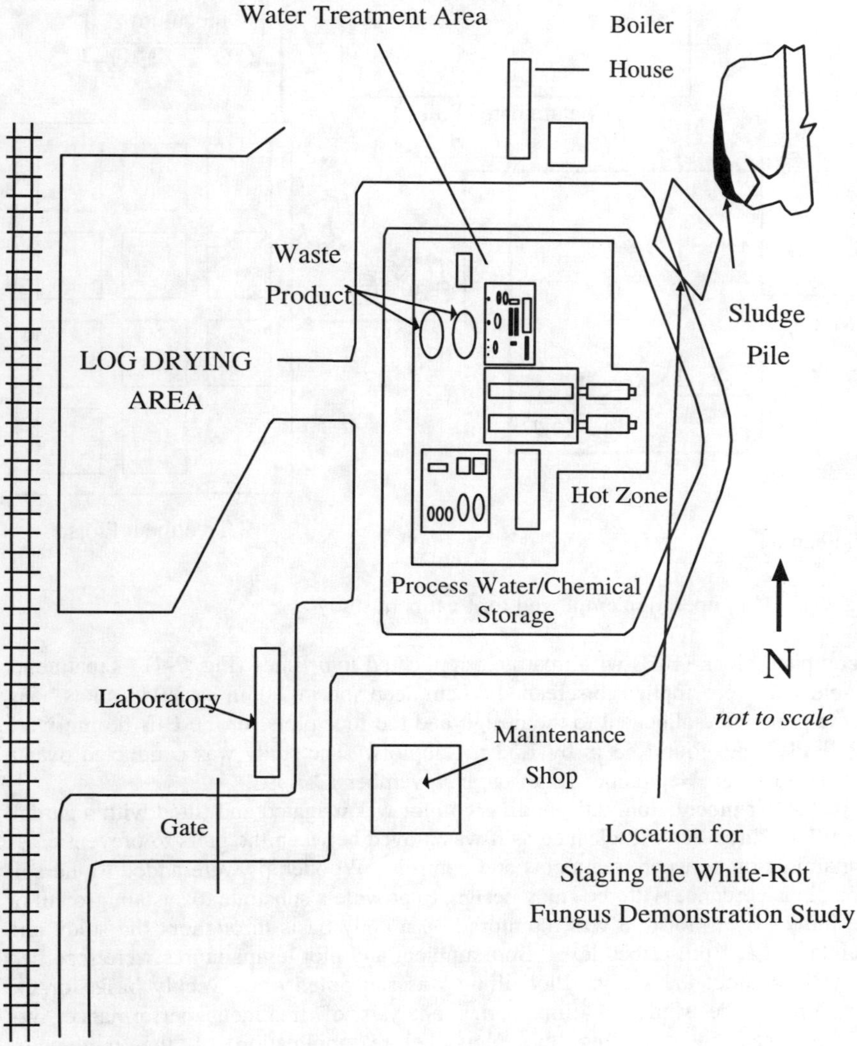

Fig. 7-2. Site detail for the Brookhaven location.

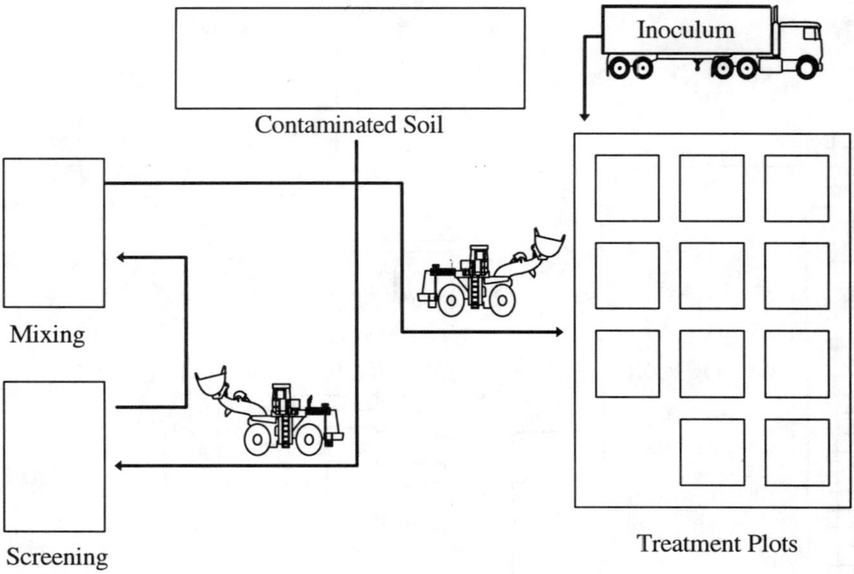

Fig. 7-3. Unit operation employed for treatment study.

incomplete block (BIB) with treatment replicated four times (Fig. 7-4). Treatments (Table 7-6) were applied to screened 25 cm deep soil in 3.05m by 3.05m plots. Six of the plots were allocated to the design and the four plots for the BIB design were subdivided into four 1.53 m by 1.53 m subplots. The study was conducted over a two month span: September 18, through November 13, 1991.

After inoculation with fungi, each plot was irrigated and tilled with a garden rototiller. The tiller was cleaned as it was moved between the plots to prevent cross contamination between treatments and controls. Woodchips were added to the soil plots, in accordance with the study design, to provide a substrate to sustain growth of the fungi. Soil moisture was monitored on a daily basis throughout the study and maintained at a prescribed level. Both ambient and plot temperatures were recorded daily throughout the study. Plot tilling was scheduled on a weekly basis for the duration of the study. A time series analysis of treatment performance was accomplished by sampling the plots before application of the treatments, immediately after treatment application and then after 1, 2, 4, and 8 weeks of operation.

Analysis of treatment data identified the *P. sordida* treatment as the most effective in biotransformation of the pentachlorophenol (Table 7-7). Transformation of the polynuclear hydrocarbon components of creosote also occurred in a similar striking fashion (Table 7-8). There is an obvious trend in the aromatic hydrocarbon data showing a removal preference for low molecular weight, as compared to higher molecular weight, hydrocarbons.

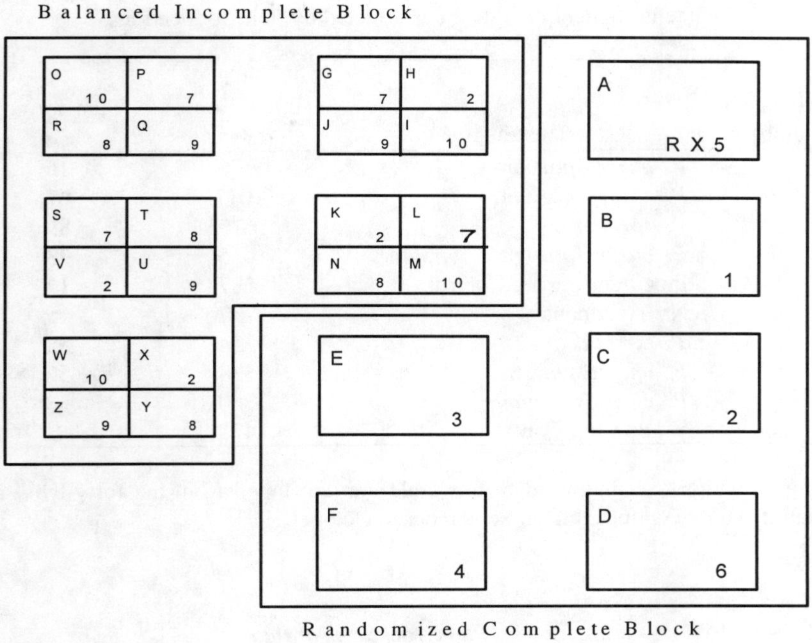

Fig. 7-4. Treatability study design.

Table 7-6. Experimental design for fungal treatment effectiveness study

Treatment Number†	Treatment	Inoculum Dry Weight (%)	Comments
1	*P. chrysosporium*	5	
2	*P. chrysosporium*	10	
3	*P. sordida*	10	
4	*P. chrysosporium/T. hirsuta*	5/5	Mixed Fungal Inoculum
5	Inoculum Carrier Control	10	Amendment Control
6	No Treatment	—	Field Control
7	*T. hirsuta*	10	
8	*P. chrysosporium*	13	
9	*P. chrysosporium*	10 (Day 0) 3 (Day 3)	Sequential Applications
10	Wood Chip Control	—	Amendment Control

†Treatment designation refers to plot layout in Fig. 7-2. Wood chips were added to all plots at a rate of 2.5% with the exception of treatment 6.

Table 7-7. Pentachlorophenol treatment as affected by field applications.†

Treatment Number†	Treatment	PCP Initial Conc. μg g^{-1}	Transformed %
1	*P. chrysosporium*	576	15
2	*P. chrysosporium*	1017	67
3	*P. sordida*	673	89
4	*P. chrysosporium/T. hirsuta*	615	23
5	Inoculum Carrier Control	687	14
6	No Treatment	737	15
7	*T. hirsuta*	334	55
8	*P. chrysosporium*	333	52
9	*P. chrysosporium*	360	55
10	Wood Chip Control	471	0

†Transformation data indicated in this table are developed from the forty-two day treatablity study conducted from September to October 1991.

Table 7-8. PAH transformation by *Phanerochaete sordida*†

PAH	Inoculum Carrier Control	No Treatment Contol	*P. sordida* 10% Inoculum
		%	
Acenapthene	68	47	95
Fluorene	57	75	95
Phenanthrene	49	59	90
Anthracene	48	57	80
Fluoranthene	42	23	72
Pyrene	22	10	52
Benzo[α]anthracene	14	11	24
Chrysene	14	6	33

†Transformation data indicated in this table are developed from the forty-two day treatablity study conducted from September to October 1991.

CONCLUSIONS

The excellent performance of *P. sordida* in biotransforming pentachlorophenol parallels previous field studies with *P. chrysosporium* at a Wisconsin site. However, at Brookhaven, *P. sordida* outperformed *P. chrysosporium*. *P. sordida* is an organism that resides in the soil under normal conditions. The soil conditions encountered at the Brookhaven site are not optimal for the cultivation of microorganisms and the concentrations of pentachlorophenol found in the Brookhaven soil are beyond the limits to sustain suitable growth for many microorganisms. All of these factors strongly suggest that fungal treatment using *P. sordida* has exceptional promise for bioremediation of soil contaminated with pentachlorophenol and other difficult to degrade pollutants. Future challenges to making the technology more cost effective include developing improved and cheaper inoculation techniques and extending the application technology to other organic pollutant classes. Incineration of pentachlorophenol waste materials continues to be an unattractive treatment option due to the expected formation of chlorinated dibenzodioxins and dibenzofurans. Viewing fungal treatment in this context provides additional incentives for the commercialization of this technology.

REFERENCES

Blondeau, R. 1989. Biodegradation of synthetic humic acids by the white-rot fungus *Phanerochaete chrysosporium.* Appl. Environ. Microbiol. 5*:*1282-1285.

Bollag, J.-M. 1979. Biochemical transformation by soil fungi. Crit. Rev. Microbiol. 2*:*35-58.

Bollag, J.-M., and S.-Y. Liu. 1985. Copolymerization of halogenated phenols and syringic acid. Pest. Biochem. and Physiol. 23*:*261.

Bumpus, J.A., and S. Aust. 1987. Biodegradation of DDT[1,1,1-trichloro-2,2-bis(4-chlorophenyl)ethane] by the white rot fungus *Phanerochaete chrysosporium.* Appl. Environ. Microbiol. 53*:*2001-2007.

Bumpus, J.A., and S.D. Aust. 1987. Biodegradation of environmental pollutants by the white rot fungus *Phanerochaete chrysosporium*: involvement of the lignin degrading system. Bioessays, 6*:*166-170.

Bumpus, J. 1993. White rot fungi and their potential use in soil bioremediation processes. p 65-100. *In* J.-M. Bollag and G. Stotzky (ed.) Soil Biochemistry. Vol 8. Dekker, New York.

Choudry, G.G., and O. Hutzinger. 1982. Mechanistic aspects of the thermal formation of halogenated organic compounds including polychlorinated dibenzo-p-dioxins. Part II: Thermochemical generation and destruction of dibenzofurans and dibenzo-p-dioxins. Toxicol. Environ. Chem. 5*:*67-93.

Cooke, W.B. 1962. The role of fungi on environmental sanitation. Develop. Indust. Microbiol. 3*:*313-318, Plenum, New York.

Deacon, J.W. 1984. Introduction to modern mycology, Blackwell Scientific Publications, Oxford.

Duncan, C.G., and F.J. Deverall. 1964. Degradation of wood preservatives by fungi. Appl. Microbiol., 12*:*57-62.

Edgehill, R.U., and R.K. Finn. 1982. Isolation, characterization, and growth kinetics of bacteria metabolizing pentachlorophenol. Eur. J. Appl. Microbiol. Biotechnol. 16:179-184.

Edgehill, R.U., and R.K. Finn. 1983. Microbial treatment of soil to remove pentachlorophenol. Appl. Environ. Microbiol. 45:1122-1125.

Goldstein, R.M., L.M. Mallory, and M. Alexander. 1985. Reasons for the possible failure to enhance biodegradation. Appl. Environ. Microbiol. 50*:*977-983.

Griffin, D.M. 1985. A comparison of the roles of bacteria and fungi. p. 221-255. *In* E.R. Leadbetter and J.S. Perlmutter (eds) Bacteria in Nature. Vol 1. Plenum, New York.

Haemmerli, S.D., M.S.A. Leisola, D. Sanglard, and A. Fiechter. 1986. Oxidation of benzo(α)pyrene by extracellular ligninases from *Phanerochaete chrysosporium.* J. Biol. Chem. 261:6900.

Hammel, K.E., and P.J. Tardone. 1988. The oxidative 4-dechlorination of polychlorinated phenols is catalyzed by the extracellular fungal lignin peroxidases. Biochemistry 27:6563-6568.

Hammel, K.E., B. Kalyanaraman, and T.K. Kirk. 1986. Oxidation of polycyclic aromatic hydrocarbons and dibenzo[p]dioxins by *Phanerochaete chrysosporium* ligninase. J. Biol. Chem. 261:16948-16952.

Hammel, K.E., and J.A. Glaser. 1990. The role of fungal lignin-degrading enzymes in aromatic pollutant biodegradation. (p. 95-97). Abstracts of EPA Symposium on Bioremedation of Hazardous Wastes: EPA's Biosystem Technology Development Program. Arlington, VA.

Haider, K.M., and J.P. Martin. 1988. Mineralization of ^{14}C-labeled humic acids and humic-acid bound ^{14}C-xenobiotics by *Phanerochaete chrysosporium.* Soil Biol. Biochem. 20:425-429.

Higgins, F.K. 1991. Degradation of xenobiotics by white rot fungi. pp 111-152. *In* G.W. Ware (ed). Revs. Environ. Contam. Toxicol. vol 122. Springer-Verlag, New York.

Kirk, T.K., and Shimada, M. 1985. Lignin biodegradation: the microorganisms involved, and the physiology and biochemistry of degradation by whiute rot fungi. p 579-605. *In* T. Higuchi (ed.) Biosynthesis and Biodegradation of Wood Components, Academic Press, New York.

Kirk, T.K. 1984. Degradation of lignin. *p.* 399-438. *In* Microbial Degradation of Organic Compounds, D.T. Gibson (ed.) Dekker, New York.

Kirk, T.K., and R.L. Farrell. 1987. Enzymatic "combustion": the microbial degradation of lignin. Ann. Rev. Microbiol. 41:465-505.

Kohler, A., A. Jager, H. Willerhausen, and H. Graf. 1988. Extracellular ligninase of *Phanerochaete chrysosporium* Burdsall has no role in the degradation of DDT. Appl. Microbiol. Biotechnol. 29:618-620.

Kozak, V.D., G.V. Simsiman, G. Chesters, D. Stensby, and J. Harkin. 1979. Reviews of the Environmental Effects of Pollutants: XI. Chlorophenols. p 260-271. U.S. Gov. Print. Off. Washington D.C. EPA-600/1-79-012.

Lamar, R.T., and D. Dietrich. 1990. In situ depletion of pentachlorophenol from contaminated soil by *Phanerochaete* sp., Appl. Environ. Microbiol 56:3093-3100.

Lamar, R.T., J.A. Glaser, and T.K. Kirk. 1989. Use of white-rot fungi to remediate soils contaminated with wood preserving waste. p. 82. Abstracts Fourth International Conference - Biotechnology in the Pulp and Paper Industry. Raleigh, N.C. May 16-19, 1989.

Lamar, R.T., J.A. Glaser, and T.K. Kirk. 1990. Fate of pentachlorophenol(PCP) in sterile soils inoculated with white-rot basidiomycete *Phanerochaete chrysosporium*: mineralization, volatilization, and depletion of PCP, Soil Biol. Biochem. 22:433-440.

Lamar. R.T. , J.A. Glaser, T.K. Kirk. 1992. White-rot fungi in the treatment of hazardous chemicals and wastes. p. 127-143. *In* G.F. Leatham (ed.) Frontiers in Industrial Mycology. Chapman & Hall, New York.

Lamar, R.T., M.J. Larsen, and T.K. Kirk. 1990. Interspecific and intraspecific variation in the pentachlorophenol-degrading abilities of white-rot basidiomycetes, Appl. Environ. Microbiol. 56*:*3519-3526.

Lyr, H. 1963. Enzymatische detoxifikation chlorieter phenole, Phytopathol. Z. 47*:*73-83.

Marks, T.S., J.D. Alpress, and A. Maule. 1989. Dehalogenation of lindane by a variety of porphyrins and corrins. Appl. Environ. Microbiol. 55*:*1258-1261.

Miskus, R.P., D.P. Blair, and J.E. Casida. 1965. Conversion of DDT to DDD by bovine rumen fluid, lake water, and reduced porphyrins. J. Agric. Food. Chem. 13*:*481-483.

Rayner, A.D.M., and C. Boddy. 1988. Fungal decomposition of wood - its biology and ecology, John Wiley & Sons, New York.

Saber, D.L., and R.L. Crawford. 1985. Isolation and characterization of *Flavobacterium* strains that degrade pentachlorophenol. Appl. Environ. Microbiol. 50*:*1512-1518.

Seitz, L.M., D.B. Sauer, R. Burroughs, H.E. Mohr, and J.D. Hubbard. 1979. Ergosterol as a measure of fungal growth. Phytopath. 69*:*1202-1203.

Sommers, L.E., C.E. Gilmour, R.E, Wildung, and S,M, Beck. 1998. The effect of water on the decomposition process in soils. p. 97-117. *In* L.F. Elliott et al. (ed.) Water potential relations in soil microbiology, SSSA Special Publication No. 9, Soil Sci. Soc. Amer., Madison, WI.

Stevenson, F.J. 1982. Humus Chemistry. Wiley, New York.

Tomke, S., M. Rundgren, and S. Eriksson. 1980. Nutritional evaluation of the white-rot fungus *Sporotrichum pulverulentum* as a feedstuff to rats, pigs, and sheep. Biotechnol. Bioeng. 22*:*2285-2303.

Unligli, H.H. 1968. Depletion of pentachlorophenol by fungi. Forest Prod. J. 18*:*45-50.

U.S. Department of Agriculture. 1980. The biological and economic assessment of pentachlorophenol, inorganic arsenicals, creosote vol 1 and 2. Technical Bulletin 1658-I and 1658-II.

Zaidi, B.R., G. Stucki, and M. Alexander. 1988. Low chemical concentration and pH as factors limiting the success of inoculation to enhance bioremediation. Environ. Toxicol. Chem. 6*:*143.

8 An Overview of Microbial Degradation in the Rhizosphere and its Implications for Bioremediation[1]

T.A. Anderson and J.R. Coats
Pesticide Toxicology Laboratory
Iowa State University
Ames, Iowa

The complex, xenobiotic nature of most synthetic chemicals encountered by microorganisms requires interaction of microbial communities at the molecular, physiological, and ecological levels to achieve biochemical transformation (Lappin et al., 1985; Anderson et al., 1995). The plant root zone fosters these types of interactions in many ways: by sustaining microbial densities an order of magnitude or more above those in nonvegetated soils, by providing a niche suitable to a diverse population of microorganisms, and/or by facilitating cometabolic transformations (Hsu & Bartha, 1979). The greater density and diversity of microorganisms commonly observed in the rhizosphere, compared with less diverse microbial communities in nonvegetated soils, often results in greater rates of xenobiotic metabolism. Vegetation has been shown to enhance microbial degradation rates of chemical residues in soils under a variety of environmental conditions (Hsu & Bartha, 1979; Sandmann & Loos, 1984; Aprill & Sims, 1990; Knaebel & Vestal, 1992; Ferro et al., 1994). Existing evidence for microbial degradation in the root zone suggests that a diverse and synergistic microbial community, rather than a single microbial strain, is responsible for the enhanced biodegradation of xenobiotics in the rhizosphere compared with nonvegetated soils (Anderson et al., 1995). Collectively, the microbial community provides the spectrum of degradative enzymes, each of which may be required for mineralization but may not all be present in a microbial strain (Bordeleau & Bartha, 1968; Gunner & Zuckerman, 1968; Senior et al., 1976; Lappin et al., 1985). These phenomena have important implications in biological remediation of waste sites where establishment on

[1]We gratefully acknowledge the participants in the 1993 ACS symposium on "Microbial Degradation of Organic Chemicals in the Rhizosphere: Implications for Bioremediation" for their contributions to the field and to the symposium. We have borrowed heavily from their presentations in developing this book chapter. Journal Paper No. J-15589 of the Iowa Agriculture and Home Economics Experiment Station, Ames, Iowa. Project No. 3187.

cultivation of selected plant species could enhance microbial degradation of organic chemical contaminants, thus providing a low-cost alternative or supplement to more expensive technologies for soil cleanup (Walton & Anderson, 1992).

Past research on microbial transformations of organic compounds in the rhizosphere has focused primarily on agricultural chemicals, such as insecticides and herbicides (Table 8-1a). A number of researchers have described an increase in pesticide degradation in the rhizospheres of a variety of plant species (for example Hsu & Bartha, 1979; Sandmann & Loos, 1984; Lappin et al., 1985). More recently, similar observations have been made for industrial chemicals (Table 8-1b) such as oil residues (Rasolomanana & Balandreau, 1987), trichloroethylene (Walton & Anderson, 1990; Anderson & Walton, 1992), polycyclic aromatic hydrocarbons (Aprill & Sims, 1990), surfactants (Knaebel & Vestal, 1992) and pentachlorophenol (Ferro et al., 1994).

Table 8-1a. Studies on the degradation of pesticides in the rhizosphere.

Pesticides	Plant	Reference	Comments
Mecoprop 2,4-D, MCPA	Wheat	Lappin et al., 1985	Microbial community isolated from wheat roots and composed of 5 species was capable of using compounds as a carbon source. However, none of the pure cultures was capable of growth on mecoprop.
2,4-D	Sugarcane Clover	Sandmann & Loos, 1984	Higher population of 2,4-D degrading microorganisms in the rhizosphere of sugarcane, but not African clover.
diazinon parathion	Bushbean	Hsu & Bartha, 1979	Increased mineralization of both compounds in the rhizosphere of hydroponic bush bean. Both compounds are suspected of being cometabolized, a process that would be expected to occur very readily in the rhizosphere.
atrazine	Corn	Seibert et al., 1981	Increase in production of atrazine metabolites by rhizosphere microorganisms, especially in the presence of decomposing roots.
atrazine metolachlor trifluralin	*Kochia*	Anderson et al., 1994	Degradation of atrazine, metolachlor, and trifluralin was greater in rhizosphere soil from *Kochia* sp., a herbicide resistant plant, than in nonvegetated soil.

As microbial degradation of organic compounds in the rhizosphere and the beneficial effects of vegetation at waste sites have been recently reviewed (Shimp et al., 1993; Anderson et al., 1993), this chapter has two primary goals: (i) Provide a brief introduction to the concept of microbial degradation in the rhizosphere by describing what is known about the interaction of microorganisms, plants, and xenobiotics; (ii) Provide an overview of the 1993 Rhizosphere Symposium sponsored by the American Chemical Society (ACS). In order to attain the first goal, appropriate information is provided throughout the text and in Table 8-1a,b. The second goal is accomplished by a section describing some of the individual presentations and themes from the ACS symposium.

Table 8-1b. Studies on the degradation of industrial chemicals in the rhizosphere.

Chemical	Plant	Reference	Comments
Oil residues	Rice	Rasolomanana & Balandreau, 1987	*Bacillus* sp. isolated from rice rhizosphere on oil residues, but only in the presence of root exudates.
VOCs	Reeds	Wolverton et al., 1983	Vegetated microbial filters increased the removal of both aromatic and aliphatic compounds.
PAHs	Prairie grasses	Aprill & Sims, 1990	Increased disappearance of polycyclic aromatic hydrocarbons in vegetated vs. nonvegetated soil columns.
Surfactants	Corn, Soybean	Knaebel & Vestal, 1992	Rhizosphere treatments significantly increased initial mineralization rates by a factor of 1.1 to 1.9.
Petroleum	Legumes	Gudin, 1978	Describes the importance of leguminous plants in reclaiming petroleum-contaminated sites.
TCE	*Lespedeza* Loblolly pine Bahia- grass Goldenrod Soybean	Walton & Anderson, 1990; Anderson & Walton, 1992	Increased degradation of TCE in rhizosphere soil and increased mineralization of ^{14}C-TCE.
Surfactants	Cattails	Federle & Schwab, 1989	Mineralization of surfactants was more rapid in the rhizosphere than in root-free sediments.
PCP	Wheatgrass	Ferro et al., 1994	Mineralization of pentachlorophenol was enhanced in soils containing wheatgrass.

THE ROOT-SOIL INTERFACE

The rhizosphere is a zone of increased microbial activity and biomass at the root–soil interface under the influence of the plant root (Curl & Truelove, 1986). Because of its importance to crop productivity, the rhizosphere has been the focus of agricultural research for many years. Several excellent and comprehensive reviews on the rhizosphere are available (Brock, 1966; Campbell, 1985; Curl & Truelove, 1986); therefore, this discussion will be limited to findings on rhizosphere microbiology and chemistry within the context of degradation of organic compounds by rhizosphere microbial communities in soil.

The overall effect of the plant–microbe interaction is an increase in microbial populations by an order of magnitude or more above microbial populations in nonvegetated soils. This "rhizosphere effect" is often expressed as the ratio of the numbers of microorganisms in rhizosphere soil to the number of microorganisms in nonrhizosphere soil, the R/S ratio (Katznelson, 1946). These R/S ratios commonly range from 5 to 20, but occasionally are as high as 100 and above (Atlas & Bartha, 1993). In a simple scenario, the presence of 100-fold more microorganisms in rhizosphere soil compared with nonvegetated soil could translate into an increase in the degradation rate for a xenobiotic compound.

The actual composition of the microbial community in the rhizosphere is dependent on root type, plant species, plant age, environmental conditions, substrate, and soil type, as well as other factors such as exposure history of the plant roots to xenobiotic compounds. Generally, the rhizosphere is colonized by a predominantly gram-negative bacterial community and leguminous plants typically exhibit a greater rhizosphere effect than nonleguminous plants (Atlas & Bartha, 1993).

Continual change at the root-soil interface, both physical and chemical, produces constant alterations in the soil matrix and microbial environment. Differences in substrate, carbon dioxide and oxygen concentrations, osmotic and redox potentials, pH, and moisture content exist between rhizosphere and nonvegetated soil. These differences, in turn, can have a dramatic influence on microbial community structure and activity (Curl & Truelove, 1986; Atlas & Bartha, 1993).

THEMES FROM THE ACS RHIZOSPHERE SYMPOSIUM

An important aspect of solving environmental problems is cooperation among numerous contributing disciplines. The current research on the rhizosphere and its potential role in biological remediation of waste sites illustrates this point. The collective breadth of expertise represented by the investigators conducting research in this area is extensive: environmental engineering, microbiology, plant sciences, environmental toxicology, and soil science all contribute to our understanding of the rhizosphere.

A symposium entitled "Microbial Degradation of Organic Chemicals in the Rhizosphere: Implications for Bioremediation" was held at the 1993 fall American Chemical Society (ACS) meeting in Chicago, IL, under joint sponsorship of the Agrochemicals and Environmental Chemistry divisions of ACS. The symposium was organized to bring together the groups conducting research on biodegradation in the rhizosphere and to summarize the current state of the science. Eighteen papers were presented from several academic, industrial, and government laboratories in the United States and two foreign countries (Table 8-2). The papers covered a variety of topics, including studies with industrial chemicals and pesticides. Several papers from the symposium will appear in a forthcoming ACS symposium series book edited by the organizers, Todd Anderson and Joel Coats. A few themes emerged from the symposium, some of which are described below. Recognition of information presented by individuals is given by enclosing names within brackets.

Table 8-2. Speakers from the 1993 American Chemical Society (ACS) symposium[†] entitled *"Microbial Degradation of Organic Chemicals in the Rhizosphere: Implications for Bioremediation."*

Speaker	Institution
Todd Anderson	Iowa State University
Andrei Barkovskii	University of Claude Bernard
Jean-Marc Bollag	Pennsylvania State University
Jay Boyle	University of Cincinnati
Michael Cole	University of Illinois
Jurich Dec	Pennsylvania State University
Paula Donnelly	University of Oklahoma
Ari Ferro	University of Utah
Robert Hoagland	USDA-Stoneville
David Knaebel	University of Idaho
Xiujin Qiu	Union Carbide Corporation
Jerome Rigot	University of California-Davis
Kyo Sato	Tohoku University
Paul Schwab	Kansas State University
Jodi Shann	University of Cincinnati
Darwin Sorensen	Utah State University
Barbara Walton	Oak Ridge National Laboratory
Robert Zablotowicz	USDA-Stoneville

[†]Held in Chicago, IL under joint sponsorship of the Agrochemicals, and Environmental Chemistry divisions of ACS.

The approaches to studying the complex interaction between plants, microorganisms, and chemicals in soil ranged from pure cultures of rhizosphere microorganisms in growth flasks [Donnelly, Dec, Rigot, Barkovskii], to rhizosphere soils [Shann, Anderson, Boyle], to more complex soil-plant systems [Knaebel, Ferro, Sorensen, Schwab, Cole, Walton], to large-scale field studies [Qiu]. Each approach brought certain benefits and limitations. Studies on pure or defined cultures of microorganisms allowed for elucidation of specific mechanisms for the transformation of organic compounds [Barkovskii]. However, the absence of competing microorganisms or the soil matrix may prevent these systems from simulating a field situation very accurately. In addition, the time between collection of rhizosphere soils and use in experiments can influence the activity of the microorganisms and the subsequent biodegradation of the xenobiotic substrate [Boyle]. Soil-plant systems enabled degradation studies to be performed under more "real world" conditions, where the rhizosphere microorganisms were being nourished by a living plant, but plants can occasionally obscure microbial degradation data [Sorensen, Knaebel]; uptake of chemicals into plant tissues, plant metabolism, and plant fixation of CO_2 can make interpretation of microbial degradation data very difficult (Aprill & Sims, 1990). The ultimate goal of such research is to apply the technology in the field [Qiu]; however, the capital- and labor-intensive nature of field research often hinders this goal.

Two intriguing and recurring aspects discussed during the symposium were the preponderance of gram-negative microorganisms in the rhizosphere and the importance of increased microbial activity in the rhizosphere compared with nonvegetated soils. There is compelling evidence from the literature that gram-negative bacteria are particularly abundant in the rhizosphere because of their ability to rapidly utilize growth substrates (Atlas & Bartha, 1993). Gram-positive species apparently do not compete as well in the rhizosphere. This idea was further supported by presentations at the symposium [Sato, Boyle, Hoagland]. In addition, gram-negative bacteria seem to have some important metabolic capabilities for degrading xenobiotic chemicals. Glutathione-S-transferase (an enzyme responsible for conjugation of xenobiotics in mammals, plants, and microorganisms) activity was particularly abundant among gram-negative bacteria including the genera *Pseudomonas* and *Enterobacter* [Zablotowicz]. These observations are consistent with previous research in the literature (LaRoche & Leisinger, 1990; Feng, 1991; Xun et al., 1992). Glutathione-S-transferase activity is also present in fungi (Cohen et al., 1986).

Additionally, increased microbial biomass did not always translate into increased microbial degradation in the rhizosphere. A better "predictor" of enhanced microbial degradation was microbial activity, which can be measured by a variety of techniques (Staley & Konopka, 1985). Two techniques which were used to measure the activity of rhizosphere microorganisms were (1) dehydrogenase assay [Cole] and (2) incorporation of acetate into microbial lipids [Knaebel].

Another theme of the symposium focused on the biodegradation endpoint being measured. Typically, these endpoints were disappearance of parent compound and/or production of $^{14}CO_2$ from degradation of ^{14}C -labeled substrates. However, another endpoint was proposed during the symposium – the incorporation of the organic compound into soil organic matter. It was speculated that humification activity of microorganisms may be particularly pronounced in the rhizosphere and could be important for immobilizing and degrading xenobiotics in soil [Bollag, Walton]. Additionally, humic substances in soils may facilitate the transformation of organic substrates [Dec] and function as electron acceptors during low-oxygen conditions [Barkovskii].

A proposal that generated much discussion among the symposium participants and attendees was the idea put forth that plants may enlist rhizosphere microorganisms to degrade xenobiotics in the root zone, thus protecting the plant from chemical injury [Walton]. A theoretical basis for this idea was presented, along with compelling experimental evidence on the response of plants and microorganisms to organic toxicants (Hoylman & Walton, 1994). Previous research seemed to provide additional evidence to support the plant protection hypothesis (Kaiser & Reber, 1970; Herring & Berring, 1988; Krueger et al., 1991). The implications of such a relationship at sites contaminated with phytotoxic chemicals (herbicides) were also discussed [Anderson]. In the case of pesticide contaminated sites, which typically have herbicide concentrations several-fold above the field application rate, establishing vegetation in an attempt to facilitate biological degradation can be difficult without plants that can survive the chemical contamination (Anderson et al., 1994).

FUTURE DIRECTIONS

Overall, the studies presented at the ACS symposium reflected the complexity of plant-microbe-toxicant interactions and the complications that may hamper elucidation of the mechanisms by which toxicants are degraded in the rhizosphere. These studies, however, add to growing experimental evidence that plant roots in conjunction with their associated microbial communities offer a potentially important treatment strategy for *in situ* biological remediation of chemically contaminated soils. Nonetheless, field demonstrations of this treatment technology have yet to be fully developed.

The relative importance of several variables, such as the morphology or surface area of the root system (e.g., tap vs. fibrous), the selective influence of the root exudates on the physiology and ecology of microorganisms, the nature of microbe and root associations present (e.g., nitrogen-fixing, mycorrhizal), or the dynamic aspects of root turnover and the release of metabolites in accomplishing biodegradation of organic compounds in the rhizosphere are still not well understood. Because an increase in microbial growth and in microbial activity in the rhizosphere seem to contribute to the enhanced metabolic degradation of certain xenobiotic compounds in the rhizosphere in some instances, an intriguing question remains. Would selection for plants with a proliferation of root hairs, or other genetically determined properties of plant roots, positively affect microbial degradation rates of specific toxicants in the rhizosphere?

CONCLUSIONS

The information presented in this chapter and at the ACS symposium on "Microbial Degradation of Organic Chemicals in the Rhizosphere: Implications for Bioremediation" illustrates the role that rhizosphere microbial communities could play in maintaining and/or remediating soil systems through metabolism of hazardous organic compounds in the root zone. Vegetation promises a means whereby legal liability for hazardous waste residues in soils may cease through total destruction (mineralization) of hazardous compounds in the rhizosphere or root zone, in contrast to technologies that merely relocate the contaminants from one place to another. A better understanding of the mechanistic interactions among plants, microorganisms, and chemicals in the root zone will help identify situations where bioremediation using rhizosphere microorganisms can be appropriate and will permit more rapid realization of this new approach to bioremediation of contaminated soils.

REFERENCES

Anderson, T.A., and B.T. Walton. 1992. Comparative plant uptake and microbial degradation of trichloroethylene in the rhizospheres of five plant species: Implications for bioremediation of contaminated surface soils. ORNL/TM-2017. Oak Ridge, TN. 186 pp.

Anderson, T.A., E.A. Guthrie, and B.T. Walton. 1993. Bioremediation in the rhizosphere. Environ. Sci. Technol. 27:2630-2636.

Anderson, T.A., D.C. White, and B.T. Walton. 1995. Degradation of hazardous organic compounds by rhizosphere microbial communities. *In* Singh, V., Ed. Biotransformations: Microbial Degradation of Health-Risk Compounds. (In press).

Anderson, T.A., E.L. Kruger, and J.R. Coats. 1994. Enhanced degradation of a mixture of three herbicides in the rhizosphere of a herbicide-tolerant plant. Chemosphere 28:1551-1557.

Aprill, W., and R.C. Sims. 1990. Evaluation of the use of prairie grasses for stimulating polycyclic aromatic hydrocarbon treatment in soil. Chemosphere. 20:253-265.

Atlas, R.M., and R. Bartha. 1993. Microbial Ecology: Fundamentals and Applications, 3rd ed. Benjamin/Cummings. Menlo Park, CA.

Bordeleau, L.M., and R. Bartha. 1968. Ecology of a pesticide transformation: Synergism of two soil fungi. Soil Biol. Biochem. 3:281-284.

Brock, T.D. 1966. Principles of Microbial Ecology. Prentice-Hall. New Jersey. Campbell, R. 1985. Plant Microbiology. Edward Arnold. Baltimore.

Cohen, E., A, Gamliel, and J. Katan. 1986. Glutathione and glutathione-S-transferase in fungi: Effect of pentachloronitrobenzene and 1-chloro-2,4-dinitrobenzene: Purification and characterization of the transferase from *Fusarium*. Pest. Biochem. Phys. 26:1-9.

Curl, E.A., and B. Truelove. 1986. The Rhizosphere. Springer-Verlag. Berlin.

Federle, T.W., and B.S. Schwab. 1989. Mineralization of surfactants by microbiota of aquatic plants. Appl. Environ. Microbiol. 55:2092-2094.

Feng, P.C.C. 1991. Soil transformation of acetochlor via glutathione conjugation. Pest. Biochem. Phys. 40:136-142.

Ferro, A.M., R.C. Sims, and B. Bugbee. 1994. Hycrest crested wheatgrass accelerates the degradation of pentachlorophenol in soil. J. Environ. Qual. 23:272-279.

Gudin, C. 1978. Interactions between oil-vegetation and soil. In Proceeding of the International Symposium on Ground Water Pollution by Oil Hydrocarbons. Prague, Czechoslovakia. pp. 411-417.

Gunner, H.B., and B.M. Zuckerman. 1968. Degradation of diazinon by synergistic microbial action. Nature. 217:1183-1184.

Herring, R., and C.L. Bering. 1988. Effects of phthalate esters on plant seedlings and reversal by a soil microorganism. Bull. Environ. Contam. Toxicol. 40:626-632.

Hoylman, A.M., and B.T. Walton. 1994. Fate of polycyclic aromatic hydrocarbons in plant-soil systems: Plant responses to a chemical stress in the root zone. ORNL/TM-12650. Oak Ridge, TN. 122 pp.

Hsu, T.S. and R.Bartha. 1979. Accelerated mineralization of two organophosphate insecticides in the rhizosphere. Appl. Environ. Microbiol. 37:36-41.

Kaiser, P. and H. Reber. 1970. Interactions entre la simazine-herbicide et les microorganismes de la rhizosphere du mais. Meded. Fac. Landbouwet. Rijksuniv. (Gent.). 35:689-705.

Katznelson, H. 1946. The rhizosphere effect of mangels on certain groups of micro-organisms. Soil Sci. 62:343-354.

Knaebel, D.B., and J.R. Vestal. 1992. Effects of intact rhizosphere microbial communities on the mineralization of surfactants in surface soils. Can. J. Microbiol. 38:643-653.

Krueger, J.P., R.G. Butz, and D.J. Cork. 1991. Use of dicamba-degrading microorganisms to protect dicamba susceptible plant species. J. Agric. Food Chem. 39:1000-1003.

Lappin, H.M., M.P. Greaves, and J.H. Slater. 1985. Degradation of the herbicide mecoprop [2-(2-methyl-4-chlorophenoxy) propionic acid] by a synergistic microbial community. Appl. Environ. Microbiol. 49:429-433.

LaRoche, S.D. and T. Leisinger. 1990. Sequence analysis and expression of the bacterial dichloromethane dehalogenase structural gene, a member of the glutathione S-transferase supergene family. J. Bact. 172:164-71.

Rasolomanana, J.L., and J. Balandreau. 1987. Role de la rhizosphere dans la biodegradation de composes recalcitrants: Cas d'une riziere polluee par des residus petroliers. Rev. Ecol. Biol. Sol. 24:443-457.

Sandmann, E.R. I.C., and M.A. Loos. 1984. Enumeration of 2,4-D-degrading microorganisms in soils and crop plant rhizospheres using indicator media: High populations associated with sugarcane (*Saccharum officinarum*). Chemosphere. 13:1073-1084.

Seibert, K.,F. Fuehr, and H.H. Cheng. 1981. Experiments on the degradation of atrazine in the maize rhizosphere. *In* Proceedings of the Theory and Practical Use of Soil Applied Herbicides Symposium. European Weed Resource Society. Paris, France. pp. 137-146.

Senior, E., A.T. Bull, and J.H. Slater. 1976. Enzyme evolution in a microbial community growing on the herbicide dalapon. Nature. 263:476-479.

Shimp, J.F., J. C. Tracy, L.C. Davis, E. Lee, W. Huang, L.E. Erickson, and J.L. Schnoor. 1993. Beneficial effects of plants in the remediation of soil and groundwater contaminated with organic materials. Crit. Rev. Environ. Sci. Technol. 23:41-77.

Staley, J.T., and A. Konopka. 1985. Measurement of *in situ* activities of nonphotosynthetic microorganisms in aquatic and terrestrial habitats. Ann. Rev. Microbiol. 39:321-346

Walton, B.T., and T.A. Anderson. 1990. Microbial degradation of trichloroethylene in the rhizosphere: Potential application of biological remediation of waste sites. Appl. Environ. Microbiol. 56:1012-1016.

Walton, B.T., and T.A. Anderson. 1992. Plant-microbe treatment systems for toxic waste. Curr. Opin. Biotechnol. 3:267-270.

Wolverton, B.C., R.C. McDonald, and W.R. Duffer. 1983. Microorganisms and higher plants for waste water treatment. J. Miss. Acad. Sci. 31:79-89.

Xun, L., E. Topp, and C.S. Orser. 1992. Confirmation of oxidative dehalogenation of pentachlorophenol by a *Flavobacterium* pentachlorophenol hydroxylase. J. Bact. 174: 5745-5747.

9 Phytoremediation: Plant-Based Remediation of Contaminated Soils and Sediments

S.D. Cunningham
DuPont
Central Research and Development
Newark, Delaware

C.R. Lee
U.S. Army Corps of Engineers
Waterways Experiment Station
Vicksburg, Mississippi

Phytoremediation is defined as the use of green plants to remove, contain, or render harmless environmental contaminants. This definition applies to all plant-influenced biological, chemical, and physical processes that aid in remediation of contaminated substrates (Cunningham & Berti, 1993). This is not a new concept. The use of plants to treat wastewater has been practiced for over 300 years (Hartman, 1975) and plant-based remediation methods for dredged material slurries (Lee et al., 1976) and metal-contaminated soils (Yamada et al., 1975) have been proposed since the mid seventies. Plant-based remediation also has been evolving in the control of both indoor air pollution (Raloff, 1989) and urban smog. The renewed interest in phytoremediation is driven by both advances in technology as well as the search for better, cost-effective solutions. Plants, along with their associated microflora, are being utilized in both containment and decontamination remediation strategies for contaminated soils as well. This chapter will focus on and discuss plant-based remediation of contaminated soils and sediments.

SITE REMEDIATION PROGRAMS

There are many technologies available to site remediation personnel. All *in-situ* site remediation programs, however, follow either a contaminant containment (also referred to as a stabilization) or contaminant reduction (often referred to as a "clean-up") strategy. Under contaminant containment, there is no actual reduction in contaminant present, but health and environmental risks are reduced to acceptable levels by a physical or chemical manipulation of the site. The construction of caps, vaults, and hydraulic isolation curtains are examples of engineering strategies in this category. A more recent example is the development of soil stabilization methods that rely on changes in soil chemistry and the growth of plants. *In-situ* contaminant reduction strategies include microbial based degradation, soil flushing, venting, heating, and electroosmosis. Plant-based decontamination strategies are among the newest techniques targeted for *in-situ* contaminant reduction.

PLANTS AS ENGINEERING STRUCTURES

It is useful to visualize and describe a plant in engineering terms in order to shift our familiar agronomic plant paradigm and to clarify the concept of phytoremediation to those professionals who are more familiar with engineering terminology. At a basic level, green plants are solar-driven pumping and filtering systems that have inherent loading, degradative, and fouling capacities. Plant roots are exploratory liquid-phase extractors that can find, alter and/or translocate elements and compounds against large chemical gradients. Root surfaces support active bacterial biofilms and fungal extensions that significantly augment soil-surface contact, metabolic capacities, and can alter most measurable soil physical and chemical parameters (Cunningham & Berti, 1993).

The plant's involvement with environmental toxicants is not new. Through evolution, plants have developed tolerance mechanisms to survive many hostile environments. Plants in most environments are continually assaulted with a wide array of organic and metallic toxins and have developed mechanisms to adapt, survive, and even thrive under these conditions. Some plant pathogens, soil microbes, and weeds produce a wide variety of toxic organic compounds, with which plants have evolved to cope. Plants growing on soil derived from outcroppings of metal-rich ore adapted survival mechanisms long before anthropomorphic soils existed. Some plants develop root mechanisms that immobilize metals in the roots and thus restrict translocation up into the plant leaves where damage could occur. Such plants are proving useful under containment strategies. Other plants have developed sophisticated mechanisms to absorb, translocate, and tolerate significant toxic quantities of heavy metals. These plants can flourish in hostile environments while all other plants die. These latter plants are beginning to prove useful in a contaminant reduction remediation strategy. These and other mechanisms that plants have developed to tolerate contaminated soil will be discussed in further detail below.

REMEDIATION BY CONTAMINANT CONTAINMENT

A critical aspect of any containment policy is to control water flux into a polluted area as well as the flux of contaminated water exiting in a containment zone. Plants excel at finding and transpiring water and have been used for both purposes (Shimp et al., 1993). Plants also control soil erosion and the subsequent off-site migration of contaminants. As such, most caps commonly use a plant cover. Additionally plants are often used to maintain the integrity ditches and berms (McAneny et al., 1985; Licht, 1993). Certain seed companies provide seed of hearty grasses for such purposes. Plant roots, especially in wetlands, also alter the ion concentrations and redox potential around their roots. In doing so they modify substrates and immobilize contaminant migration in percolating water through wetland areas (Wenzel et al., 1992). This phenomenon is exploited in many wastewater remediation schemes involving natural and constructed wetlands as well as in reed beds for treatment. In many remediation schemes plants can also absorb

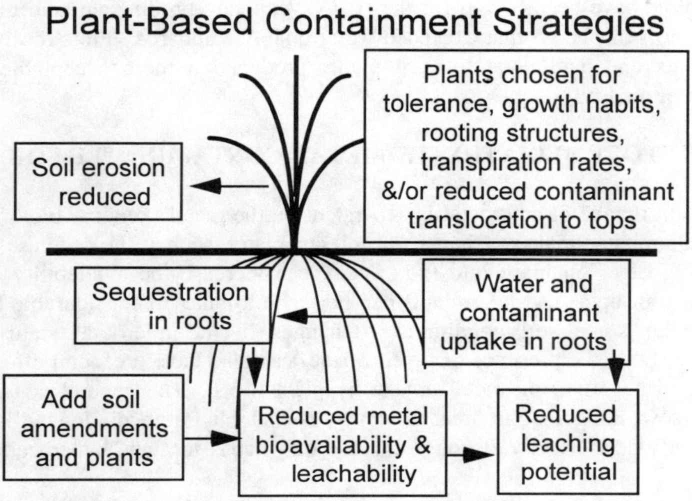

Fig. 9-1. A schematic of phytoremediation as a pollutant containment system.

bioavailable contaminants into roots and choosing plants with restricted translocation into plant leaves aids in a containment strategy (Oyler, 1988; Bradshaw & Chadwick, 1980). Fig. 9-1 provides an overview of such a containment strategy. Under containment remediation, the mobility of contaminants can be reduced by the addition of soil amendments that reduce contaminant solubility. These amendments have included alkalizing agents, phosphates, and organic materials all designed to render the toxic materials more insoluble and unavailable to leaching and plant uptake. In addition, plants with certain mechanisms for dealing with excessive concentrations of contaminants in soils can be planted to further immobilize the pollutants. For example, heavy metals and lipophilic organics can be immobilized in roots or on root surfaces with little mobility in the plant or soil system (Baker, 1981; Bell, 1992). Furthermore, most tolerant plants will stabilize the contaminated matrix against wind and rainfall erosion.

Phytoremediation as a containment technique borrows from much previous research in revegetation of restored strip mine lands. Metal-tolerant plants, such as *Festuca rubra* (red fescue), have been used to vegetate and control soil erosion on metal mine tailing and waste piles (Bradshaw & Chadwick, 1980). These plants do not translocate large amounts of metals into their leaves and consequently can be used in grazing management strategies for these highly contaminated soils in the United Kingdom. Animal grazing on these restored sites is limited to short time periods and animals are rotated to noncontaminated land for the rest of their feeding requirements.

Another example of this strategy was described by Brandon et al., (1991) for acidified, metal-contaminated dredged material placed in an upland disposal environment. Soil amendments of lime, coarse limestone gravel, horse manure and acid/metal tolerant plants were used to control metal solubility and to totally vegetate the contaminated sediment after placement in an upland environment.

Control plots have been barren for the past 6 years and produce an acid metal rich surface runoff of pH 3.5 that exceeded water quality standards. Plots receiving soil amendments and acid/metal tolerant plants produced a more acceptable surface runoff water quality.

PHYTOREMEDIATION OF METAL-CONTAMINATED SOILS

Plant uptake of any metal is dependent on the metal's relative bioavailability in the contaminated matrix. Changes in soil chemistry, such as the decrease of pH or the limiting of anion input into the system, can increase the availability of many metals for root uptake. Many plants can be made to take up considerable levels of metals under some soil conditions. Changes in the microbial status of the rhizosphere (e.g., the presence of mycorrhizae) can also have profound effects (both positive and negative) on metal uptake by plant roots. The general consensus of researchers working in this area, however, is that phytoremediation will only be economically cost effective through the use of what is termed "hyperaccumulator" plants.

There are plants that tolerate excessive concentrations of metals in soils by absorbing, translocating and storing the metals in a nontoxic way. Many of these plants have evolved on metal-rich soils. Research over the last two decades has shown that certain specialized plants have the ability to accumulate up to 3% (by dry weight) of heavy metals and up to 25% (by dry weight) in sap/latex with no apparent plant damage (Brooks & Jaffre, 1974; Farago et al., 1977; Bradshaw et al., 1978; Ernst, 1978; Morrison et al., 1980; Brooks & Jaffre, 1981; Baker et al., 1983; Reeves et al., 1983; Nishizono et al., 1987; Baker & Brooks, 1989; Baker et al., 1990; and Baker & Walker, 1990). Geobotantists have located similar types of plants in numerous countries around the world such as the United Kingdom, Germany, Switzerland, Spain, France, Italy, Bulgaria, Greece, Czechoslovakia, Zaire, The Philippines, N. Caledonia, New Zealand, Japan and the United States. These authors document that the accumulation of these elements is dependent on plant, metal, and environmental conditions. The mechanisms governing this amazing tolerance and uptake of excessive concentrations of metals into plant leaves have been the subject of active research (Ernst & Weinert , 1972; Farago et al., 1975; Lee et al., 1977; Ernst, 1978; Morrison et al., 1981; Grill et al., 1989; Verkleij et al., 1990; and Gabbrielli et al., 1991). Tolerance mechanisms include zinc accumulation in cell walls; nickel associated with the pectin in large cells; nickel, cobalt and zinc being chelated by malic acid; zinc associated with phytochelatin; nickel chelation by citrate; and cobalt associated with calcium oxalate crystals in plant tissues. Knowledge of the tolerance mechanisms will assist in the identification of the genetic characteristics necessary for transferring metal tolerance to plants capable of larger biomass production and with deeper rooting characteristics. It has been suggested that metal contaminated soils can be decontaminated by using hyperaccumulator plants (Yamada et al., 1975, Ernst 1988; and Baker et al., 1991). The latter reference also suggested that in some cases the resultant metal-rich biomass (bio-ore) could be ashed and the metals recycled economically. This "biomining" of metals also may be applicable as a mining technique for metals with significant economic value (Fig. 9-2). The coupling of hyperaccumulation of metals with larger biomass and potentially deeper rooting systems through genetic engineering may result in a powerful phytoremediation technology. It is the hope of many that this technology may provide an economical, permanent, site remediation technique for many sites which have only more expensive engineering cleanup technologies available. Plant and soil scientists are developing the technologies, that if combined and perfected, will enable genetically engineered phytoremediation crops to be a reality in the near future.

PHYTOREMEDIATION OF ORGANIC CONTAMINATED SOILS

Plants are continuously assaulted by a wide array of biologically active organic compounds. Many plant pests (fungal, bacterial, and weed) make a wide array of biocidal agents to which plants have developed tolerance mechanisms. More recently, with the development of the pesticide industry, we have begun to appreciate the metabolic capacity of the plant system. Most modern herbicides rely on crop selectivity based on metabolic differences between plant species (Hatway, 1989). This capacity is the backbone of the multibillion dollar selective-herbicide

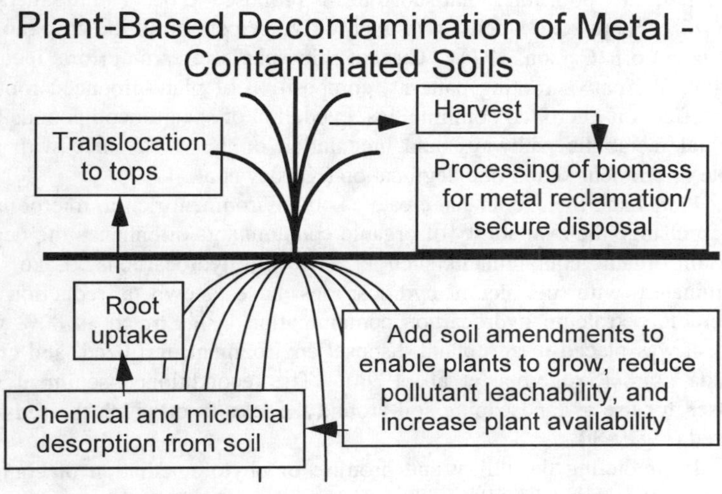

Fig 9-2. A schematic of plant -based soil decontamination methods.

market. With these herbicides, desirable plants rapidly metabolize the pesticide into a nontoxic compound, whereas undesirable weeds do not, and are consequently killed. These metabolic capacities of plants are wide spread, and are both constitutive and inducible (Hatzios & Hoagland, 1989). They have also begun to prove potentially exploitable in the remediation of contaminated soils. Under this remediation strategy, plants absorb pollutants from the soil and either metabolize them into nontoxic materials or incorporate them into a stable cell component (e.g., lignin). This capacity of plants to detoxify xenobiotics is widely recognized. In parallel to mammalian detoxification systems, plants have even been described as planetary "green livers" (Sandermann, 1992). As our capacity to genetically engineer plants has increased, many researchers have begun to incorporate both prokaryotic and eukaryotic metabolic genes into the plant genome. One currently popular industrial target is to engineer plants for increased tolerance to currently registered herbicides, thus increasing the herbicide's utility under a wider variety of agronomic conditions. Many concepts, techniques, and constructs used in producing plants with increased herbicide metabolism may also find a niche in the remediation market. Patents have been issued in this area (Gordon et al., 1989).

Despite this capacity to metabolize certain xenobiotics, plants are generally "constructive" and not "destructive" organisms. As compared to microbial metabolism, plants act on a narrower range of substrates, perform simpler degradative steps, and generally do not reduce the xenobiotic to a molecularly simple end point. It makes both intuitive and logistical sense to combine the engineering infrastructure and autotrophic nature of the plant with the degradative capacity of the microbe. One such plant-microbe arrangement exists in the biologically active area around the roots of a plant (rhizosphere). Initial research in this field shows promise (Anderson & Walton, 1992) and is the subject of another paper in this symposium (Anerson & Coats, 1995). Co-metabolism of pollutants that are normally poor microbial substrates is proposed to occur in the metabolically active rhizosphere zone where up to twenty-five percent of all the crop biomass can be sloughed off (Carson, 1974). Current rhizosphere research efforts include plant selection, alternative rooting patterns, composition of plant-produced root exudate and its effect on microbial communities, exudation of specific compounds to induce microbial metabolic pathways, and inoculation of the rhizosphere with microbes capable of efficient xenobiotic degradation (Kinsley et al., 1993).

Plants and their roots can create a soil environment rich in microbial activity that can change the availability of organic contaminants or enhance the degradation of certain organic contaminants such as petroleum hydrocarbons. Lake sediments contaminated with petroleum hydrocarbons have shown a reduction in total recoverable petroleum hydrocarbon contamination by as much as 80% when the sediment was placed in an upland disposal environment, fertilized, and cropped to bermuda grass (Skogerboe et al., 1994). The reconditioned sediment has been proposed for use as landscaping soil around the lake from which the sediment was removed (Fig. 9-3.).

In predicting the utility and promise of phytoremediation of contaminated soils, we have drawn heavily on the study, development, and registration of soil-applied pesticides. Using models of the availability, uptake, and flux of xenobiotics in the plant (Hanks & Ritchie, 1991), we can make certain generalizations. The pesticide industry often attempts to predict the fate and effect of a pesticide from relatively simple measurable parameters. One such useful parameter is the octanol-water partitioning coefficient (K_{OW}).

Contaminants with log K_{OW} ($\leq$1) are very water-soluble and would be predicted to cause groundwater contamination without a hydrological containment system or restricted rainfall pattern. Additionally, plants do not generally accumulate these water-soluble compounds at a rate surpassing passive influx in the transpiration stream. Compounds with this high water solubility are generally considered mobile in both plant xylem and phloem. In some cases this can lead to substantial accumulation; however the botany, chemistry, physics, and regulatory constraints suggest that these compounds would provide a relatively poor target for phytoremediation.

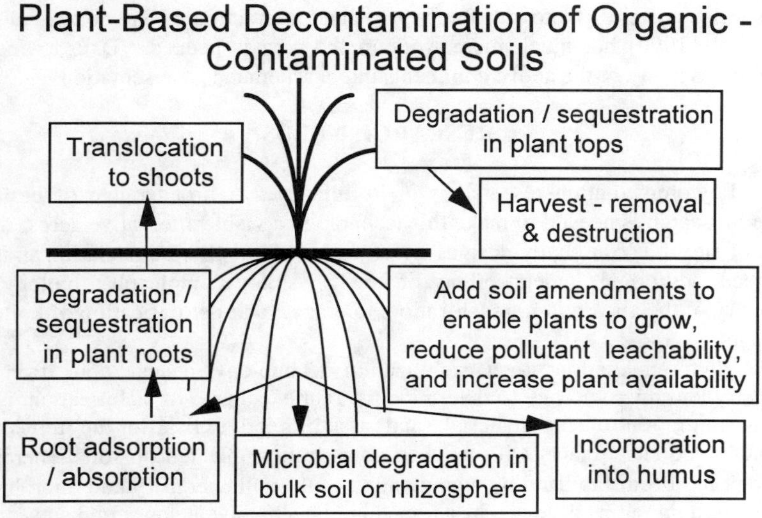

Fig. 9-3. A schematic of plant-based decontamination of soil contaminated with organics.

Slightly more lipophilic compounds (log K_{OW}'s approximately 1 to 4) are taken up by roots and are considered mobile in the xylem but generally phloem-immobile. It is in this range (2.5 to 4) that many soil-applied herbicides are found. This is perhaps the best evidence that they have the physical properties which permit translocation in the soil zone, root uptake, and good interaction with the plant metabolic machinery. These types of compounds would be expected to be good targets for phytoremediation and many priority pollutants are found in this range. Toward the low end of this scale, compounds again would be expected to cause groundwater problems. The suitability of phytoremediation as a remediation alternative would therefore be dependent on soil type, rainfall, how the material entered the environment, toxicology, and regulatory status of the site.

Lipophilic compounds (log K_{OW}'s > 4) are adsorbed to roots as well as soils (Bell, 1992). These compounds are not substantially translocated to the shoot. Remediation technologies that harvest roots (Bell, 1992; McMullin, 1993) are being explored. The main problem with such an approach seems to be the relative immobility of the pollutant in the soil matrix and the potential for high worker-exposure, yet solutions for these objections are being explored. The other likely approach for degradation of these type of compounds is to rely on rhizosphere and root degradative capacities (Schwab & Banks, 1994). Most priority pollutants are not ionizable at physiologically reasonable pH's. Mobility guidelines in both the soil and the plant change when the material is an ion (Bromilow et al., 1987; Rigitano et al. 1987). Certain volatile compounds (e.g., the soil applied herbicide Treflan) should be nonmobile under these guidelines, however in a volatile state they can travel in the soil void space to be absorbed by roots. Parallels may be appropriate in the phytoremediation area as well and can be predicted to some degree based on the Henry's law constant of the pollutant in question. Many soil, plant, and plant-root characteristics are also known to influence plant interaction of any compound (e.g., Hsu et al., 1990) but much more work in the area is needed. Data in Fig. 9-3 summarizes many of the above comments into a schematic representation.

RESEARCH NEEDS

Phytoremediation research is in its infancy. A broad range of basic and applied research is needed to make this technology a viable alternative across a wide range of sites. To rationally design, select, and develop phytoremediation strategies we need additional understanding of the physiology, molecular biology, and chemistry of the absorption, translocation, tolerance, and detoxification of pollutants by plants.

Phytoremediation as a stabilization technology needs input from soil chemistry and microbiology to determine the relative degree of stabilization, predict off-site migration of the material, and assign relative risk to the remediation technique and site. Plants with reduced translocation, planted on soils wherein the pollutant has been stabilized by manipulation of the soil chemistry and microbiology is just beginning to be accepted as a remediation technique at superfund sites, yet the fundamentals are nearly 50 years old. Work in establishing protocols, standards, and application techniques for the research, engineering remediation, and regulatory authorities is needed.

Phytoremediation as a decontamination technique, needs researchers in finding plants with high metal uptake rates, large biomass, and tolerance to a wide array of environmental conditions and constraints. Despite the best efforts of geobotantists, these plants currently may not exist and the ideal plant may require breeding, selection, or molecular biological strategies to develop. To this end, molecular biologists, botanists, agronomists, and soil scientists will need to cooperate closely to develop applicable phytoremediation systems. The best cooperative efforts of those experts in such disparate fields as irrigation science and molecular biology will be needed to fully realize the potential in plant-based remediation.

The decontamination of soils contaminated with organics requires a better fundamental understanding of plant-soil interactions. To accomplish this, the disciplines of biology, physiology, microbiology, and microbial and plant genetics should be focused on the interactions of plants and their microbial communities in contaminated soil. Increasing the degradative capacity of plants is in its early stages and much effort is needed. Degradative genes, cloned from both prokaryotic and eukaryotic systems have begun to be expressed in plant tissue. Many major agricultural chemical companies are incorporating herbicide degradation genes into crop germplasm. Finding, cloning, and expressing degradative genes targeted toward industrial pollutants should not be far behind. Some of these systems may already function in such a capacity, but most are untested in this role. Exploiting the ancillary enzyme systems, cofactors, and altered plant tissues is a remarkable opportunity well within the grasp of many laboratories.

Plant structures have never been optimized for remediation. Rooting depth, structure and density can no doubt be altered to increase degradative efficiency. Again, the theorectical, practical, and applied sciences have fairly established techniques that have not yet been brought to bear in the phytoremediation of contaminated soils.

CONCLUSIONS

Our corporate and governmental experiences in remediation have shown that current methods of cleaning surface soils of metals and non-volatile organics are clumsy, costly and physically disruptive. Phytoremediation is an emerging technology that promises a viable remediation when pollutants: i) are near the surface, ii) are relatively nonleachable, and iii) pose little imminent risk to health or the environment. Few polluted sites have been remediated using plants. The engineering community has its doubters. The agronomic community; however, has established a firm technical basis suggesting that plants do indeed pick up metals, absorb toxins, and detoxify certain organics. Coupling the hyperaccumulation of metals and the enhanced degradation or sequestration of organics with larger biomasses and deeper rooting systems may result in a powerful phytoremediation technology that will provide cheaper, permanent, and less invasive remediation options. Plant and soil scientists have developed the technologies, that if combined and perfected, will enable crops specifically designed for phytoremediation to be a reality in the near future. We believe that exploiting and extending our knowledge base in the area of plant-environmental interactions is an essential element in developing this technology. The technical, economic, and regulatory climate for the perfection of this research is excellent.

REFERENCES

Anderson, T.A., and B.T. Walton. 1992. Comparative plant uptake and microbial degradation of trichloro-ethylene in the rhizospheres of five plant species-implications for bioremediation of contaminated surface soils. OakRidge Nat. Labs. Environmental Sciences Division Publication #3809. ORNL/TM-12017.

Anderson T.A. and J.R. Coats. 1995. An overview of microbial degradation in the rhizosphere and its implications for bioremediation. p. 133-141. *In* H.D. Skipper and R.F. Turco (ed.) Bioremediation: Science and applications. SSSA Spec. Publ. 43. ASA,CSSA and SSSA, Madison WI.

Baker, A. J. M. 1981. Accumulators and excluders - strategies in the response of plants to heavy metals. J. Plant Nutr. 3:643-654.

Baker, A. J. M., R.R. Brooks, A.J. Pease, and F. Malaisse. 1983. Studies of copper and cobalt tolerance in three closely related taxa within the genus *Silene L. (Caryophyllaceae)* in Zaire. Plant Soil 73:377-385.

Baker, A.J.M., and R.R. Brooks. 1989. Terrestrial higher plants which hyperaccumulate metal elements-A review of their distribution, ecology, and phytochemistry. Biorecovery. 1:81-126.

Baker, A.J.M., K. Ewart, G.A.F. Hendry, P.C. Thorpe, and P.L. Walker. 1990. The evolutionary basis of cadmium tolerance in higher plants. p. 23-29. *In* J. Barcelo (ed.) Proceedings of 4th international conference on environmental contamination, Oct. 1990. CEP Consultants Ltd., Edinburgh, UK.

Baker, A.J.M., and P.L. Walker. 1990. Ecophysiology of metal uptake by tolerant plants. *In* A.J. Shaw (ed.) Heavy metal tolerance in plants: evolutionary aspects CRC Press Inc., Boca Raton, FL.

Baker, A.J.M., R.D. Reeves, and S.P. McGrath. 1991. *In situ* decontamination of heavy metal polluted soils using crops of metal-accumulating plants - A feasibility study. *In* R. E. Hinchee and R. F. Olfenbutter (ed). *In Situ* bioreclamation: applications and investigations for hydrocarbon and contaminated site remediation. Butterworth-Heinemann, Boston.

Bell, R.M. 1992. Higher plant accumulation of organic pollutants from soils. Risk Reduction Engineering Laboratory, ORD, USEPA. Technical Report. EPA/600/R-92/138.

Bradshaw, A. D., M.O. Humphreys, and M.S. Johnson. 1978. The value of heavy metal tolerance in the revegetation of metalliferous mine wastes. p. 311-34. *In* G.T. Goodman, and M.J. Chadwick (ed.) Environmental management of mineral wastes. Sijthoff and Nordhoff. The Netherlands.

Bradshaw, A.D., and M.J. Chadwick. 1980. The restoration of land: The geology and reclamation of derelict and degraded land. Univ. California Press, Berkeley.

Brandon, D.L., C.R.Lee, J.W. Simmers, and J.G Skogerboe. 1991. Long-term evaluation of plants and animals colonizing contaminated estuarine dredged material placed in both upland and wetland environments. Miscellaneous paper D-91-5 US Army Engineer Waterways Experiment Station, Vicksburg MS.

Bromilow, R.H., R.L.O. Rigitano, G.G. Briggs, and K. Chamberlain. 1987. Phloem translocation of non-ionised chemicals in *Ricinus communis*. Pestic. Sci. 19:85-99.

Brooks, R.R., and T. Jaffre. 1974. Some New Zealand and New Caledonian plant accumulators of nickel. J. Ecol. 62:493-499.

Brooks, R.R., S. Shaw, and A.A. Marfil. 1981. Some observations on the ecology, metal uptake, and nickel tolerance of *Alyssum serpyllifolium* subspecies from the Iberian Peninsula. Vegetation 45:183-188.

Carson, E.W. 1974. The Plant Root and Its Environment, University Press of Va.

Cunningham, S.D., and W.R. Berti. 1993. The remediation of contaminated soils with green plants: an overview. *In vitro* cellular and developmental biology-plant. 29:207-212.

Ernst, W.H.O., and H. Weinert. 1972. Localization von zink in den blatternvon *Silene cucubalus* Wib. Z. Pflanzenphysiol. 66:258-264.

Ernst, W.H.O. 1978. Physiology of Heavy Metal Resistance in Plants. *In* Proceedings of International Conference on Heavy Metals in the Environment Vol II: Part I: 121-136. Sept. 1978. CEP Consultants Ltd., Edinburgh, UK.

Ernst, W.H.O. 1988. Decontamination of mine sites by plants: An analysis of the efficiency. p 305-310. *In* Proceedings of international conference on environmental contamination. CEP Consultants, Ltd., Edinburgh, UK.

Farago, M.E., A.J. Clark, and M.J. Pitt. 1975. The chemistry of plants which accumulate metals. Coord. Chem. Rev. 16:1-8.

Farago, M.E., A.J. Clark, and M.J. Pitt. 1977. Plants which accumulate metals. I. The metal content of three Austrialian plants growing over mineralized sites. Inorg. Chim. Acta 24:53-56.

Gabbrielli, R., C. Mattioni, and O. Vergnano. 1991. Accumulation mechanisms and heavy metal tolerance of a nickel hyperaccumulator. J. Plant Nutr. 14:1067-1080.

Grill, E., E.L. Winnacker, and M.H. Zenk. 1989. Occurrence of heavy metal binding phytochelatins in plants growing in a mining refuse area. Experientia. 44:539-540.

Gordon, I., S.A. Sojka, and M.P Gordon. 1989. Plant transformation construct. US Patent application 369886.

Hanks, J., and J.T. Ritchie (ed). 1991. Modeling plant and soil systems. Am. Soc. Agron. Agronomy series, Number 31. Madison, WI.

Hartman, W.J., Jr. 1975. An evaluation of land treatment of municipal wastewater and physical siting of facility installations, US Department of the Army.

Hatway, D.E. 1989. Molecular mechanism of herbicide selectivity, Oxford Univ Press. New York.

Hatzios, K.K., and R.E. Hoagland. 1989. Crop safeners for herbicides: development, uses and mechanisms of action. Academic Press. New York.

Hsu, F.C., R.L. Marxmiller, and A.Y.S. Yang. 1990. Study of root uptake and xylem translocation of cinmethylin and related compounds in detopped soybeans using a pressure chamber technique. Plant Physiol. 93, 1573-1578.

Kinsley, M.T, F.B. Metting, J.K. Fredrickson, and R.J. Seidler. 1993. In-situ stimulation vs bioaugmentation: can plant inoculation enhance biodegradation of organic compounds. Proceedings of the Air and Waste Management Association's 86th Annual meeting and conference. July 14 -18, 1993 Denver CO -WA-89.04. AWMA, Pittsburgh, PA .

Lee, C.R., R.E. Hoeppel, P.G.Hunt, and C.A. Carlson. 1976. Feasibility of functional use of vegetation to filter, dewater, and remove contaminants from dredged material. Technical Report D-76-4 (June). U.S. Army Engineer Waterways Experiment Station, Vicksburg, MS.

Lee, J., R.D. Reeves, R.R. Brooks, and T. Jaffre. 1977. Isolation and identification of a citrate-complex of nickel from nickel-accumulating plants. Phytochem. 16:1503-1505.

Licht, L. A. 1993. Ecolotree Cap: Densely Rooted Trees for Water Management on Landfill Covers. *In* Proceedings of the Air and Waste Management Association's 86th Annual meeting and conference. July 14 -18, 1993 Denver CO -WA-89. Paper A1549. AWMA, Pittsburgh, PA.

McAneny, C. C., P. G. Tucker, J. Morgan, C. R. Lee, M. F. Kelly, and R. C. Horz. 1985. Covers for uncontrolled hazardous waste sites. U.S. Gov. Print. Office. EPA/540/2-85/002.

McMullin, E. 1993. Absorbing idea. California Farmer Feb. p 20-24.

Morrison, R.S., R.R. Brooks, R. D. Reeves. 1980. Nickel uptake by *Alyssum* species. Plant Sci. Lett. 17:451-457.

Morrison, R. S., R.R. Brooks, R. D. Reeves, F. Malaisse, M. Aronson, and G. Merriam. 1981. The diverse chemical forms of heavy metals in tissue extracts of some metallophytes from Shaba Province, Zaire. Phytochem. 20:255-458.

Nishizono, H., S. Sukuki, and F. Ishii. 1987. Accumulation of heavy metals in the metal tolerant fern, *Athyrium yokoscense*, growing on various environments. Plant Soil 102:65-70.

Oyler, J.A. 1988. Remediation of metals-contaminated site near a zinc smelter using sludge/fly ash amendments:herbaceous species. *In* D.D. Hemphill, (ed.) Trace substances in environmental health-XXII, A Symposium, University of Missouri, Columbia, MO.

Raloff, J. 1989 Greenery filters out indoor air pollution. Science News 136:212.

Rigitano, R.L.O., R.H. Bromilow, G.G. Briggs, and K. Chamberlain. 1987. Phloem translocation of non-ionised chemicals in *Ricinus communis*. Pestic. Sci. 19:113-133.

Reeves, R.D., R.R. Brooks, and T.R. Dudley. 1983. Uptake of nickel by species of *Alyssum*, *Bornmuellera* and other genera of old world Tribus Alysseae. Taxon 32:184-192.

Sandermann, H. Jr. 1992. Plant metabolism of xenobiotics. TIBS 17:82-84.

Schwab, A.P. and M.K. Banks. 1994. Biologically mediated dissipation of polyaromatic hydrocarbons in the root zone. *In* T.J. Anderson, and J.R. Coats (ed.) Microbial degradation of organic chemicals in the rhizosphere: Implications for bioremediation. Amer. Chem. Soc. Symp. Ser.

Shimp, J.F., J.C. Tracey, L.C. Davis, E. Lee, W. Huang, L.E. Erickson, and J.L. Schnoor. 1993. Beneficial effects of plants in the remediation of soil and groundwater contaminated with organic compounds. CRC Critical Reviews in Environ. Scien. and Tech. 23: 41-77.

Skogerboe, J.G., H.E. Tatem, C.R. Lee, J.W. Simmers, D.L. Brandon, J.M. Brannon, and M.R. Palermo. 1994. Evaluation of upland disposal: reconditioning and bioremediation tests of Hamlet City lake sediment, NC. Miscellaneous Paper EL-94-US Army Engineer Waterways Experiment Station, Vicksburg, MS.

Verkleij, J. A.C., P. Koevoets, J. van't Riet, J.A. de Knecht, and W.H.O. Ernst. 1990. *In* H. Rennenberg et al. (ed.) The role of metal-binding compounds (Phytochelatins) in the cadmium-tolerance mechanism of Bladder Campion (*Silene vulgaris*). Sulfur nutrition and sulfur assimilation in higher plants. SPB Academic Publishing, Inc. The Hague.

Wenzel, W.W., M.A. Pollak, and W.E.H. Blum. 1992. Dynamics of heavy metals in soils of a reed bed system. Intern. J. Environ. Anal. Chem. 46:41-52.

Yamada, K., K. Miyahara, and T. Kotoyori. 1975. Studies on soil pollution caused by heavy metals. Part IV: Soil purification by plants that absorb heavy metals. Gamm Ken Nogyo Shienjo Hokoku. 15: 39-54.

10 Contaminants at U.S. DOE Sites and Their Susceptibility to Bioremediation[1]

R.J. Lenhard, R.S. Skeen and T.M. Brouns
Pacific Northwest Laboratory
Richland, Washington

Over the past 40 years of defense nuclear production operations, the U.S. Department of Energy (U.S. DOE), its predecessor agencies, and its contractors have generated large volumes of hazardous and radioactive wastes. Discharge of these wastes to the soil column has resulted in widespread soil and groundwater contamination at U.S. DOE sites. Although many of the contaminants also occur at public, commercial, and federal hazardous waste sites, U.S. DOE contaminated sites are unique in the number and diversity of contaminants and the presence of complex mixtures of contaminants. In addition, many of U.S. DOE hazardous waste sites consist of large volumes of soil and groundwater in which contamination from organics, metals, and radionuclides is dilute. Existing remediation technologies are too costly, too slow, and ineffective for meeting cleanup standards. As alternatives to existing remediation technologies, U.S. DOE is investing in basic and applied research to identify, develop, and demonstrate *in situ* technologies that would solve these problems.

In situ bioremediation is one technology that has the potential to address many of the contaminants and contaminant mixtures found at U.S. DOE sites. However, much of the research and most field applications to date have been limited to single contaminants, single contaminant classes, or very simple mixtures of contaminants. Therefore, additional research and development will be needed to address the complex mixtures of degradable and non-degradable contaminants found at most U.S. DOE sites.

[1]This paper was supported in part by the Subsurface Science Program, Office of Health and Environmental Research, and in part by the VOC-Arid Integrated Demonstration, Office of Technology Development, U.S. Department of Energy (U.S. DOE). Pacific Northwest Laboratory is operated for U.S. DOE by Battelle Memorial Institute under Contract DE-AC06-76RLO 1830.

The U.S. Department of Energy (U.S. DOE) operates sites throughout the United States that encompass a land mass approximately equal to the combined areas of the states of Rhode Island and Delaware. At these sites, U.S. DOE has engaged in multidisciplinary research; production of nuclear materials; development, testing, and fabrication of nuclear and non-nuclear weapons; construction, testing, and operation of nuclear reactors; and management of various radioactive and non-radioactive wastes from its operations. Although much of the wastes generated at U.S. DOE facilities has been stored in tanks (e.g., high-level radioactive wastes) or stored in the ground in a retrievable form (e.g., transuranic waste), by far the largest volume has been disposed of via ponds, cribs, basins, injection wells, or landfills or to the ground surface. Originally, such disposal of wastes was believed to have limited long-term consequences for health and the environment. However, in fact significant subsurface contamination problems have resulted. Many of the chemicals are now of health concern and regulated under Federal and State statutes (Federal Register, 1985a, b, 1989, 1990).

Over 3,000 contaminated sites have been identified at U.S. DOE installations (U.S. GAO, 1988a,b); in the current economic and technologic environment the estimated cost of restoring these sites is prohibitive. In many cases, existing remediation technologies are too costly, slow, or ineffective. In other cases, the necessary technologies simply do not exist.

The extent and complexity of contamination at U.S. DOE sites vary widely. Contamination problems at some sites are confounded by the disposal of large amounts of different inorganic and organic chemicals. Such sites present especially difficult problems because remediation strategies cannot be focused on individual contaminants, but must address all of the contaminants that were disposed of at the site and how they interact with each other and the environment. Furthermore, the remediation strategies must comply with regulatory statutes, which can be restrictive for sites contaminated with radionuclides. Unfortunately, the full extent and complexity of subsurface contamination at U.S. DOE facilities is largely unknown; records are commonly incomplete, and there are sometimes disparities between disposal records and subsequent chemical analyses.

To make cleanup of subsurface contamination at its sites by direct remediation/restoration activities possible, U.S. DOE is sponsoring environmental technology development and basic research on subsurface contaminant behavior. A primary objective of this research and development is to provide a new generation of environmental restoration technologies that will reduce the cost, time, and risk associated with cleanup.

Currently, above-ground treatment of contaminated soils and groundwater is the primary method for remediation. Unfortunately, above-ground treatment requires expensive excavation or extraction of large volumes of contaminated media. Above-ground treatment also has other shortcomings, which center on the difficulties and costs associated with recovering and handling contaminants from soil and groundwater. For example, the slow recovery of nonaqueous phase liquids (NAPLs) from aquifer solids (Keely et al., 1987; Powers et al., 1991) can require pump-and-treat operations for tens of years before cleanup goals can be achieved. NAPLs are a particular problem for U.S. DOE because of the large volumes of associated radioactive and mixed wastes at U.S. DOE sites. For mixed waste sites, pump-and-treat systems cannot simply employ air stripping to remove volatile organic compounds (VOCs). Instead, an expensive system of metal, radionuclide, and anion treatment operations will also be needed to meet the discharge requirements for treated water.

Bioremediation is one innovative technology that may have widespread applicability to U.S. DOE's wastes because it can operate *in situ* and may address a suite of contaminants. Bioremediation involves the use of microorganisms or other living species to destroy, immobilize, or otherwise transform contaminants to less hazardous forms.

The feasibility of bioremediation for many organic and inorganic chemical species has been demonstrated; however, there have been very few full-scale applications of bioremediation at U.S. DOE sites. As with conventional pump-and-treat systems, bioremediation applications have been limited because so many of the contaminants often occur as mixtures. For example, although above-ground biological systems are proven for treatment of groundwater and soil contaminated with nitrates, fuels, and chlorinated solvents, when these degradable compounds are mixed with difficult-to-degrade or nondegradable contaminants, such as fully chlorinated hydrocarbons, metals, or radionuclides, bioremediation may not be effective. However, unlike conventional above-ground treatment, bioremediation does not have the potential to treat a wider range of contaminants through degradation, sorption, and precipitation processes.

To lower costs and avoid many of the problems associated with conventional above-ground treatment, U.S. DOE's research and technology development efforts have focused on *in situ* treatment technologies, especially *in situ* bioremediation. *In situ* treatment circumvents the high costs associated with recovering large volumes of contaminated material (e.g, groundwater and soil) and with treating very low levels of secondary contaminants that contribute little to the overall health risk of a waste site. In addition, it has been shown that the time required for remediation at many sites can be reduced using *in situ* technologies (Skeen et al., 1993).

Fundamental microbiological and applied *in situ* bioremediation research is being supported by U.S. DOE. Fundamental microbiological research is being supported by the Subsurface Science Program (SSP), Office of Health and Environmental Research, U.S. DOE. The SSP also conducts fundamental research on a wide variety of other topics related to the environment, such as contaminant chemistry, multiphase fluid flow, deep microbiology, and spatial heterogeneity. Applied *in situ* bioremediation research and development is being supported by the Office of Technology Development (OTD), U.S. DOE. OTD's Integrated Programs and Integrated Demonstrations are designed to accelerate the development and testing of new technologies for cleaning up hazardous and radioactive wastes and to evaluate an array of technologies for all phases of remediation under actual field conditions. Bioremediation has played a key role in two integrated demonstrations for cleanup of chlorinated hydrocarbons at arid and nonarid sites and in the In Situ Remediation Integrated Program (ISRIP).

This paper reviews the susceptibility to bioremediation of the most frequently reported contaminants and contaminant mixtures at 91 waste sites at 18 U.S. DOE facilities. Riley and Zachara (1992) identified these contaminants and determined generic chemical mixtures appropriate for use in basic research into the subsurface geochemical and microbiological behavior of mixed contaminants. The publication analyzed data found in 89 documents from 18 U.S. DOE facilities to identify the most prevalent contaminants. Our review emphasizes organics (especially chlorinated hydrocarbons), metals, and radionuclides. Ongoing basic research, development, and field-scale bioremediation tests within U.S. DOE are also summarized.

MOST COMMON CHEMICALS

Riley and Zachara (1992) considered the groundwater and vadose zones separately and then grouped similar compounds into 13 contaminant classes (Table 10-1). The 91 waste sites reviewed by Riley and Zachara represent only a small subset of the total number identified at U.S. DOE facilities because detailed data on the chemical composition of the U.S. DOE waste sites are limited, particularly when both the groundwater and vadose zones are considered.

Table 10-1. Contaminant Classes at U.S. DOE Sites†.

Contaminant Class	Representative Constituents
Metals	Lead, chromium, mercury
Anions	Nitrate, fluoride, cyanide
Radionuclides	Tritium, plutonium, uranium
Chlorinated hydrocarbons	Trichloroethylene
Fuel hydrocarbons	Benzene, toluene, xylenes
Phthalates	Bis-2-ethylhexylphthalate
PCBs	Arochlor mixtures
Explosives	Trinitrotoluene
Ketones	Acetone, methyl ethyl ketone
Pesticides	Lindane, DDT
Alkyl phosphates	Tributyl phosphate
Complexing Agents	Nitriloacetic acid
Organic Acids	Oxalic acid, citric acid

†Proposed by Riley and Zachara (1992).

Vadose Zone

Figure 10-1 shows the frequencies of occurrence of the 13 contaminant classes at the 91 sites. The most common contaminant class in the vadose zone was radionuclides, followed by metals and chlorinated hydrocarbons. Approximately 45% of the sites were contaminated with radionuclides and 35% were contaminated with chlorinated hydrocarbons. Approximately 25% of the sites were contaminated with polychlorinated biphenyls (PCBs) and 25% with fuel hydrocarbons. Anions were present at approximately 20% of the sites, and all other contaminant classes were present at less than 10% of the sites.

Of the radionuclides, uranium was reported in the vadose zone at 12 of the 18 U.S. DOE facilities. Plutonium and cesium were found at 10 of the facilities, and tritium and strontium were found in the vadose zone of 6 facilities. Thorium, cobalt, technetium, and iodine were reported for only a few facilities. Some radionuclides

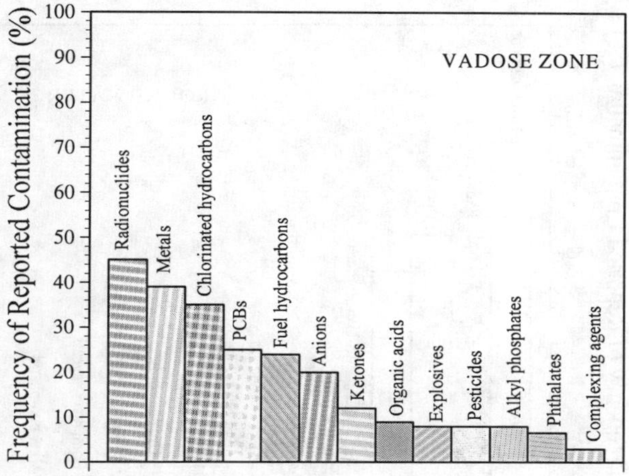

Fig. 10-1. Frequency of occurrence of the contaminant classes in the vadose zone of the 91 contaminated U.S. DOE sites (Riley & Zachara, 1992).

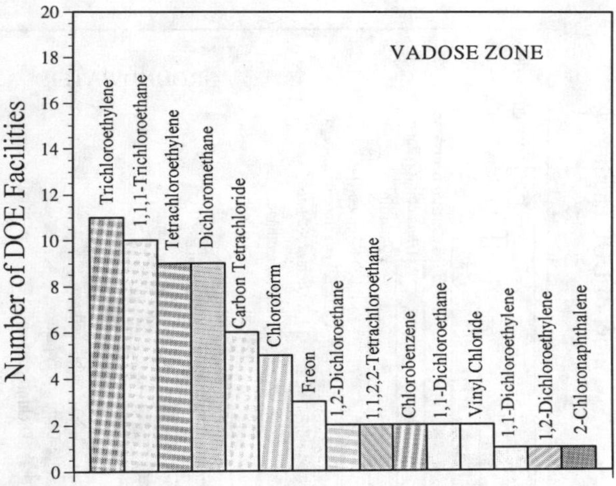

Fig. 10-2. Frequency of occurrence of the chlorinated hydrocarbons in the vadose zone at the 18 U.S. DOE facilities (Riley & Zachara, 1992).

were found only at facilities with specific processing operations. For example, iodine-129 was reported for only the Hanford Site.

The most common metals in the vadose zone were copper, chromium, zinc, mercury, arsenic, and cadmium. All were reported with similar frequencies of occurrence for the 18 facilities.

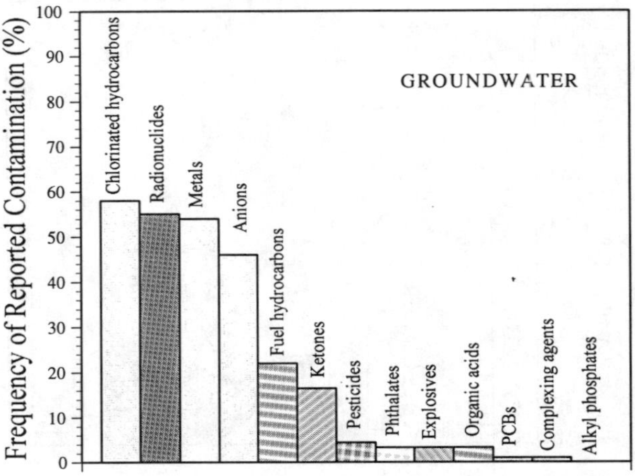

Fig. 10-3. Frequency of occurrence of the contaminant classes in the groundwater of the 91 contaminated U.S. DOE sites (Riley & Zachara, 1992).

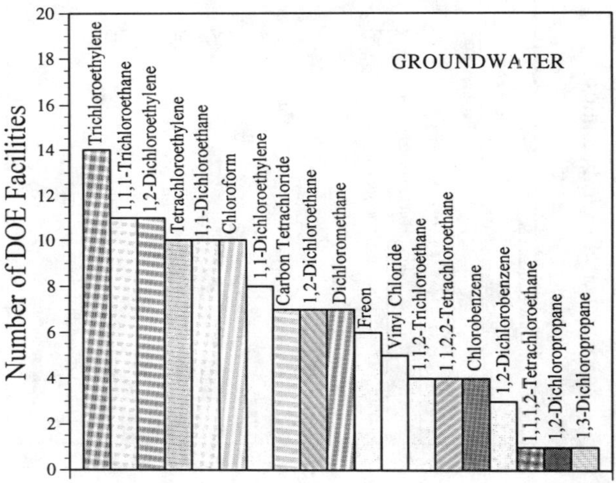

Fig. 10-4. Frequency of occurrence of the chlorinated hydrocarbons in the groundwater at the 18 U.S. DOE facilities (Riley & Zachara, 1992).

Fig. 10-2 represents the frequency of occurrence of specific chlorinated solvents at the 18 U.S. DOE facilities. The most common chlorinated hydrocarbons at the U.S. DOE facilities were trichloroethylene (TCE), 1,1,1-trichloromethane, tetrachloroethylene, and dichloromethane. Carbon tetrachloride (CCl_4) and chloroform occurred at only 6 and 5 of the facilities, respectively. Other chlorinated hydrocarbons were found to a lesser extent.

Groundwater

Fig. 10-3 shows the frequency of occurrence of contaminant classes in groundwater at the 91 U.S. DOE waste sites. The most frequently reported contaminant class was chlorinated hydrocarbons, followed by radionuclides, metals, anions, and, to a lesser extent, fuel hydrocarbons. Approximately 60% of the sites were contaminated with chlorinated hydrocarbons, 55% with radionuclides and metals, 45% with anions, and 20% with fuel hydrocarbons. The least common contaminant classes in groundwater were pesticides, PCBs, and complexing agents. Only four sites at one U.S. DOE facility were contaminated with pesticides, only one site was contaminated with PCBs, and only one site was contaminated with complexing agents. No alkyl phosphates were reported to be in the groundwater at the U.S. DOE waste sites; however, these compounds may not have been measured because they are not currently listed by the U.S. Environmental Protection Agency as a priority pollutant.

Nineteen chlorinated-hydrocarbon compounds were identified in the groundwater. The frequency of occurrence of these compounds at the 18 U.S. DOE facilities is represented in Fig. 10-4. The most commonly occurring chlorinated hydrocarbon was TCE, which appeared in the groundwater at 14 of the 18 facilities. Tetrachloroethylene, 1,1,1-trichloromethane, 1,2-dichloroethylene, 1,1-dichloroethane, and chloroform appeared at more than half of the facilities.

Of the radionuclides, tritium, uranium, and strontium appeared in the groundwater in over half of the facilities. The most common metals in groundwater were lead, chromium, arsenic, and zinc. The most common anion was nitrate, which is consistent with extensive use of nitric acid and nitrate salts in nuclear fuels reprocessing and fabrication (Stenner et al., 1988). The most frequently occurring fuel hydrocarbons in the groundwater were benzene, toluene, ethylbenzene, and xylene.

MOST COMMON CONTAMINANT MIXTURES

Because of past practices at U.S. DOE facilities, few sites are contaminated by just one set or class of compounds. Instead, sites are typically contaminated with multiple compound classes. Remediation of these sites poses special problems, because the contaminant types may interact synergistically or antagonistically in a way that significantly affects remediation strategies. For example, even though one compound may be readily biodegraded by indigenous microorganisms, bioremediation may be ineffective at a site because the site is also contaminated with another compound that is toxic to the microorganisms. Furthermore, the presence of co-contaminants may either stabilize or mobilize certain compounds, thereby significantly changing environmental risks.

Vadose Zone

The most common two-compound class mixtures in the vadose zone contained either metals or radionuclides. In fact, mixtures of metals and radionuclides were the most prevalent contaminant mixture. Of the 91 sites, 25 sites reported mixtures of metals and radionuclides. Metal-PCB mixtures were reported at 18 of the sites, metal-chlorinated hydrocarbon mixtures were reported at 16 of the sites, and chlorinated hydrocarbon-fuel hydrocarbon mixtures were reported at 15 of the sites. Contaminant mixtures consisting of either PCBs, anions, or chlorinated hydrocarbons with radionuclides were reported for approximately 15 of the sites.

In the three-compound class mixture, the most common mixture was metals-radionuclides-PCBs, which occurred at 13 of the sites. The second most common three-compound class mixture was metals-radionuclides-chlorinated hydrocarbons, which was reported for 11 of the sites. At 10 of the sites, metal-chlorinated hydrocarbon-PCB mixtures were reported. Other three-compound mixtures that combined either metals, radionuclides, or chlorinated hydrocarbons with anions, PCBs, or fuel hydrocarbons occurred at 6 to 9 of the sites. All other combinations of three-compound mixtures occurred at 6 or fewer of the sites.

Four-compound mixtures were less common; the highest frequency of any four-compound class mixtures was at seven sites. The mixtures metals-anions-radionuclides-chlorinated hydrocarbons and metals-radionuclides-chlorinated hydrocarbons-PCBs both occurred at seven sites. All other four-compound mixtures were reported at five or fewer of the waste sites.

Groundwater

The most frequent two-compound class mixture in groundwater was metals and chlorinated hydrocarbons, which was reported at 38 sites. The metal-radionuclide mixture was reported at 36 sites, the metal-anion mixture at 33 sites, the anion-radionuclide mixture at 33 sites, and the radionuclide-chlorinated hydrocarbon mixture at 32 sites. The anion-chlorinated hydrocarbon mixture was reported for 26 sites, and all other two-compound mixtures occurred at fewer than 18 sites. Because pesticides, phthalates, explosives, organic acids, PCBs, and complexing agents were rarely reported in the groundwater, they occurred in mixtures at 4 or fewer sites.

In the three-compound mixtures, groupings of metals and radionuclides with either anions or chlorinated hydrocarbons were the most frequent; they occurred at 29 sites each. Mixtures of anions and chlorinated hydrocarbons with either metals or radionuclides were reported to be at 25 and 23 sites, respectively. All other three-compound mixtures occurred at 16 or fewer sites.

The most frequent four-compound class, metals-anions-radionuclides-chlorinated hydrocarbons, was reported at 23 of the sites. Four-compound class mixtures containing metals with either radionuclides or chlorinated hydrocarbons with other contaminant classes were each reported at approximately 10 of the sites.

IN SITU BIOREMEDIATION OF U.S. DOE CONTAMINANTS

Biological technologies have the potential to treat the commonest organic materials at U.S. DOE sites, including fuel hydrocarbons, PCBs, and chlorinated solvents (Eaton, 1985; Vogel et al., 1987; Hill et al., 1989; Abramowicz, 1990; Atlas, 1991). Some less common organics can also be removed using biological processes. These compounds include some ketones, organic acids, explosives, and pesticides. In addition, some anions, such as nitrate and cyanide, can be degraded biologically (Knowles, 1982; Richards & Shieh, 1989).

Overall, these organic and inorganic compounds are attractive targets for bioremediation because microorganisms can readily and completely degrade them. However, most research efforts and field applications of *in situ* bioremediation have addressed only aqueous-phase treatment of organics. This is important because a larger volume of organic chemicals may exist as a NAPL or sorbed phase or may be in the gaseous phase. Except for bioventing technologies for petroleum hydrocarbons (Hoeppel et al., 1991) and some recent work with vapor-phase bioreactors for off-gas streams, very little has been done in the area of bioremediation for vadose zone contamination. Most field applications of vadose-zone remediation have focused on soil flushing to transfer contaminants into the aqueous phase for subsequent biodegradation.

Some of U.S. DOE's metal and radionuclide contamination could be treated using such biological technologies as bioaccumulation or bioprecipitation. For example, a biological system can remove uranium and chromium from soil (Gorby and Lovley, 1992). In this process, the valence state of a soluble metal is biologically reduced, which results in the formation of insoluble compounds that can be precipitated and recovered. Biological leaching of metals and radionuclides has also been studied. For instance, biological leaching of uranium from ores and tailings is a viable commercial recovery method (Francis et al., 1989). However, biological processes have not ben applied for mobilization of metals and radionuclides for environmental purposes. Sorption of heavy metals by microbial cells, a process referred to as "biosorption," has been recognized as a potential alternative to existing technologies for the removal and perhaps recovery of heavy metals from industrial waste streams and natural water (e.g., groundwater, streams, ponds).

Except for biological leaching, most bioremediation processes for metals and radionuclides have been researched and tested only for *ex situ* applications, in which wastewater streams or groundwaters are processed through treatment vessels. The applicability of these processes for *in situ* treatment has not been adequately explored. Even in the case of biological leaching, most applications involve *ex situ* heap leaching rather than *in situ* applications. However, biosorption of uranium, thorium, and radium has been studied extensively because of the radionuclides' usefulness and adverse health effects (Horikoshi et al., 1979; Tsezos & Volesky, 1981, 1982; Tsezos & Keller, 1983; Tsezos et al., 1986, 1987; Gadd & White, 1989; Tsezos & Deutschmann, 1990). The susceptibility of lead, chromium, and a variety of other metal contaminants to separation from aqueous streams by means of biosorption has also been studied (Beveridge and Murray, 1976; Friis and Meyers-Keith, 1986; Siegel et al., 1986; Galun et al., 1987; Macaskie & Dean, 1987; Kong et al., 1992a, b, 1993).

Very little research has been conducted to evaluate or develop processes for treating mixtures of contaminants, despite the importance of mixtures at U.S. DOE sites that include metals and radionuclides in conjunction with organics. Much of the research that has combined organics or metals and radionuclides has focused attention on only a single contaminant or single classes of contaminants and not on all of the chemical and/or biological interactions. By using this approach, the interaction of organic contaminant biodegradation with metal or radionuclide mobilization, sorption, or precipitation cannot be evaluated. The lack of research is in part due to the added complexity associated with mixtures and the difficulty of attempting to understand their complex interactions.

Technical Barriers to *In Situ* Bioremediation

One of the primary technical barriers to widespread application of *in situ* bioremediation at U.S. DOE sites is not unique to U.S. DOE or to bioremediation. It stems from the complex environment of the subsurface and the current abilities to access, monitor, and control subsurface processes. Specifically, the success of *in situ* treatment and bioremediation depends on the ability to deliver nutrients or reagents, monitor the process, maintain acceptable operating conditions, and confirm that the process worked. Each of these steps is difficult given the heterogeneity of the subsurface environment, the complexity of geochemical, biological, and contaminant interactions, and limited access through small boreholes. For U.S. DOE, the process is further complicated by the presence of complex contaminant mixtures whose natural biochemical and geochemical transformations and reactions are relatively unknown.

From the perspective of chemical or environmental engineers, the ideal process for biological treatment uses an above-ground well-mixed continuous flow reactor that offers relatively easy design and scale-up relationships, monitoring, and process control. Unfortunately, the subsurface environment provides very few of these attributes. Recent bioremediation efforts have focused on these technical barriers, including development and testing of better mixing and fluid transport methods, simulation tools for scale-up and performance evaluation, and monitoring systems for near-real-time process feedback.

The development of methods to treat soils that are contaminated with recalcitrant compounds, such as chlorinated solvents, is another technical barrier that will require more research and development. Techniques are needed to introduce gas-phase nutrients, to control water content without mobilizing contaminants, and to cost-effectively induce anaerobic conditions in the subsurface environment.

For biosorption to be a viable option, future work must demonstrate that an engineered system can be used for large-scale *in situ* remediation. Current research has focused on demonstrating that biomass can effectively remove metals from simple solutions; efforts to evaluate *in situ* applications and actual groundwater have been limited. A better understanding of the applicability of biosorption for environmental problems will develop only as the technology is demonstrated on actual aqueous streams of environmental concern. In addition, biosorption may only temporarily immobilize contaminants; therefore, the regulatory and public acceptance of biosorption prior to large-scale application will be critical.

Methods are also needed to treat mixtures of biodegradable compounds that require different microbial populations and/or media. This is particularly true for mixtures of co-metabolically degraded compounds, such as chlorinated solvents. For example, TCE, dichloroethylene, trichloromethane, dichloroethane, chloroform, and dichloromethane are destroyed by microorganisms in aerobic environments (Fogel et al., 1986; Henson et al., 1988; Strand et al., 1990), but more halogenated compounds, such as CCl_4 and perchloroethylene, are degraded only under anaerobic conditions (Fogel et al., 1986; Vannelli et al., 1990; Criddle et al., 1990). However, anaerobic conditions have also been shown to destroy less halogenated compounds, suggesting that a purely anaerobic process might be effective in treating a mixture of chlorinated contaminants (Bouwer & McCarty, 1983; Freedman & Gossett, 1989). The process is complicated, however, by the fact that rates of biological transformation under anaerobic conditions decrease with less halogenated molecules (Freedman & Gossett, 1989). Hence, less halogenated species are removed more slowly, which can result in a significant transient buildup of hazardous intermediates. For example, during anaerobic TCE biodegradation, even more hazardous vinyl chloride may form as an intermediate (Freedman & Gossett, 1989). Contrarily, in aerobic systems, less halogenated species are degraded faster and hazardous products are not produced (Fogel et al., 1986; Henson et al., 1988; Phelps et al., 1990; Arvin, 1991; Semprini et al., 1991). Hence, several options for *in situ* destruction of a mixture of chlorinated contaminants in groundwater are possible. However, too little information is available to determine the most practical and effective mode of operation.

Another technical hurdle to making bioremediation more widely applicable to U.S. DOE sites is integrating organic destruction processes with other biotic and abiotic processes that treat metals and radionuclides. This integration is necessary because many of the waste sites with biodegradable materials are also contaminated with radionuclides and metals. Applying *in situ* bioremediation at such sites will require consideration of how the nondegradable species are to be removed and of how increased biological activity will affect the treatment scheme. For example, if nondegradable species are treated by enhancing their mobility through electroosmosis or the addition of chelating agents, then it may be necessary to delay biological treatment, because increasing numbers of microorganisms could both reduce the permeability of the subsurface and make the metal inaccessible by increasing its bioaccumulation. Alternatively, if *in situ* biosorption is being considered for immobilization of metals or radionuclides, a treatment technology for organic co-contaminants that might thermally or chemically destroy the indigenous microbial population would be undesirable.

U.S. DOE RESEARCH AND DEVELOPMENT ACTIVITIES

Current research and technology development efforts within U.S. DOE are focused on resolution of many of the technical issues identified above. These research and development efforts encompass the range from basic science through full-scale demonstration and deployment. The purpose of this research and development is to (i) develop and demonstrate technologies in the near term that can address primary contamination problems, (ii) develop the detailed scientific basis for predicting how mixtures of contaminants behave in the subsurface, and (iii) develop and demonstrate technologies in the long term that can address the complex waste mixtures by building on the success of earlier, less complex field efforts and the enhanced knowledge produced from the fundamental research. The development and demonstration efforts within U.S. DOE's Office of Environmental Restoration and Waste Management have so far focused on single compounds or single classes of compounds rather than the complex mixtures of contaminants that actually occur. The basic science programs conducted by U.S. DOE's Office of Energy Research have focused most of their efforts on understanding the complex biological and geochemical processes that occur with specific classes of compounds.

Basic Research

The Office of Energy Research's SSP conducts long-term fundamental research in the physical, chemical, and biological sciences. A significant effort is being directed toward investigation of how subsurface microorganisms interact with complex contaminant mixtures in varying subsurface environments. The natural variations in the distributions of microorganisms in the subsurface are also being studied. Knowledge gained from this research will help lay scientific foundations for i) developing improved remediation concepts, ii) accurately assessing potential risks to human health and the environment posed by contaminated U.S. DOE sites, and iii) characterizing the nature and extent of subsurface contamination.

Ongoing SSP research is investigating the biological transformation and transport of chelated metal/radionuclide complexes. The chelated metal/radionuclide complexes currently being investigated are nitrilotriacetic-cobalt and citrate-uranium, in which the chelate portion of the complex may be subject to aerobic degradation and the metal/radionuclide portion may be reduced as an electron acceptor by the microorganisms. Each of these transformation mechanisms changes the thermodynamic stability of the complex as well as kinetic properties of sorption.

In another project, localized changes in microbial population dynamics and function are being investigated in porous media containing distinct fine-grained organic inclusions. Upscaling transport and microbial dynamics from the controlled intermediate-scale laboratory flow cell to the prediction of field-scale phenomena is one focus of this work. In physically and chemically heterogeneous systems, it is hypothesized that reactive transport can be scaled up by superimposing the scaling of microbial kinetic reactions onto the distinct scaling of the physical transport process.

A third project is investigating the effects of microorganisms on subsurface NAPL behavior. A major question being addressed in these studies is how and to what extent microorganisms modify NAPL flow and chemical transport, with emphasis on vadose-zone contamination. Other investigations focus on the effects of microbial growth, i.e., pore clogging and biosurfactant production, on NAPL flow. Physical and biological submodels will be coupled to predict NAPL retention and conduction in porous media.

Technology Development

As a developing technology, bioremediation has gained most acceptance as a treatment for single-class compounds, such as petroleum hydrocarbons (fuels and oils), that are directly metabolized by growing microorganisms. Current development efforts within U.S. DOE's Office of Environmental Restoration and Waste Management are applying bioremediation technology to more complex contaminants, such as chlorinated hydrocarbons, radionuclides, and PCBs. A full field demonstration of *in situ* TCE bioremediation by methanotrophic bacteria was just completed at U.S. DOE's Savannah River Site in South Carolina (Hazen, 1992). Another TCE bioremediation project under way at Lawrence Livermore National

Laboratory is evaluating the use of passive microbial filters for *in situ* treatment of a migrating VOC plume.

As part of the VOC-Arid Integrated Demonstration Program, an *in situ* bioremediation technology is being developed to destroy nitrate and CCl_4 in the groundwater at U.S. DOE's Hanford Site. The field test of this technology will begin in the spring of 1994 and continue for approximately one year. The goal of this test is to stimulate native denitrifying microorganisms to degrade the contaminants, by mixing an electron donor (acetate) and an electron acceptor (nitrate) into the contaminated aquifer using a dual-screen well developed at Stanford University. The upper screen of this well is used as an injection point, and the lower screen serves as an extraction point. In addition to demonstrating biological destruction of CCl_4 and nitrate, the effectiveness of scale-up techniques developed in the petrochemical industry for bioremediation is being evaluated. The rigorous methodology being used to design the bioremediation test site and to determine the best operating and monitoring strategy is based on combining fluid mixing and transport predictions with numerical descriptions for biological transport and reaction kinetics (Skeen et al., 1993).

Treatment of uranium-contaminated soils is the focus of a new development effort at the Idaho Nuclear Engineering Laboratory, utilizing acid-producing microorganisms to help solubilize uranium for use at U.S. DOE's Fernald Plant. This technique is based on processes used in the mining industry for mobilizing uranium from ore. Other development efforts using uranium have focused on bioaccumulation (sorption) and anaerobic transformations of uranium complexes (bioprecipitation) at Oak Ridge National Laboratory and Brookhaven National Laboratory (Francis et al., 1989), respectively. Bioremediation of PCBs by indigenous microorganisms has also been a topic of ongoing field research at Oak Ridge National Laboratory since 1988 (Hill et al., 1989).

CONCLUSIONS

Contaminants at U.S. DOE sites encompass a range of common industrial pollutants. However, the prevalence of contaminant mixtures including organics, metals, and radionuclides is relatively unique to U.S. DOE's facilities. Bioremediation has been shown to be effective for destruction of many of the organic pollutants. The technology also has promise for application to many of the metals and radionuclides; however, field demonstrations for these applications have not yet been attempted. Because of the complexity of biodegradation of even a single-compound class, little has been done to develop or demonstrate *in situ* bioremediation technologies for multicompound combinations. The current bioremediation demonstration on CCl_4 and nitrates within the VOC-Arid Integrated Demonstration is one the first efforts to address inorganic and organic co-contaminants simultaneously. Additional research, technology development, and field demonstrations are needed to evaluate the applicability of *in situ* bioremediation to U.S. DOE's most common contaminant mixtures.

REFERENCES

Abramowicz, D.A. 1990. Aerobic and anaerobic biodegradation of PCBs: A review. Crit. Rev. Biotechnol. 3: 241-251.

Arvin, E. 1991. Biodegradation kinetics of chlorinated aliphatic hydrocarbons with methane oxidizing bacteria in an aerobic fixed biofilm reactor. Water Res. 25:873-881.

Atlas, R.M. 1991. Bioremediation of fossil fuel contaminated soils. p. 14-32. *In* R.E. Hinchee and R.F. Olfenbuttel (ed.) *In situ* Bioreclamation. Butterworth-Heinemann, Boston, MA.

Beveridge, T.J., and R.G.E. Murray. 1976. Uptake and retention of metals by cell walls of Bacillus subtilis. J. Bacteriol. 127:1502-1518.

Bouwer, E.J., and P.L. McCarty. 1983. Transformations of 1- and 2-carbon halogenated aliphatic organic compounds under methanogenic conditions. Appl. Environ. Microbiol. 45:1286-1294.

Criddle, C.S., J.T. DeWitt, and P.L. McCarty. 1990. Reductive dehalogenation of carbon tetrachloride by *Escherichia coli* K-12. Appl. Environ. Microbiol. 56: 3247-3254.

Eaton, D.C. 1985. Mineralization of polychlorinated biphenyls by Phanerochaete crysosporium: A ligninolytic fungus. Enzyme Microb. Technol. 7:194-196.

Federal Register. 1985a National primary drinking water regulations; Volatile synthetic organic chemicals. Federal Register, 50 FR No. 219, 46880.

Federal Register. 1985b. National primary drinking water regulations; Synthetic organic chemicals, inorganic chemicals and microorganisms. Federal Register, 50 FR No. 219, 46936.

Federal Register. 1989. National primary and secondary drinking water regulations. Federal Register, 54 FR No. 97, 22062.

Federal Register. 1990. National primary and secondary drinking water regulations: Synthetic organic chemicals and inorganic chemicals. Federal Register, 55 FR No. 143, 30370.

Fogel, M.M., A.R. Taddeo, and S. Fogel. 1986. Biodegradation of chlorinated ethenes by a methane-utilizing mixed culture. Appl. and Environ. Microbiol. 51:720-724.

Francis, A.J., C.J. Dodge, J.B. Gillow, and J.E. Cline. 1989. Microbial transformations of uranium in wastes. *In* Proc. of the 2nd International Conference on Chemistry and Migration Behavior of Actinides and Fission Products in the Geosphere, Monterey, CA, November 6-10, 1989. BNL-43483, Brookhaven National Laboratory, Upton, NY.

Freedman, D.L.., and J.M. Gossett. 1989. Biological reductive dechlorination of tetrachloroethylene and trichloroethylene to ethylene under methanogenic conditions. Appl. Environ. Microbiol. 55:2144-2151.

Friis, N., and P Meyers-Keith. 1986. Biosorption of uranium and lead by Streptomyces longwoodensis. Biotechnol. Bioengr. 28:21-28.

Gadd, G.M., and C. White. 1989. Removal of thorium from simulated acid process streams by fungal biomass. Biotechnol. Bioengr. 33:592-597.

Galun, M., E. Galun, B. Siegel, P. Keller, H. Lehr, and S. Siegel. 1987. Removal of metal ions from aqueous penicillium biomass: Kinetic and uptake parameters. Water, Air, Soil Pollut. 33:359-371.

Gorby, Y.A., and D.R. Lovley. 1992. Enzymatic uranium precipitation. Environ. Sci. Technol. 26:205-207.

Hazen, T.C. 1992. Test plan for *in situ* bioremediation demonstration of the Savannah River Integrated Demonstration Program. DOC/OTD TTP No.: SR 0566-01 (U), WSRC-RD-91-23, Westinghouse Savannah River Company, Aiken, SC.

Henson, J.M., M.V. Yates, J.W. Cochran, and D.L. Shackleford. 1988. Microbial removal of halogenated methanes, ethanes, and ethylenes in an aerobic soil exposed to methane. FEMS Microbiol. Eco. 53:193-201.

Hill, D.L., T.J. Phelps, D.C. White, A.V. Palumbo, T.L. Donaldson, and G.W. Strandberg. 1989. Bioremediation of polychlorinated biphenyls: Degradation capabilities in field lysimeters. Appl. Biochem. Biotechnol. 20/21:233-243.

Hoeppel, R.E., R.E. Hinchee, and M.F. Arthur. 1991. Bioventing soils contaminated with petroleum hydrocarbons. J. Industrial Microbiol. 8:141-146.

Horikoshi, T., A. Nakajima, and T. Sakaguchi. 1979. Uptake of uranium by Chlorella regularis. Agric. Biological Chem. 43:617-623.

Keely, J.F., C.D. Palmer, R.L. Johnson, M.D. Piwoni, and C.D. Enfield. 1987. Optimizing recovery of contaminant residuals by pulsed operation of hydraulically dependent remediations. p. 91-104. *In* Proc. of the NWWA/API Conference on Petroleum Hydrocarbons and Organic Chemicals in Ground Water-Prevention, Detection and Restoration. National Water Well Association, Dublin, OH.

Knowles, R. 1982. Denitrification. Microbiol. Rev. 46:43-70.

Kong, S, D.R. Yonge, D.L. Johnstone, J.N. Petersen, and T.M. Brouns. 1992a. Chromium distribution in subcellular components between fresh and starved subsurface bacterial consortium. Biotechnol. Letters. 14:521-524.

Kong, S, D.R. Yonge, D.L. Johnstone, J.N. Petersen, and T.M. Brouns. 1992b. Comparison of chromium adsorption to starved and fresh subsurface bacterial consortium. Biotechnol. Techniq. 6:143-148.

Kong, S, D.R. Yonge, D.L. Johnstone, J.N. Petersen, and T.M. Brouns. 1993. Competing ion effect on chromium adsorption with fresh and starved subsurface bacterial consortium. Biotechnol. Letters. 15:87-92.

Macaskie, L.E., and A.C.R. Dean. 1987. Use of immobilized biofilm of *Citrobacter* sp. for the removal of uranium and lead from aqueous flows. Enzyme Microbial Technol. 9:2-4.

Phelps, T.J., J.J. Niedzielski, R.M. Schram, S.E. Herbes, and D.C. White. 1990. Biodegradation of trichloroethylene in continuous-recycle expanded-bed bioreactors. Appl. Environ. Microbiol. 56:1702-1709.

Powers, S.E., C.O. Loureiro, L.M. Abriola, and W.J. Weber, Jr. 1991. Theoretical study of the significance of nonequilibrium dissolution of nonaqueous phase liquids in subsurface systems. Water Resour. Res. 27:463-477.

Richards, D. J., and W. K. Shieh. 1989. Anoxic-oxic activated-sludge treatment of cyanide and phenols. Biotech. Bioeng. 33:32-38.

Riley, R.G., and J.M. Zachara. 1992. Chemical contaminants on U.S. DOE lands and selection of contaminant mixtures for subsurface science research. U.S. Gov. Print. Office, Washington D.C U.S. DOE/ER-0547T.

Semprini, L., G.D. Hopkins, P.V. Roberts, D. Grbic-Galic, and P.L. McCarty. 1991. A field evaluation of *insitu* biodegradation of chlorinated ethenes: Part 3, Studies of competitive inhibition. Ground Water 29:239-250.

Siegel, S., P. Keller, M. Galun, H. Lehr, B. Siegel, and E. Galun. 1986. Biosorption of lead and chromium by penicillium preparations. Water, Air, Soil Pollut. 27:60-75.

Skeen, R.S., S.P. Luttrell, B.S. Hooker, and J.N. Petersen. 1993. *In situ* bioremediation of Hanford groundwater. Remediation 3:353-67.

Stenner, R.D., K.H. Cramer, K.A. Higley, S.J. Jette, D.A. Lamar, T.J. McLaughlin, D.R. Sherwood, and N.C. van Houten. 1988. Hazard ranking system evaluation of CERCLA inactive waste sites at Hanford: Evaluation methods and results. PNL-6456, vol. 1, Pacific Northwest Laboratory, Richland, WA.

Strand, S.E., M.D. Bjelland, and H.D. Stensel. 1990. Kinetics of chlorinated hydrocarbon degradation by suspended cultures of methane-oxidizing bacteria. Res. J. WPCF 62:124-9.

Tsezos, M., and A.A. Deutschmann. 1990. An investigation of engineering parameters for the use of immobilized biomass particles in biosorption. J. Chem. Technol. Biotechnol. 48:29-39.

Tsezos, M., and D.M. Keller. 1983. Adsorption of radium-226 by biological origin adsorbents. Biotechnol. Bioengr. 25:201-215.

Tsezos, M., and B. Volesky. 1981. Biosorption of uranium and thorium. Biotechnol. Bioengr. 23:583-604.

Tsezos, M., and B. Volesky. 1982. The mechanism of thorium biosorption by *Rhizopus arrhizus*. Biotechnol. Bioengr. 24:955-969.

Tsezos, M., M.H.I. Baird, and L.W. Shemilt. 1986. The kinetics of radium biosorption. Chem. Engr. J. 33:B35-B41.

Tsezos, M., M.H.I. Baird, and L.W. Shemilt. 1987. The elution of radium adsorbed by microbial biomass. Chem. Engr. J. 34:B57-B64.

U.S. Department of Energy (U.S. DOE). 1990. Basic research for environmental restoration. U.S. Gov. Print. Office, Washington D.C U.S. DOE/ER-0482T.

U.S. Government Accounting Office (U.S. GAO). 1988a. Nuclear waste problems associated with U.S. DOEO's inactive waste sites. U.S. Gov. Print. Office, Washington D.C. GAO/RCED-88-169, U.S.

U.S. Government Accounting Office (U.S. GAO). 1988b. Nuclear waste, supplemental information on problems at U.S. DOE's inactive waste sites. U.S. Gov. Print. Office, Washington D.C. GAO/RCED-88-229FS.

Vannelli, T., M. Logan, D.M. Arciero, and A.B. Hooper. 1990. Degradation of halogenated aliphatic compounds by the ammonia-oxidizing bacterium Nitrosomonas europaea. Appl. Environ. Microbiol. 56:1169-1171.

Vogel, T.M., C.S. Criddle, and P.M. McCarty. 1987. Transformations of halogenated aliphatic compounds. Environ. Sci. Technol. 21:722-736.

11 Applications of Bioremediation in the Cleanup of Heavy Metals and Metalloids

W.T. Frankenberger, Jr. and M.E. Losi
Department of Soil and Environmental Sciences
University of California
Riverside, California

A major emphasis has been placed on bioremediation of organic compounds and their fate/transport throughout the environment. Another important class of chemicals polluting our environment includes the inorganics, particularly metals and metalloids such as chromium (Cr), mercury (Hg), arsenic (As) and selenium (Se). Recent work has revealed that microbial transformations may have several applications in the cleanup of heavy metals and metalloids in soil as well as in water.

Heavy metals/metalloids and inorganic compounds are the second and third most frequently reported categories of hazardous substances just behind the organics/volatile organic compounds (VOCs). Approximately 50% of the Superfund sites as specified by the EPA Records of Decisions (RODs) between 1982 and 1986 contain inorganic compounds which are designated as a threat or problem (Palmer et al., 1988). These contaminants include acids (18 sites), arsenic (30 sites), asbestos (2 sites), chromium (27 sites), heavy metals (79 sites), and various inorganics (50 sites).

Whereas the goal of remediating heavy metal/metalloid-laden water is the removal of the contaminant, this is not necessarily the case with contaminated soil. The most commonly used remedial technology to deal with metals and metalloids in soil is solidification. Solidification, or fixation, involves a process in which the contaminated matrix is stabilized, fixated, solidified or encapsulated into a solid material via the addition of a resin or other chemical compound such as cement or pozzolans. This technology is designed to reduce the production of leachate in the environment. There are a variety of commercial pozzolanic materials available today, which are proprietary and offer cementitious properties. Although the contaminants are chemically and/or physically bound in the resulting matrix, they are not destroyed. The issue of stability in solidification technology is of major concern. Other technologies used to treat soils contaminated with heavy metals and metalloids include vitrification, soil washing and particle size separation.

Remedial technologies accepted in the detoxification of heavy metal/metalloid-contaminated soils employ chemical treatments involving additions

---

 Bioremediation: Science and Applications. SSSA Special Publication 43.

of compounds such as: limestone or apatite to precipitate metals; clays and/or organic materials, which results in sorption and precipitation reactions; and zeolite, giving rise to unexchangeable interactions. Recent efforts have demonstrated the feasibility of using crop exudators and plant hyperaccumulators to control the mobilization of trace metals. Mechanical and physical means currently used include soil cover barriers and deep tillage which results in redistribution and dilution.

Novel applications in bioremediation have recently been designed for aquatic systems, but relatively fewer applications are available for contaminated soils, which in many cases represents a more complex problem. The fate and transport of inorganic solutes in soils can be controlled by chemical and biochemical processes such as mineral precipitation, dissolution, complexation, oxidation-reduction and ion adsorption. Volatilization also occurs, but is generally limited to a few elements. Some of these reactions are abiotic and some are under biotic control and thus, both will be subject to discussion in this review. The focus of this review article is to discuss bioremediation approaches to detoxify soils and water contaminated with Cr, Hg, As and Se, making use of state-of-the-art techniques which are being developed today.

BIOREMEDIATION OF CHROMIUM CONTAMINATION

Chromium (Cr) has many industrial uses and, as a result, anthropogenic discharge into the environment often occurs. Hexavalent chromium (Cr(VI)) forms chromates and dichromates which are toxic and mutagenic, soluble over a wide pH range, and generally mobile in soil/water systems. The trivalent form Cr(III), is considerably less toxic and is virtually nonmobile, largely because it precipitates as oxides and hydroxides at pH 5 and above (McGrath and Smith, 1990; Cary, 1982). Thus, the reduction of Cr(VI) to Cr(III) represents a viable remediation/ detoxification strategy.

Traditional techniques for remediating chromate-contaminated water involve reduction of Cr(VI) to Cr(III) by chemical (usually Fe^{2+}) or electrochemical means at pH >5 followed by precipitation and finally filtration or sedimentation (Eary and Rai, 1988). The discovery of microorganisms that preferentially reduce Cr(VI) to Cr(III) has led to applications in the bioremediation field which are potentially more cost effective (Table 11-1). There are currently two strategies for immobilizing Cr through microbial transformations; biosorption/reduction and gaseous bioreduction.

Biosorption/Reduction of Chromium

Bioreduction of chromate can occur as a direct result of microbial metabolism. It takes place under both aerobic and anaerobic conditions, and has been shown to be enzymatic in nature (Bopp et al., 1983; Horitsu et al., 1987). Chromate can be the terminal electron acceptor in the oxidation of organic matter, or can be reduced as a detoxification mechanism (Wood & Wang, 1983). Russian researchers first proposed using Cr(VI)-reducing bacterial isolates in removal of chromates from various industrial effluents (Romanenko et al., 1976; Romanenko & Koren'kov, 1977). Since then, a wide variety of bacterial strains with accelerated Cr(VI) reducing capabilities have been isolated from chromate-contaminated water, soils and sediments (Table 11-1). It is evident that these organisms can mediate

Table 11-1. Bioremediation of chromium.

Mechanism	Organism	Description and Effectiveness	Source
Bioreduction	*Pseudomonas ambigua* G-1	Cr(VI) concentration was lowered from 150 to ~35 mg L^{-1} over 36 hours in liquid media. Shown to be enzymatic.	Horitsu et al., 1987
Bioreduction	*Pseudomonas* K-21	Reduction rates were proportional to the glucose concentration in the medium. [Cr(VI)] dropped from 200 to 100 mg L^{-1} in the presence of 0.2% glucose and to ~0 mg L^{-1} with 1% glucose after 40 hours.	Shimada & Matsushima, 1983
Bioreduction	*Pseudomonas fluorescens* LB300	Reduced about half of Cr(VI) in media spiked with 40 mg L^{-1} Cr(VI) over 48 hours.	Bopp & Ehrlich, 1988
Biosorption	*Oscillatoria* sp.	Algal cultures removed 20% of Cr from water spiked at levels from 1-20 mg L^{-1}	Filip et al., 1979
Biosorption	*Arthrobacter* sp. *Agrobacter* sp.	Accumulated Cr with increasing concentration gradient of Cr(VI) up 400 mg L^{-1} (*Arthrobacter* sp.) and 100 mg L^{-1} (*Agrobacter* sp.)	Coleman & Paran, 1983

(continued on next page)

Table 11-1. Continued.

Mechanism	Organism	Description and Effectiveness	Source
Bioreduction/ biosorption	*Pseudomonas aeruginosa* S128	Removed 15-50% of Cr from liquid media spiked with 1000 mg L^{-1} Cr(VI) over 72 hours. Glucose enhanced the removal rate.	Marquez et al., 1982
Bioreduction	Sulfate reducing bacteria	[Cr(VI)] in water was lowered from 11 to <0.21 mg L^{-1} with addition of sulfate and acetate in an anaerobic bioreactor which produced H_2S, reducing the Cr(VI).	DiFilippi & Lupton, 1992
Bioreduction	*Enterobacter cloacae* HO1	Cr(VI) at 182 mg L^{-1} in wastewater was completely reduced at high cell densities with appropriate nutrient additions.	Ohtake et al., 1990
Bioreduction	*Enterobacter cloacae* HO1	90% of Cr was removed from water with an initial [Cr(VI)] of 208 mg L^{-1}. Cultures were contained within dialysis tubes submerged in the contaminated water. Chromate diffused into the tube, was reduced, precipitated and unable to diffuse out.	Komori et al., 1990

(continued on next page)

Table 11-1. Continued.

Mechanism	Organism	Description and Effectiveness	Source
Biosorption/ bioreduction	Sulfate-reducing bacteria	100% Cr removal was achieved in water with [Cr(VI)] up 150 mg L^{-1}.	Roda & Smirnova, 1989
Biosorption/ bioreduction	*Chlamydomonas* sp.	An average of 30% Cr removed from water with initial concentrations of 0.2 mg L^{-1} in an aerobic bioreactor system.	Ayoub & Sayigh, 1987
Bioreduction	*Pseudomonas putida* PRS2000	The isolate reduced the [Cr(VI)] in liquid media from 1.04 to about 0.36 mg L^{-1} over 90 hrs. Reduction was shown to be associated with soluble protein in cells. Process was unaffected by SO_4^{2-} or NO_3^-	Ishibashi et al., 1990
Biosorption	*Chlorella vulgaris*; *Zoogloea ramigera*	Maximum adsorption was seen at pH 2 and at 25-50°C for both organisms. Adsorption increased with [Cr(VI)] up to 200 mg L^{-1} (*C. vulgaris*) and 75 mg L^{-1} (*Z. ramigera*).	Zumriye et al., 1990

(continued on next page)

Table 11-1. Continued.

Mechanism	Organism	Description and Effectiveness	Source
Bioreduction	Consortium of bacteria	Reduced all CrO_4^{2-} in liquid media at 30 mg L^{-1} after 48 hours, and less as [Cr(VI)] was increased.	Eliseeva et al., 1991
Bioreduction	*Aeromonas dechromatica*	Effective reduction observed with appropriate carbon sources (propyl alcohols and ethyl acetate), $Cu(OH)_2$, Zn^o and $Zn(OH)_2$ promoted bioreduction, Cd had no effect, and Ni, Fe(III) and Cu inhibited the transformation.	Kvasnikov et al., 1986, 1987
Bioreduction	Unidentified pure bacterial cultures	Reduced 40-60% of Cr(VI) added to liquid media at 10 and 25 mg L^{-1}; approximately 50% at 60 mg L^{-1}, and 10% at 120 mg L^{-1} over 18 days with no additional nutrient supplement.	Losi et al., 1994b
Bioreduction/ Immobilization in surface soils (Biofilter)	Indigenous soil organisms	Removed up to 98% of Cr(VI) added in irrigation water with a [Cr(VI)] of 1.0 mg L^{-1}. Important factors shown to be organic matter loadings, O_2 status and residence time in bioactive zone.	Losi et al., 1994a

bioreduction of chromates. Research is presently underway to develop commercially viable bioremediation techniques exploiting these organisms.

The major focus of bioreduction of Cr(VI) is aimed towards the development of bioreactors, which consist of a reduction phase, with Cr(VI)-reducing bacteria immobilized on inert matrices within the reactor, followed by a settling, or filtration phase to remove Cr(III) precipitates. Chromate-contaminated water is pumped through the reactor along with various carbon sources and nutrient supplements, and the Cr(VI) is reduced, precipitated and removed. The main advantages of this system are considered to be its cost effectiveness, and that no reagents containing other heavy metals (reductants) are added. A potential disadvantage is that batch experiments in the laboratory have suggested that the lowest achievable effluent Cr concentrations are probably around 1 mg L^{-1} (Apel & Turick, 1991; Mattison, 1992a), which is twenty times higher than the national EPA drinking water standard of 0.05 mg L^{-1}. Also, reaction rates may be slow, since CrO_4^{2-} must diffuse into direct contact with the cells, which maximizes its toxicity effects as well. Since optimum bioreduction has been correlated with the exponential growth phase of the bacteria (Apel & Turick, 1991), conditions within the bioreactor must be adjusted to sustain high growth rates. Bioreactors such as these are probably best suited to treating wastewater prior to discharge rather than *in situ* applications because flow rates are expected to be relatively slow, and pumping of groundwater can significantly raise the cost.

We recently investigated a land application method (biofilter) for remediating Cr(VI)-contaminated groundwater (Losi et al., 1994a). The method consists of using contaminated water high in Cr(VI) to irrigate an alfalfa field (supplemented with an organic amendment) where reduction, precipitation and immobilization would take place. A glasshouse experiment was conducted which involved passing the Cr(VI)-contaminated water through soil columns which had been amended with dried cattle manure to stimulate microbial activity. This led to lower O_2 concentrations in the effluent of the amended treatments compared to unamended controls. Alfalfa was grown in some of the soil columns as a treatment.

In this experiment, we were able to treat relatively large volumes of water spiked with 1.0 mg L^{-1} Cr(VI), yielding outflow Cr levels consistently below 0.02 mg L^{-1} (Losi et al., 1994a). We determined that reduction followed by precipitation was the dominant removal mechanism, and that lower O_2 levels favored the reaction. Subsequent experiments provided evidence that indigenous microorganisms, in conjunction with a readily available carbon source (cattle manure) were largely responsible for the excellent removal rates (Losi et al., 1994b). Fig. 11-1 illustrates the chemical and biological reactions that are responsible for this transformation. In light of this approach, it appears that reactive constituents in soils may be useful (and economical) in optimizing bioreactor systems designed to treat Cr(VI)-contaminated water. Perhaps more importantly, our study indicated that biofilters show promise for remediation of Cr-contaminated aquifers which underlie open land. This method is potentially very cost effective requiring only animal manure as an additive, and has an additional advantage of producing alfalfa with slightly elevated Cr levels, which could be useful for supplementing the Cr content of animal feed (Bartlett & James, 1988). The potential to rapidly and efficiently treat

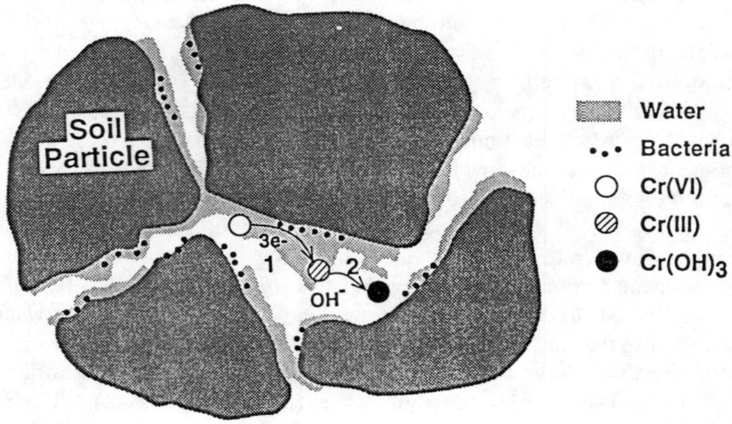

Fig. 11-1. Direct, enzymatic reduction of Cr(VI) followed by chemical precipitation. 1. Biological transformation of Cr(VI) -- contact between bacterial cell and Cr(VI) results in enzymatic reduction to Cr(III) (this reaction also occurs chemically). 2. Non-biological precipitation of Cr(III) -- Cr(III) reacts with OH^- to form insoluble/immobile $Cr(OH)_3$.

enormous quantities of contaminated water could be expected to offset the cost of pumping from the aquifer.

Another application of direct biosorption/reduction was demonstrated by Komori et al. (1990). In this study, anaerobic Cr-reducing bacterial cultures contained within dialysis tubes were submerged in contaminated water. Chromate diffused into the tubes, was reduced, precipitated and was thus sequestered within the tubes. Laboratory studies employing this system demonstrated that 90% of Cr was removed from water with an initial Cr(VI) concentration of 208 mg L^{-1}. Advantages of this type of system are again cost and reagent efficiency. A major disadvantage is that the process is gradient-driven and thus diffusion into the tube will slow as Cr(VI) concentrations are lowered. This could lead to relatively long residence times needed to reach acceptable removal rates. Another disadvantage might be a low suitability for *in situ* applications.

Gaseous Bioreduction of Chromium

Gaseous bioreduction involves microbial reduction of Cr(VI) by a metabolic byproduct (in this case, H_2S), which is produced in anaerobic environments by sulfate-reducing bacteria (Fig. 11-2). This technology has shown some promise for both anaerobic bioreactor systems and possibly for *in situ* applications.

DiFilippi and Lupton (1992) designed an anaerobic bioreactor utilizing marine derived sulfate-reducing bacteria immobilized as a biofilm on gravel. Their experiments have suggested that with additional clarification treatment, effluent Cr levels as low as 0.01 mg L^{-1} may be attainable (Mattison, 1992a). An advantage of this system over biosorption/reduction is that the CrO_4^{2-} need not come into direct contact with the bacterial cells for reduction to occur. Instead, the H_2S diffuses out into the medium, which can have several important benefits, including enhancement of the reaction rate, and protection of the cells from the toxic effects of the Cr(VI). This could promote more efficient reduction, and may explain why effluent Cr(VI)-concentrations are lower with this system than the concentrations reported by Apel & Turick (1991) utilizing direct (enzymatic) bacterial reduction. A potential disadvantage of this system is that anaerobicity (requiring energy and costs) must be maintained within the reactor.

A version of this technique is being evaluated as a possible *in situ* treatment for immobilization of Cr in chromate-contaminated soils and groundwater. Such an application would involve production of H_2S in groundwater or deep within the soil profile by *in situ* stimulation of sulfate-reducing bacteria with additions of sulfate and other nutrients. This process was previously shown to be effective in the reduction and sedimentation of Cr(VI)-containing tannery wastes entering Otago Harbor in New Zealand (Smillie et al., 1980).

In addition to the obvious cost advantages (over methods such as pump and treat), this system would have other benefits as well. First, H_2S can also reduce Mn(IV), which can reoxidize Cr(III) to Cr(VI), thus stabilizing Cr(III) precipitates. Second, H_2S has the potential to diffuse through soil/water systems, reducing adsorbed chromates that may otherwise not come into contact with bacterial cells, providing a more complete treatment (Lupton, F., see Mattison, 1992a). The need for maintaining anaerobicity may be a limiting factor in this technique.

Mattison (1992a) reports that of the methods discussed, gaseous bioreduction technology is the most promising at present, and is furthest along in terms of development. The final test will be to determine if these techniques can compete with chemical methods in efficiency as well as cost-effectiveness. If so, they would certainly be environmentally preferred, and could be operational in the near future.

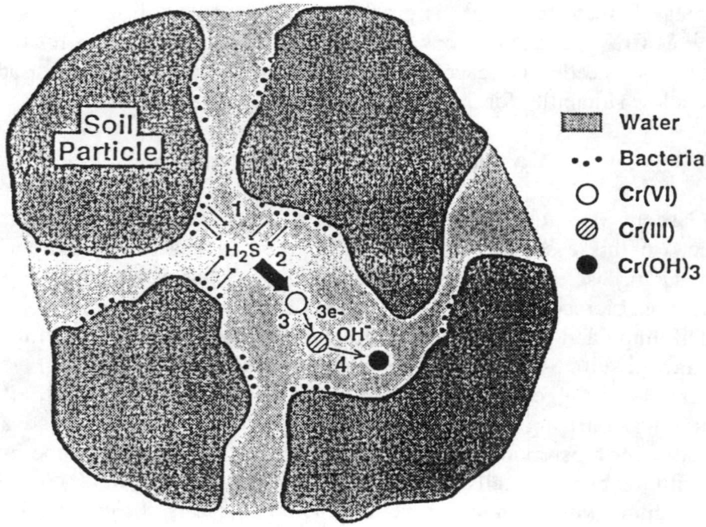

Fig. 11-2. Indirect, gaseous reduction of Cr(VI) followed by chemical precipitation. 1. Hydrogen sulfide is produced by sulfate-respiring bacteria in anaerobic systems. 2. Hydrogen sulfide gas diffuses into contact with Cr(VI). 3. Cr(VI) is chemically reduced to Cr(III) with H_2S serving as the electron donor. 4. Cr(III) reacts with OH^- to form insoluble/immobile $Cr(OH)_3$.

BIOREMEDIATION OF MERCURY CONTAMINATION

Mercury pollution has received a great deal of publlicity since the infamous Minamata Bay incident in Japan, where direct discharge of Hg-containing industrial waste led to extremely high Hg levels in fish, which, when eaten by humans, caused physical impairments and death. Aquatic or ganisms found in surface waters in the United States also have elevated levels of Hg, particularly in Florida (Everglades), believed to be a result of bioaccumulation within the food chain. Biomagnification can result when Hg concentrations in lakes and streams may be undetectable. Concentrations in fish tissues can be far greater than levels in surrounding water, particularly when methylmercury is present. Another concern has involved contaminated grains treated with Hg compounds (used as fungistatic agents), which have posed health threats in Iraq, Pakistan, Guatemala and the U.S. (Ehrlich, 1990).

Table 11-2. Bioremediation of mercury.

Mechanism	Organisms	Description and Effectiveness	Source
Biosorption	Hg resistant bacteria	Accumulated up to 2-4% by weight of non-volatile Hg from aerated solutions containing 50-100 mg Hg^{2+} L^{-1}. Carbon source-methanol or acetic acid. Biomass was separated from solution by thermal decomposition (400-500°C) to recover the distillate.	Glombitza et al., 1984a, b
Biosorption	AlgaSORB (*Chlorella vulgaris*) is heated to 300-500°C and immobilized on a silica-based matrix	Strong adsorption of Hg^{2+} over pH range of 2-6 due to ligands present in cell walls. Feed range of 0.14 to 1.6 mg L^{-1} yielded effluent <10 µg L^{-1}.	Dennis Darnall marketed by Bio-Recovery (See Mattison, 1992b)
Bioreduction	*Chlorella emersonii* (CCAP 211/8a) as free culture or Ca-alginate immobilized beads	Rapid rates of removal (99% of Hg removed from a 1 mg L^{-1} solution in 12 days with immobilized system)	Wilkinson et al., 1989

(continued on next page)

Table 11-2. Continued.

Mechanism	Organisms	Description and Effectiveness	Source
Bioreduction	Mercury resistant bacteria (*Xanthomonas maltophilia, Aeromonas hydrophilla, Microbacterium* sp., *Alcaligenes eutrophus, Pseudomonas paucimobilis*	Porous support coated with Bakers yeast and exposed to nonsterile Hg feed. Hg resistant bacteria colonize particles. Loading rates of 2-48 mg Hg^{2+} h^{-1} resulted in effluent of 50-100 µg L^{-1}.	Gesellschaft für Biotechnologische Forschung (GBF, Institute for Biotechnology Research) in Braunschweig Germany. K. Timmis (see Mattison, 1992b)
Bioreduction	Consortium from municipal activated sludge	Hg^{2+}- containing feed solution was added to fluidized bed. Hg^{2+} reduced to Hg^0 as it contacts biofilm-coated sand. 100 mg L^{-1} Hg^{2+} feed → 1 mg L^{-1}. 2 mg L^{-1} Hg^{2+} → <0.01 mg L^{-1}.	Conly Hansen, Dept. of Nutrition and Food Science, Utah State Univ. (See Mattison, 1992b)
Methylation	Methanogens	$Hg^{2+} \xrightarrow{CH_3B_{12}} \rightarrow (CH_3)Hg^+ \xrightarrow{CH_3B_{12}} (CH_3)_2Hg$	Wood et al., 1968; DeSimone et al., 1973

(continued on next page)

Table 11-2. Continued.

Mechanism	Organisms	Description and Effectiveness	Source
Methylation	*Clostridium cochlearium*	HgO, $HgCl_2$, $Hg(NO_3)_2$ $Hg(CN)_2$, $Hg(SCN)_2$ $Hg(OOCCH_3)_2 \rightarrow (CH_3)_2Hg$	Yamada & Tonomura, 1972a, b
Methylation	*Desulfovibrio desulfuricans*	Active Hg^{2+} methylation in anoxic estuarine sediment.	Compeau & Bartha, 1985
Methylation	*Neurospora crassa*	Hg^{2+} complexes with homocysteine or cysteine in the presence of a methyl donor (betaine or choline) and transmethylase cleaves $(CH_3)Hg$	Landner, 1971

Current studies are now focusing on biosorption, bioreduction and methylation as a bioremedial approach to immobilize Hg (Table 11-2).

Biosorption of Mercury

A number of naturally occurring Hg-biotransformations may have applications in bioremediation of Hg-contaminated soil and/or water. Mercury resistant bacteria have been isolated by F. Glombitza at the Akademie der Wissenschaften (Table 11-2). These Hg-resistant bacteria can accumulate up to 2-4% by weight of non-volatile Hg from aerated solutions containing 50-100 mg L^{-1} Hg^{2+} (Mattison, 1992b). Biomass can be suspended in a feed solution within a bioreactor, and provided with a carbon source such as methanol or acetic acid plus nutrients. The Hg-laden biomass is separated from solution and thermally decomposed at 400-500°C to recover the mercury distillate.

Another technology in Hg biosorption employs an immobilized algae marketed by Bio-Recovery Systems under the name Alga-SORB (Mattison, 1992b). This product is prepared by heating the algae (*Chlorella vulgaris*) to 300-500°C and immobilizing it on a silicate based matrix. Alga-SORB shows strong adsorption for Hg^{2+} independent of pH over the pH range of 2-6. Mercury adsorption occurs as a result of interactions with "soft" ligands (forms covalent complexes with "soft" functional groups which contain N or S) present in the cell wall. In an EPA sponsored SITE demonstration, Alga-SORB columns treated with 0.14-1.6 mg L^{-1} resulted in an effluent consistently below 10 μg L^{-1} Hg.

Unfortunately, biosorption will not remove organically bound Hg compounds such as phenylmercuric acetate or methyl mercury (Mattison, 1992b). These biosorption systems are only effective on Hg in its cationic form (Hg^{2+}). Therefore, Hg covalently bound to organic carbon such as methyl mercury must be hydrolyzed to the free Hg ion prior to removal by biosorption.

Bioreduction of Mercury

Another naturally occurring process with potential applications is microbial reduction of Hg^{2+} to the elemental form (Hg^0) followed by volatilization. To be used as a method to remove Hg, volatile Hg^0 must be captured, requiring a contained system such as a bioreactor (Fig. 11-3). This process would obviously be limited in the volume of water which could be treated. Some of the more well-characterized systems applicable to bioremediation have been developed with *Chlorella emersonii* (Wilkinson et al., 1989), Hg-reducing bacteria (Institute of Biotechnological Research [GBF], Braunschweig, Germany, see Mattison, 1992b) and a bacterial consortium from municipal activated sludge (Hansen, see Mattison, 1992b) (Table 11-2).

Ionic mercury (Hg^{2+}) can be reduced to volatile metallic Hg^0 by *Pseudomonas* spp., enteric bacteria, *Staphylococcus* sp. and *Cryptococcus* sp. (Ehrlich, 1990). Often the ability to reduce Hg^{2+} correlates with Hg resistance. The enzyme involved is mercuric reductase, a soluble, NADPH-dependent cytoplasmic flavoprotein. This enzyme is active in the presence of excess thiols (R-SH) such as mercaptoethanol, dithiothreitol, glutathione or cysteine.

Methylation of Mercury

Microbial methylation of Hg has been observed in methanogenic bacteria, and is also a potentially useful reaction. In this process, $(CH_3)_2Hg$ is formed from Hg^{2+} with the involvement of methylcobalamin (Wood et al., 1968). Methylation of Hg can occur under both anaerobic and aerobic conditions. In addition to methanogenic bacteria, *Clostridium cochlearium* can methylate HgO, $HgCl_2$, $Hg(NO_3)_2$, $Hg(CN)_2$, $Hg(SCN)_2$, and $Hg(OOCCH_3)_2$ (Yamada & Tonomura, 1972a, b). Sulfate reducing bacteria such as *Desulfovibrio desulfuricans* can methylate Hg^{2+} in anoxic estuarine sediments (Compeau & Bartha, 1985). Although these processes certainly show potential for detoxifying mercury-contaminated soils and water, none have been used commercially. Until methods can be developed, tested, and demonstrated to be effective in field experiments, their feasibility as possible bioremediation techniques will remain in question.

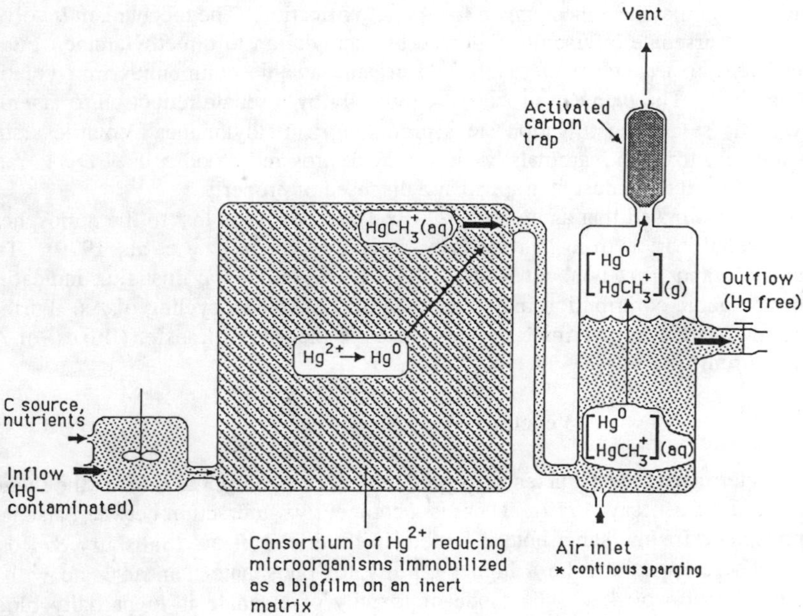

Fig. 11-3. Bioreactor design for treatment of Hg-contminated water.

Aerobic bacteria such as *Pseudomonas* spp., *Bacillus megaterium, Echerichia coli, Enterobacter aerogenes* and fungi including *Aspergillus niger, Neurospora crassa, Scopulariopsis,* sp., *Drevicaulis* sp. and *Saccharomyces cerevisiae* are also active in methylating Hg^{2+}.

BIOREMEDIATION OF ARSENIC CONTAMINATION

Arsenic (As) has both metallic and nonmetallic properties and is a member of the nitrogen family. It occurs naturally in the environment as well as through anthropogenic discharge in a variety of chemical states. Arsenic forms alloys with various metals, and covalently bonds with carbon, hydrogen, oxygen, and sulfur (Ferguson & Gavis, 1972). Arsenate (AsO_4^{3-}) can replace phosphate in energy transfer phosphorylation reactions, and arsenite (AsO_2^-) has a high affinity for thiol groups of proteins, inactivating many enzymes. The poisonous character of As made it an effective herbicide and insecticide. The ubiquity of As in the environment, its biological toxicity, and its redistribution are factors evoking public concern.

Table 11-3 lists a composite of studies which suggest the potential for bioremediation strategies involving As-contaminated soils and water. Both oxidation and methylation are microbial transformations involved in the redistribution and global cycling of As. Oxidation involves the conversion of toxic arsenite to less toxic arsenate. Bacterial methylation of inorganic arsenic is coupled to the methane biosynthetic pathway in methanogenic bacteria under anaerobic conditions and may be a mechanism for As detoxification. The mechanism involves reduction of arsenate to arsenite followed by methylation to dimethylarsine. Fungi are also able to transform inorganic and organic arsenic compounds into volatile methylarsines. The pathway proceeds aerobically by arsenate reduction to arsenite followed by several methylation steps producing trimethylarsine. Volatile arsine gases are very toxic to mammals because they destroy red blood cells (LD_{50} in rats; 3.0 mg kg^{-1}) and thus must be trapped and disposed of properly.

It is estimated that as much as 2.10 x 10^7 kg of As is lost to the atmosphere through volatilization from land surfaces annually (MacKenzie et al., 1979). The continental vapor flux is about 8 times that of the continental dust flux indicating that the biogenic contribution may play a significant role in cycling of As. Further studies are needed to determine mass balances in the rate of transfer (fluxes) of As in the environment.

Vegetative Uptake of Arsenic

In terrestrial plants, arsenate is preferentially taken up 3 to 4 times the rate of arsenite (Asher & Reay, 1979). There is a competitive interaction between arsenate and phosphate for the same uptake system in terrestrial plants (Salisbury & Ross, 1969). There appears to be a higher affinity for phosphate than arsenate with a discriminate ratio of 4:1. The mode of toxicity of arsenate is to partially block protein synthesis and interfere with phosphorylation, but the presence of excess phosphate prevents this mode of action. It has not been established whether volatile arsenic can be released by plants.

Bacterial Oxidation of Arsenic

A number of microbially mediated oxidation reactions are currently being studied in the interest of developing bioremediation techniques for detoxifying As-contaminated soil and water. *Bacillus* and *Pseudomonas* spp. have been isolated which can oxidize arsenite to the less toxic arsenate. In addition, a strain of *Alcaligenes faecalis* obtained from raw sewage was capable of oxidizing arsenite (Phillips & Taylor, 1976). Energy is not recovered from this reaction but rather,

Table 11-3. Bioremediation of arsenic.

Mechanism	Organisms	Description and Effectiveness	Source
Bio-oxidation	Mixed bacterial culture	Electronic industry uses gallium arsenide (GaAs) in light emitting diodes and integrated circuits (computer chips). Rejected chips and toxic fragments are treated with inoculum at 75°C at pH 1.5 transforming GaAs to Ga solution and AsO_4^{3-}	Technical Research Inc., Salt Lake City, Utah, Patent, US 5030425 (1991) (see Mattison, 1992c)
Bio-oxidation	*Alcaligenes faecalis*	Energy is not obtained, but oxidation was proposed to be a detoxification mechanism.	Phillips & Taylor, 1976
Bio-oxidation	*Alcaligenes* sp.	Induced by arsenite and consumes O_2	Osborne & Ehrlich, 1976
Methylation	Methanogenic bacteria. (*Methanobacterium* strain M.O.H.)	Methylates arsenate, arsenite and methylarsonic acid to dimethylarsine under anaerobic conditions.	McBride & Wolf, 1971

(continued on next page)

Table 11-3. Continued.

Mechanism	Organisms	Description and Effectiveness	Source
Methylation	*Flavobacterium* sp.	Methylated dimethylarsinic acid to trimethylarsine oxide.	Chau & Wong, 1978
Methylation	*Penicillium brevicaule* (*Scopulariopsis brevicaulis*)	Produced trimethylarsine in the presence of methylarsonic acid or dimethylarsinic acid.	Challenger, 1945
Methylation	*Candida humicola, Gliocladium roseum* and *Penicillium* sp.	Converted methylarsonic and dimethylarsinic acid to trimethylarsine. *C. humicola* used AsO_2^- and AsO_4^{3-} as substrates to produce trimethylarsine.	Cox & Alexander, 1973a
Methylation	*Candida humicola*	Methylated a wood preserving fungicide, chromated copper arsenate into a volatile As species.	Cullen et al., 1984
Methylation	*Alcaligenes* sp. *Pseudomonas* sp.	Methylated AsO_2^- or AsO_4^{3-} into arsine under aerobic conditions	Cheng & Focht, 1979

(continued on next page)

Table 11-3. Continued.

Mechanism	Organisms	Description and Effectiveness	Source
Methylation	*Penicillium* sp.	Isolated from an agricultural evaporation pond. Capable of producing trimethylarsine from methylarsonic and dimethylarsinic acids. Optimum conditions for trimethylarsine production included: (i) a medium in which 100 mg L^{-1} methanearsonic acid served as the As source; (ii) pH 5 to 6; (iii) temperature of incubation, 20° C. The addition of carbohydrates and sugar acids to the minimal medium suppressed trimethyl-arsine production. The amino acids, phenylalanine, isoleucine and glutamine promoted trimethylarsine production with an enhancement ranging from 10.2- to 11.6-fold over control without amino acid enhancement.	Huysmans & Frankenberger, 1991

oxidation is proposed to be a detoxification mechanism (Osborne & Ehrlich, 1976). Biochemical studies on resting cells revealed that the oxidation process is induced by arsenite. The use of respiratory inhibitors prevented further oxidation of arsenite, indicating that oxygen served as the terminal electron acceptor. In extreme environments such as acid mine waters, As concentrations are as high as 2.0 to 13 mg L^{-1} and the major inorganic species is arsenite. Oxidation of arsenite by heterotrophic bacteria play an important role in detoxifying such an environment, catalyzing as much as 78 to 96% of the arsenite to arsenate (Wakao et al., 1988).

Bacterial Methylation of Arsenic

Bacterial methylation of inorganic As, has been studied extensively using methanogenic bacteria. Methanogenic bacteria are a morphologically diverse group consisting of coccal, bacillary, and spiral forms, but are unified by the production of methane as their principal metabolic end product. They are present in large numbers in anaerobic ecosystems, such as sewage sludge, freshwater sediments, and composts where organic matter is decomposing (McBride et al., 1978). It has been shown that at least one species of *Methanobacterium* is capable of methylating inorganic As to produce volatile dimethylarsine. Arsenate, arsenite, and methanearsonic acid can serve as substrates in dimethylarsine formation. Inorganic As methylation is coupled to the CH_4 biosynthetic pathway and may be a widely occurring mechanism for As removal/detoxification. A bioreactor design which makes use of the ability of methanogenic bacteria to produce volatile dimethylarsine from inorganic As is shown in Fig. 11-4. *Methanobacterium* strain M.O.H. cell-free extracts, when incubated under anaerobic conditions with $[^{74}As]Na_2AsO_4$, a methyl donor (methylcobalamin, CH_3CoB_{12}), H_2, and ATP, produced a volatile $[^{74}As]$-dimethylarsine (McBride & Wolfe, 1971). The pathway involves the reduction of arsenate to arsenite with subsequent methylation by a low molecular weight cofactor, Coenzyme M (CoM). Methanearsonic acid added to cell-free extracts is not reduced to methylarsine but requires an additional methylation step before reduction. However, dimethylarsinic acid is reduced to dimethylarsine even in the absence of a methyl donor (McBride & Wolfe, 1971).

Whole cells of methanogenic bacteria under anaerobic conditions also produce dimethylarsine as a biomethylation end product of arsenic but not heat-treated cells, indicating that this is a biological reaction. Furthermore, samples collected from a number of different anaerobic ecosystems (anaerobic sewage digestor sludge and rumen from cattle) which produced methane also transformed arsenate into dimethylarsine.

Under anaerobic conditions biomethylation of As proceeds only to dimethylarsine, which is stable in the absence of oxygen, but is rapidly oxidized under aerobic conditions. Dimethylarsine in anaerobic environments can react with disulfide bonds present on particulates, thus reducing the concentration of soluble As.

Interestingly, another study indicated that resting cell suspensions of *Pseudomonas* sp. and *Alcaligenes* sp. incubated with either arsenite or arsenate under anaerobic conditions, produced arsine but no other intermediates (Cheng & Focht, 1979). *Aeromonas* sp. methylated dimethylarsinic acid to trimethylarsine oxide. Methylation of As is pH-dependent with the highest rates occurring at pH 3.5

to 5.5, suggesting that As mobilization from the sediments to the overlying water phase is enhanced by acidification (Baker et al., 1983). Freshwater algae also produce methanearsonic and dimethylarsinic acids; however, there is some question at to whether biomethylation of As in freshwater is a widespread, common process.

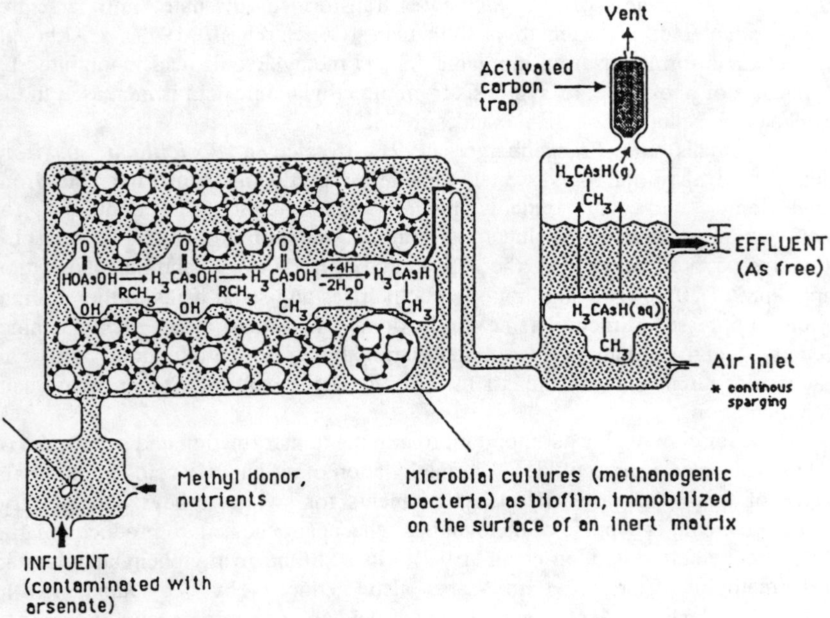

Fig. 11-4. Bioreactor design for treatment of As-contaminated water.

Fungal Methylation of Arsenic

In addition to bacteria, several fungi have demonstrated the ability to transform arsenic. It is well established that fungi are able to volatilize As as methylarsine compounds, which are derived from inorganic and organic As species. The volatilized As dissipates from the cells, effectively reducing the arsenic concentration to which the fungus is exposed. The importance of fungal metabolism of As dates back to the early 1800s where a number of poisoning incidents in Germany & England were caused by trimethylarsine gas (Challenger, 1945). The victims lived in musty rooms with a characteristic garlic-like odor. Molds growing on wallpaper decorated with arsenical pigments (Scheele's green and Schweinfürter green) produced the toxic trimethylarsine gas. Since then, several species of fungi have been identified that are able to volatilize As (Cox & Alexander, 1973a). The fungus, *Penicillium brevicaule* (*Scopulariopsis brevicaulis*) produces trimethylarsine when grown on bread crumbs containing either methylarsonic acid or dimethylarsinic acid.

In another study, three different fungal species, *Candida humicola, Gliocladium roseum*, and *Penicillium* sp. were capable of converting methylarsonic and dimethylarsinic acids to trimethylarsine (Cox & Alexander, 1973a). In addition, *C. humicola* used arsenate and arsenite as substrates to produce trimethylarsine. Cell-free homogenates of *C. humicola* transformed arsenate into arsenite, methylarsonic acid, and dimethylarsinic acid (Cullen et al., 1979). Although trimethylarsine formation from inorganic As and methylarsonic acid is inhibited by the presence of phosphate, its synthesis from dimethylarsinic acid is increased in the presence of phosphate (Cox & Alexander, 1973b).

Huysmans and Frankenberger (1991) isolated a *Penicillium* sp. from agricultural evaporation pond water capable of producing trimethylarsine from methylarsonic acid and dimethylarsinic acid. Optimum conditions for trimethylarsine production included: a minimal medium in which 100 mg L^{-1} methylarsonic acid served as the As source; pH 5 to 6; and an incubation temperature of 20°C. The addition of carbohydrates and sugar acids to the minimal medium suppressed trimethylarsine production. The amino acids, phenylalanine, isoleucine, and glutamine promoted trimethylarsine production with an enhancement ranging from 10.2- to 11.6-fold over the control without amino acid supplementation.

Soil fungi may play an important role in the transformation and movement of As chemicals used in agriculture. The methylation of arylarsonic acids is important because of their wide use as food supplements for swine, turkeys, and poultry. *Candida humicola* is capable of methylating benzenearsonic acid to produce volatile dimethylphenylarsine (Cullen et al., 1983). In addition, methylphenylarsinic acid and dimethylphenylarsine oxide are also reduced by *C. humicola* to dimethylphenylarsine. Arsanilic acid, which contains an amino group at the *para* position of phenylarsonic acid was not converted to a volatile arsine, but it has been reported that soils treated with arsanilic acid release their As component.

The adaptiveness of *C. humicola* in methylating As is evident by the fact that dilute solutions of the highly effective wood preserving fungicide, chromated copper arsenate (CCA), is depleted of arsenic through volatilization (Cullen et al., 1984). It has been demonstrated that a variety of soils have the potential to produce alkylarsines (Woolson, 1977). Soils amended with inorganic and methylated arsenic herbicides produce dimethylarsine and trimethylarsine (Baker et al., 1983; Hassler et al., 1984; Woolson, 1977; Woolson & Kearney, 1973). The organisms responsible for volatilization come from diverse environments suggesting that a number of species have the capacity to produce alkylarsines. For example, mixed communities of soil microorganisms produced dimethylarsine and trimethylarsine, which was trapped in bell jars over soil and lawns that were previously treated with methylarsenicals (Braman & Foreback, 1973).

BIOREMEDIATION OF SELENIUM CONTAMINATION

Selenium is classified as a metalloid having properties of both a metal and a nonmetal. It is markedly similar to sulfur in its chemistry, with its primary oxidation states being 6^+, 4^+, 0 and 2^-. Elemental Se (Se^0) is often associated with sulfur in compounds such as selenium sulfide (SeS_2) and polysulfides. Selenate (SeO_4^{2-}) and selenite (SeO_3^{2-}) are common ions in natural waters and soils. Reduced Se compounds include volatile methylated species such as dimethylselenide (DMSe, $[CH_3]_2Se$), dimethyldiselenide (DMDSe, $[CH_3]_2Se_2$) and dimethylselenone ($[CH_3]_2SeO_2$). Nonvolatile, reduced Se species include seleno-amino acids such as selenomethionine, selenocysteine, and selenocystine, in which Se is substitut for S. Inorganic reduced Se forms include mineral selenides and hydrogen selenide (H_2Se). Many instances of Se contamination in soils and natural waters have been noted, and have been directly linked to the deformation and death of higher organisms. Bioremediation strategies for Se are further along in terms of implementation than methods for Cr, Hg, and As, which are still largely in the experimental stage. Techniques discussed here will include vegetative uptake, bioreduction and volatilization of Se (Table 11-4).

Vegetative Uptake of Selenium

The accumulation and removal of Se from drainage water by vegetation is a viable bioremediation approach. Uptake is generally restricted to the soluble fraction of Se. In order to be effective, the vegetation must rely on the soil microflora to mineralize the complexed, insoluble organic Se into a soluble form. Vegetative uptake holds some promise because the Se is incorporated into the plant tissue, thus facilitating permanent removal upon harvesting of the plants. A possible drawback of this type of system could be that the seleniferous plant material would need special handling. The plant material could be used, however, to augment animal feed in Se-deficient areas.

The accumulation of Se in plants is highly variable and can favorably or adversely affect their growth, survival, and reproduction (Trelease, 1945; Eisler, 1985). The extent of Se accumulation depends on the plant, speciation of Se, pH, salinity, and the calcium carbonate content of the soil (Gissel-Nielsen et al., 1984; McNeal & Balistrieri, 1989). Soils of Hawaii may contain 6-15 mg Se^{4+} kg^{-1} yet seleniferous vegetation is not found, while plants growing in alkaline soils containing <1 mg Se^{6+} kg^{-1} may accumulate Se to potentially toxic concentrations (Lakin, 1973; U.S. Environmental Protection Agency, 1975). The type of soils which give rise to high concentrations of Se in plants -- hence potential animal toxicosis -- are usually alkaline and contain free calcium carbonate (Lakin, 1961; Rosenfeld & Beath, 1964). Seleniferous vegetation has been found in Australia, Ireland, Israel, South Africa, U.S.A. and Venezuela.

Based on their relationship to Se uptake, plants may be roughly divided into three groups: primary Se accumulators containing Se concentrations in excess of 100 mg kg^{-1} dry weight (e.g., many species of *Astragalus, Haplopappus, Machaeranthera* and *Stanleya*); secondary Se accumulators, which rarely concentrate more than 50 to 100 mg kg^{-1} (e.g., *Aster,* and some spcies of *Astragalus, Atriplex, Castilleja, Grindelia, Gutierrezia,* and *Mentzela*); and plants that usually

Table 11-4. Bioremediation of selenium.

Mechanism	Organisms	Description and Effectiveness	Source
Vegetative uptake	*Astralagus, Haplopappus, Machaeranthera* and *Stanleya*	Se accumulators	Rosenfield & Beath, 1964
Vegetative uptake	*Atriplex patula, Bassia hyssopifola, Melilotus indica* and *Salsola kali*	Use of native plant species at Kesterson Reservoir to remove Se in soils	Wu, 1994
Agroforestry	*Eucalyptus*	Propagation of salt-tolerant trees with uptake of drainage water high in Se	Cervinka, 1994
Deselenification	*Acremonium falciforme, Penicillium citrinum, Ulocladium tuberculatum, Alternaria alternata*	Volatilization through methylation in the conversion of soluble Se to a gaseous species	Karlson & Frankenberger, 1988; Thompson-Eagle et al., 1989
Bioreduction	*Clostridium, Desulfovibrio, Desulfotomaculum*	An anaerobic soil column reducing SeO_4^{2-} and SeO_3^{2-} to Se^0	Kerr-McGee Patent US 4519912 (1985) (see Mattison, 1992d)

(continued on next page)

Table 11-4. Continued.

Mechanism	Organisms	Description and Effectiveness	Source
Bioreduction	*Desulfovibrio desulfuricans* subsp. *aestuarri*	Reduction of SeO_4^{2-} and SeO_3^{2-} to Se^0	Zehr & Oremland, 1987
Bioreduction	Algae and SeO_4^{2-} reducing bacteria	Algae removed NO_3^- and water was treated with anaerobic bacteria to further reduce $[NO_3^-]$ by denitrification and SeO_4^{2-} to SeO_3^{2-}. SeO_3^{2-} was precipitated by adding $FeCl_3$	Gerhardt et al., 1991
Bioreduction	Anaerobic granular sludge inoculum	Methanol was added as a C source. NO_3^- must be consumed through denitrification before SeO_4^{2-} is reduced Final product is Se^0.	Owens, personal communication

(continued on next page)

Table 11-4. Continued.

Mechanism	Organisms	Description and Effectiveness	Source
Bioreduction	*Thauera selenatis*	Reduction of SeO_4^{2-} in the presence of NO_3^-. Macy's system uses two bioreactors in series. The first is an anaerobic sludge blanket reactor and the second is a fluidized bed reactor. In this system, large amounts of nitrate are required as a nitrogen source, while acetate is used as an organic reductant. Under these conditions, concentrations of both nitrate (250-380 mg L^{-1}) and selenate/selenite (350-380 mg L^{-1}) in the feed are decreased by 98% in the effluent.	Macy et al., 1993a

do not absorb more than 50 mg Se kg^{-1} (e.g., grasses) while growing in seleniferous soils (Rosenfeld & Beath, 1964; Shrift, 1973). Aquatic plants such as algae may absorb enormous quantities of Se from water.

Selenium can also be removed through agroforestry farming practices (Cervinka, 1994). Agroforestry operates on the principle that drainage water containing salts, including Se, are resources that have an economic value. Agroforestry provides farming practices that facilitate management of Se and other salts through alternative cropping systems which rely on salt-tolerant trees (e.g., *Eucalyptus*) and halophytic plants to reduce the volume of contaminated drainage water and thereby facilitate the removal of excess Se and other salts. Further work is being carried out to determine the tolerance of eucalyptus trees and shrubs to the extreme salinities and high concentrations of boron found in the drainage water and their ability to selectively take up SeO_4^{2-} in the presence of large quantities of other oxyanions.

Bioreduction of Selenium

Bioreduction has been extensively investigated as a technique for immobilizing Se in soils, and for removing it from contaminated water. In living systems, Se tends to be reduced rather than oxidized (Shamberger, 1983). Anaerobic cell extracts of *Micrococcus lactilyticus, Clostridium pasteurianum*, and *Desulfovibrio desulfuricans* have been shown to use molecular hydrogen to reduce SeO_3^{2-}, but not SeO_4^{2-}, to Se^{2-} (Woolfolk & Whiteley, 1962). However, Zehr and Oremland (1987) reported that washed cells of *D. desulfuricans* subsp. *aestuarii* were capable of reducing SeO_4^{2-} to Se^{2-}. Upon an acetate- SeO_4^{2-} enrichment of a seleniferous soil, Macy et al. (1989) were able to isolate two anaerobic bacteria: one which reduced SeO_4^{2-} to SeO_3^{2-} and the other reduced SeO_3^{2-} to Se^0. *Pseudomonas maltophilia* O-2, an isolate from a toxic waste site, reduced SeO_3^{2-} as well as salts of Hg, Pb, Cd, Ag, Au, Cr, and Sn to their elemental forms (Latinwo et al., 1990). Bacterial resistance to Hg^{2+}, SeO_3^{2-} and Pb^{2+} was found to be associated with a single large plasmid (Latinwo et al., 1990). Selenite is reduced to Se^0 by aerobically grown *Salmonella heidelberg* (McCready et al., 1966) as well as by cell-free extracts of *Streptococcus faecalis* and *S. faecium* (Tilton et al., 1967). Oremland et al., (1989) demonstrated *in situ* (anoxic sediments) and in pure culture, the anaerobic bacterial conversion of seleno-oxyanions to Se^0. Their results indicate that dissimilatory SeO_4^{2-} reduction to Se^0 may be a major biological transformation for reduction of Se oxyanions in anoxic sediments.

Microbial reduction of Se to the elemental (insoluble) form represents an important mechanism for immobilizing or removing seleniferous compounds from contaminated systems by filtration, sedimentation, bioadsorption and coprecipitation. A variety of processes (mostly bioreactors) for removing Se from wastewater have been developed using these or similar principles. An example of this technique is the Kerr-McGee soil column precipitation process which utilizes an anaerobic soil column colonized with *Clostridium* sp. capable of reducing soluble Se^{4+} and Se^{6+} to the elemental (Se^0) form (Mattison, 1992d). In field tests with 3-ft diameter soil columns, reduction occurred from over 1,000 mg L^{-1} Se to under 50 mg L^{-1} at a flow rate of 100-175 L day^{-1} m^{-2} of column area. The performance of this column was improved by the addition of a bacterial consortium including sulfate-reducing bacteria. This soil column technique has the benefits of low cost and maintenance, but the matrix requires regeneration at some point during its operation.

Oremland et al. (1991) has reported that selenate-respiring bacteria are ubiquitous in nature, functioning even in highly saline soils and sediments. However, if nitrate is present, SeO_4^{2-} reduction does not occur. Oremland holds a patent which describes a process involving a 2-stage reduction using algae in the aerobic first stage to deplete nitrate concentrations below 1 mM. The water is then fed to an anoxic bioreactor containing selenate respiring bacteria where SeO_4^{2-} is reduced to insoluble Se^0 and deposited within the biomass. Approximately 56 mg L^{-1} as SeO_4^{2-} solution was reduced to 0.14 mg L^{-1} in seven days of incubation.

The Oswald Process (Gerhardt et al., 1991) is similar to Oremland's technology in that the system uses algal growth to take up nitrate, lowering its concentration to under 10 mg L^{-1}. This is a critical step since, as mentioned, Se reduction is inhibited by high nitrate concentrations. The biomass suspension is then transferred to an anoxic unit where denitrification and selenate reduction are carried out by bacteria which are able to convert selenate to selenite rather than Se^0. After the anoxic reduction step, $FeCl_3$ is added to precipitate the inorganic selenite, thereby reducing soluble selenium levels from 200-400 mg L^{-1} to 7-12 mg L^{-1}.

An alternate approach, reported by Macy and coworkers (Macy et al., 1993a, Macy et al., 1993b; Macy, 1994) consists of a system using the newly identified bacterial species, *Thauera selenatis*, which is able to reduce both nitrate and selenate simultaneously. The presence of nitrate interferes with SeO_4^{2-} reduction in other bacteria because both SeO_4^{2-} and NO_3^- reduction are catalyzed by the same enzyme, nitrate reductase. To obtain complete reduction, Macy's system uses two bioreactors in series. The first is an anaerobic sludge blanket reactor and the second is a fluidized bed reactor. In this system, large amounts of nitrate are required as a nitrogen source, while acetate is used as an organic reductant. Under these conditions, concentrations of both nitrate (250-380 mg L^{-1}) and selenate/selenite (350-380 mg L^{-1}) in the feed are decreased by 98% in the effluent.

Lastly, the Owens Process (Owens, see Mattison, 1992d) uses a technology based upon an anaerobic mechanism to reduce SeO_4^{2-} to Se^0. Methanol is added as a carbon source. Selenium reduction will not take place until nitrate is consumed, but enough methanol is added to support both denitrification as well as Se reduction. After the nitrate is consumed, selenium reduction takes place stepwise, from selenate to selenite to insoluble Se^0.

Deselenification by Microbial Volatilization

Another naturally occuring process which can be accelerated, and thus useful in removal of Se from contaminated soils and water is volatilization through methylation. In natural systems, this is thought to be a protective mechanism used by microorganisms to detoxify their surrounding environment. The process permanently removes Se from soil and water (Fig. 11-5). The major groups of Se methylating organisms isolated from soils and sediments are bacteria and fungi (Abu-Eirreish et al., 1968; Barkes & Fleming, 1974; Challenger, 1945; Doran, 1982; Karlson & Frankenberger, 1988), while in water, bacteria are thought to play a more dominant role (Thompson-Eagle & Frankenberger, 1991). Deselenification of toxic Se species including, SeO_4^{2-} and various organoselenium compounds into a less toxic, volatile form (DMSe) is apparently a widespread transformation in seleniferous environments (Chau et al., 1976; Doran, 1982; Francis et al., 1974). Dimethylselenide is the major metabolite of Se volatilization (Barkes & Fleming, 1974; Doran & Alexander, 1975; Doran, 1982; Francis et al., 1974; Karlson & Frankenberger, 1988, 1989; Thompson-Eagle et al., 1989), although other Se compounds such as DMDSe, methaneselenone ($[CH_3]_2SeO_2$), methaneselenol (CH_3SeH), and dimethylselenenyl sulfide (CH_3SeSCH_3) may also be produced (Shrift, 1973; Chau et al., 1976; Reamer & Zoller, 1980; Chasteen et al., 1990). Characterization of this naturally occurring microbial Se transformation and removal process led to the discovery that volatilization can be accelerated to the point where there is a significant decline in the initial Se inventory within a matter of months.

Since Se volatilization permanently removes Se from contaminated soil and water, it has potential applications for the successful maintenance and operation of evaporation pond disposal as well as for drainage water itself. Research in our laboratory has shown that microorganisms naturally present in saline, alkaline drainage water, and soil methylate Se into DMSe. This naturally occurring transformation can be dramatically accelerated by the addition of specific amendments (e.g. pectin and proteins). The advantages of biostimulation of the indigenous microflora over bioaugmentation (adding preselected microorganisms) are twofold: first, survival of alien organisms is not required; and second, the indigenous microbiota have adapted to their surrounding environment, and are capable of multiplying rapidly and making a substantial contribution to deselenification in a relatively short period of time. When the Se levels decline below the acceptable level, amendments are no longer necessary and the microbial population quickly dies back to its original level.

Microbial volatilization of Se has been tested as a bioremediation approach to reduce toxic levels of Se at Kesterson Reservoir (Merced County, California). We conducted a field investigation with the goals of identifying the most effective practices for accelerating volatilization and to obtain information necessary for the determination of time and factors affecting this technology (Frankenberger, 1989; Frankenberger et al., 1990). A variety of nutrient additives and environmental conditions were evaluated in contaminated areas containing Se concentrations ranging from 10 to 750 mg kg^{-1}.

Seasonal variation of gaseous Se emission was evident with the highest emission recorded in the late spring and summer months. The volatilization rates were well correlated with soil temperature, with greater evolution of Se observed under warmer conditions. The greatest emission of gaseous Se with all treatments occurred at the initiation of the project when the Se inventory was high. As time proceeds, the Se inventory available for methylation appears to be decreasing with each season of warmer temperature. The most stimulatory soil treatment for Se volatilization (citrus peel, $[NH_4]_2SO_4$ plus Zn [added as a micronutrient]) yielded emission rates as high as 808 μg Se m^{-2} h^{-1} which was approximately 42-fold greater than the background level. The application of casein, a milk protein, also promoted volatilization of Se. Approximately 60% of the total Se inventory was removed with these treatments after 2 years (Frankenberger & Karlson, 1989).

Another field experiment was initiated at the Sumner Peck Ranch (near Fresno, CA) in October of 1987 to assess microbial volatilization from a sediment comprised mainly of clay (Frankenberger et al., 1990). The volatilization/temperature relationship observed at Kesterson held, but in this experiment, the most effective amendment was cattle manure. After 21 months of treatment, the application of water removed approximately 32% of the Se inventory, while treatment with cattle manure plus wheat straw removed 58%.

In summary, the parameters which enhance volatilization of Se in the field are an available C source, aeration, moisture, and high temperatures. Rototilling promotes volatilization as long as the soil is kept moist. Frequent tillage is needed to support the aerobic methylating organisms, enhance soil porosity (this facilitates the diffusion of the alkylselenide gas) and break the crust that may form as a result of sprinkler irrigation, or rainfall. Also recommended is irrigation, with wetting and drying cycles, in order to release the organically bound Se to the methylating organisms. Water should only be applied to moisten the upper few centimeters of soil otherwise water-soluble Se may be translocated out of the surface layer, thus making it unavailable for volatilization.

Fate of Selenium Gas

Once the Se is methylated into a gaseous species, it is released into the atmosphere, and diluted and dispersed by air currents away from the contaminated source. DMSe reacts with OH^- and NO_3^- radicals and O_3 within a few hours to yield products which are as yet uncharacterized (Atkinson et al., 1990). However, it is likely that these oxidized products may be scavenged by aerosols or sorbed onto particulates which have a relatively long residence time (7 to 9 days) in the atmosphere (Mosher & Duce, 1989), and can travel considerable distances (Mayland et al., 1989). The changes in speciation that occur in the atmosphere are as yet unknown due to the problems of collection and analysis. At the present time, the fate of DMSe in the atmosphere is subject to much debate among Se researchers. The controversy is based on the concentration of Se in remote regions of the world, DMSe emission rates, and interactions in relation to the sulfur cycle.

Recently an acute toxicity study by O. G. Raabe and M. A. Al-Bayati (see Frankenberger & Karlson, 1988) was conducted on the inhalation of DMSe by rats. This study consisted of 85 adult rats exposed to four concentrations of DMSe (0, 1607, 4499, and 8037 ppm) for 1 h. It was concluded that the effect of DMSe was one of irritation rather than injury. The half-life of DMSe appears to be very short and the compound was eliminated mainly via the lung. The data indicate that inhaled DMSe vapor is nontoxic to the rat at concentrations of up to 8037 ppm (Frankenberger & Karlson, 1988).

CONCLUSION

Bioremediation is a proven technology in the removal/detoxification of many environmental contaminants. There are a number of advantages to biotreatment systems including cost effectiveness, potential for *in situ* applications, and energy/reagent efficiency. The major emphasis of bioremediation has been on degradation of organic compounds, but other potential uses are currently being assessed. Application of bioremediation principles to heavy metal/metalloid contamination is largely in the experimental stage, but a growing body of evidence suggests that new biotechnologies for treating soils and water containing toxic forms of these elements may be in use in the near future.

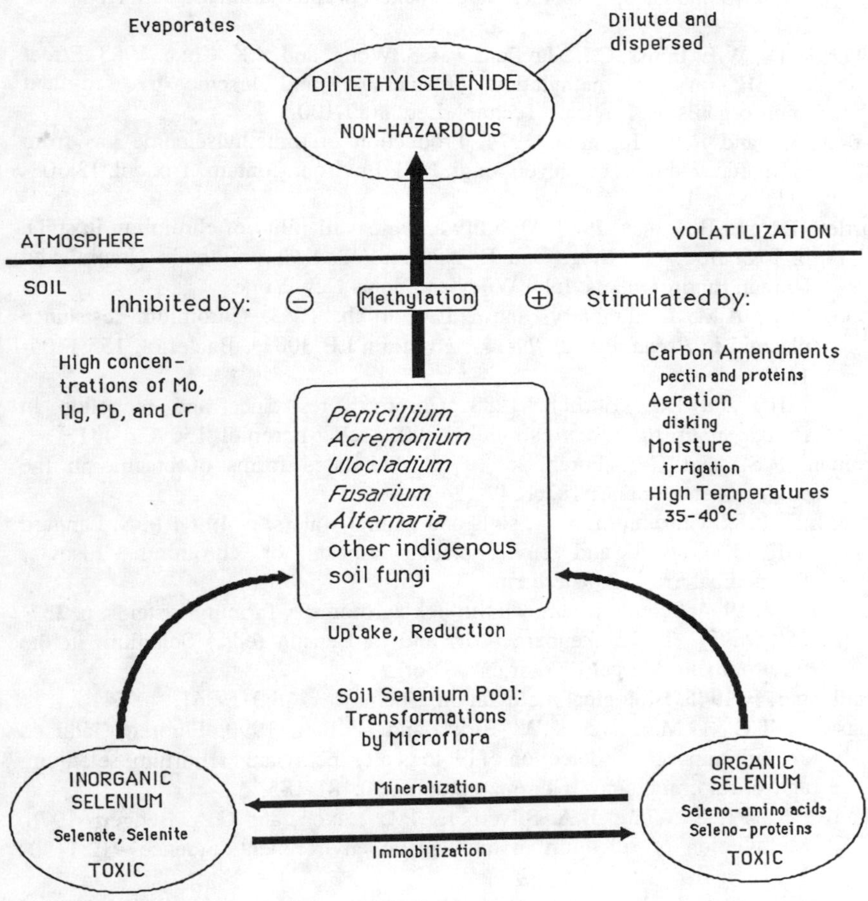

Fig. 11-5. Accelerated microbial volatilization used as a technology to dissipate selenium.

REFERENCES

Abu-Eirreish, G.M., E.I. Whitehead, and O.E. Olson. 1968. Evolution of volatile selenium from soils. Soil Sci. 106:415-420.

Apel, W.A., and C.E. Turick. 1991. Bioremediation of hexavalent chromium by bacterial reduction. p. 377-387. *In* R. W. Smith (ed.) Mineral Bioprocessing. The Minerals, Metals & Materials Society. Warrendale, PA.

Asher, C.J., and P.F. Reay. 1979. Arsenic uptake by barley seedlings. Aust. J. Plant Physiol. 6:459-466.

Atkinson, R., S.M. Aschmann, D. Hasagawa, E. T. Thompson-Eagle, and W. T. Frankenberger, Jr. 1990. Kinetics of the atmospherically-important reactions of dimethylselenide. Environ. Sci. Technol. 24:1326-1332.

Ayoub, G.M., and B.A. Sayigh. 1987. The effects and removal of chromium in *Chlamydomonas* sp. Toxicity Assessment: An International Quarterly 2:253-264.

Baker, M.D., W.E. Inniss, C.I. Mayfield, P.T.S. Wong, and Y.K. Chau. 1983. Effect of pH on the methylation of mercury and arsenic by sediment microorganisms. Environ. Technol. Lett. 4:89-100.

Barkes, L., and R.W. Fleming. 1974. Production of dimethylselenide gas from inorganic selenium by eleven fungi. Bull. Environ. Contam. Toxicol. 12:308-311.

Bartlett, R., and B. James. 1988. Mobility and bioavailability of chromium in soils. p. 276. *In* J. O. Nriagu and E. Nieboer (eds.) Chromium in Natural and Human Environments. John Wiley and Sons, New York.

Bopp, L.H., A.M. Chakrabarty, and H.L. Ehrlich. 1983. Chromium resistance plasmid in *Pseudomonas fluorescens* strain LB 300. J. Bacteriol. 155:1105-1109.

Bopp, L.H., and H.L. Ehrlich. 1988. Chromate resistance and reduction in *Pseudomonas fluorescens* strain LB 300. Arch. Microbiol. 150:426-431.

Braman, R.S., and C.C. Foreback. 1973. Methylated forms of arsenic in the environment. Science 182:1247-1249.

Cary, E.E. 1982. Chromium in air, soils and natural waters. p. 49-64 *In* S. Langard (ed.) Biological and environmental aspects of chromium. Elsevier Biomedical Press, Amsterdam.

Cervinka, V. 1994. Selenium removal through agroforestry farming systems. p. 237-250. *In* W. T. Frankenberger, Jr. and S. Benson (eds.) Selenium in the Environment. Marcel Dekker, New York.

Challenger, F. 1945. Biological methylation. Chem. Rev. 36:315-361.

Chasteen, T.C., G.M. Silver, J.W. Birks, and R. Fall. 1990. Fluorene induced chemiluminescence detection of biologically methylated tellurium, selenium, and sulfur compounds. Chromatographia 30:181-185.

Chau, Y.K., P.T.S. Wong, B.A. Silverberg, P.L. Luxon, and G.A. Bengert. 1976. Methylation of selenium in the aquatic environment. Science 912:1130-1131.

Chau, Y.K., and P.T.S. Wong. 1978. Occurrence of biological methylation of elements in the environment. Am. Chem. Soc. Symp. Ser. 82:39-53.

Cheng, C.N., and D.D. Focht. 1979. Production of arsine and methylarsines in soil and in culture. Appl. Environ. Microbiol. 38:494-498.

Coleman, R.N., and J.H. Paran. 1983. Accumulation of hexavalent chromium by selected bacteria. Environ. Technol. Lett. 4:149-156.

Compeau, G.C., and R. Bartha. 1985. Sulfate-reducing bacteria: Principal methylators of mercury in anoxic estuarine sediments. Appl. Environ. Microbiol. 50:498-502.

Cox, D.P., and M. Alexander. 1973a. Production of trimethylarsine gas from various arsenic compounds by three sewage fungi. Bull. Environ. Contam. Toxicol. 9:84-88.

Cox, D.P., and M. Alexander. 1973b. Effect of phosphate and other anions on trimethylarsine formation by *Candida humicola*. Appl. Microbiol. 25:408-413.

Cullen, W.R., B.C. McBride, and A.W. Pickett. 1979. The transformation of arsenicals by *Candida humicola*. Can. J. Microbiol. 25:1201-1205.

Cullen, W.R., A.E. Erdman, B.C. McBride, and A.W. Pickett. 1983. The identification of dimethylphenylarsine as a microbial metabolite using a simple method of chemofocusing. J. Microbiol. Method. 1:297-303.

Cullen, W.R., B.C. McBride, A.W. Pickett, and J. Regalinski. 1984. The wood preservative chromated copper arsenate is a substrate for trimethylarsine biosysthesis. Appl. Environ. Microbiol. 47:443-444.

DeFilippi, L.J., and F.S. Lupton. 1992. Bioremediation of soluble Cr(VI) using sulfate reducing bacteria. p. 138-141. Allied Signal Research: National R&D Conference on the Control of Hazardous Materials..

DeSimone, R.E., M.W. Penley, L. Charbonneau, S.G. Smith, J.K.M. Wood, H.A. O. Hill, J.M. Pratt, S. Ridsdale, and J.P. Williams. 1973. The kinetics and mechanism of cobalamin-dependent methyl and ethyl transfer to mercuric ion. Biochem. Biophys. Acta. 304:851-863.

Doran, J.W. 1982. Microorganisms and the biological cycling of selenium. Adv. Microbiol. Ecol. 6:1-32.

Doran, J.W., and M. Alexander. 1975. Microbial formation of dimethylselenide and the biological cycling of selenium. Abstracts of the annual meeting of the Amer. Society for Microbiology, New York, NY. N22, p. 188.

Eary, L.E., and D. Rai. 1988. Chromate removal from aqueous waste by reduction with ferrous iron. Environ. Sci. Technol. 22:972-977.

Ehrlich, H.L. 1990. Geomicrobiology. p. 267-282. Second Edition. Marcel Dekker, Inc., New York.

Eisler, R. 1985. Selenium hazard to fish and wildlife. Serv. Biol. Rep. 85(1.5). U. S. Gov. Print Office. Washington D.C.

Eliseeva, G.S., T.M. Klyushnikova, T.P. Kasatkina, and N.S. Serpokrylov. 1991. Reduction of Cr(VI) by microorganisms in media with inedible plant raw material. Khimiya i Tekhnologiya Vody. 13:72-75.

Ferguson, J.F., and J. Gavis. 1972. A review of the arsenic cycle in natural waters. Water Res. 6:1259-1274.

Filip, D., S.T. Peters, V.D. Adams, and E.J. Middlebrooks. 1979. Residual heavy metal removal by an algae-intermittent sand filtration system. Water Res. 13:305-313.

Francis, A.J., J.M. Duxbury, and M. Alexander. 1974. Evolution of dimethylselenide from soils. Appl. Microbiol. 24:424-429.

Frankenberger, W.T., Jr., and U. Karlson. 1988. Dissipation of soil selenium by microbial volatilization at Kesterson Reservoir. U.S. Dept. of Interior, Bureau of Reclamation, December. Contract No. 7-FC-20-05240.

Frankenberger, W.T., Jr., and U. Karlson. 1989. Dissipation of soil selenium by microbial volatilization at Kesterson Reservoir. U.S. Dept. of Interior, Bureau of Reclamation, December. Contract No. 7-FC-20-05240.

Frankenberger, W.T., Jr., U. Karlson, and K.E. Longley. 1990. Microbial volatilization of selenium from sediments of agricultural evaporation ponds. California Water Resources Control Board under interagency agreement No. 7-125-250-1.

Gerhardt, M.B., F.B. Green, R.D. Newman, T.J. Lundquist, R.B. Tresan, and W.J. Oswald. 1991. Removal of selenium using a novel algal bacterial process. Res. J. Water Pollut. Control Fed. 63:799-805.

Gissel-Nielson, G., U.C. Gupta, M. Lamand, and T. Westermarck. 1984. Selenium in soils and plants and its importance in livestock and human nutrition. Adv. Agron. 37:397-460.

Glombitza, F., U. Iske, and C. Gwenner. 1984a. Biosorption of mercury by microorganisms. Acta Biotechnol. 4:281-284.

Glombitza, F., U. Iske, and C. Gwenner. 1984b. Cultivation of mercury-resistant microorganisms. Acta Biotechnol. 4:285-288.

Hassler, R. A., D.A. Klein, R.R. Meglen. 1984. Microbial contribution to soluble and volatile arsenic dynamics in retorted oil shale. J. Environ. Qual. 13:466-470.

Horitsu, H., S. Futo, and Y. Miyazawa. 1987. Enzymatic reduction of hexavalent chromium by hexavalent chromium tolerant *Pseudomonas ambigua* G-1. Agric. Biol. Chem. 51:2417-2420.

Huysmans, K.D., and W.T. Frankenberger, Jr. 1991. Evolution of trimethylarsine by a *Penicillium* sp. isolated from agricultural evaporation pond water. Sci. Total Environ. 105:13-28.

Ishibashi, Y., C. Cervantes, and S. Silver. 1990. Chromium reduction in *Pseudomonas putida*. Appl. Environ. Microbiol. 56:2268-2270.

Karlson, U., and W. T. Frankenberger, Jr. 1988. Effects of carbon and trace element addition on alkylselenide production by soil. Soil Sci. Soc. Am. J. 52:1640-1644.

Karlson, U., and W. T. Frankenberger, Jr. 1989. Accelerated rates of selenium volatilization from California soils. Soil Sci. Soc. Amer. J. 53:749-753.

Komori, K., A. Rivas, K. Toda, and H. Ohtake. 1990. A method for removal of toxic chromium using dialysis-sac cultures of a chromate-reducing strain of *Enterobacter cloacae*. Appl. Microbiol. Biotechnol. 33:117-119.

Kvasnikov, N. S., N. S. Serpokrylov, T. M. Klyushnikova, T. P. Kasatkina, I. M. Zukov, and L. L. Tokareva. 1986. Optimization of a nutrient medium for *Aeromonas dechromatica* reducing Cr(VI). Khimiya i Tekhnologiya Vody. 8:64-66.

Kvasnikov, N. S., N. S. Serpokrylov, T. M. Klyushnikova, T. P. Kasatkina, I. M. Zukov, and L. L. Tokareva. 1987. Reduction of Cr(VI) by a culture of *Aeromonas dechromatica* KS-11 in the presence of certain heavy metals. Khimiya i Tekhnologiya Vody. 9:159-162.

Lakin, H. W. 1961. Geochemistry of selenium in relation to agriculture. p. 3-12. *In* M. S. Anderson et al. (eds.) Selenium in Agriculture. USDA handbook 200. U. S. Gov. Print. Office, Washington D.C.

Lakin, H. W. 1973. Selenium in our environment. Adv. Chem. 123:96-111.

Landner, L. 1971. Biochemical model for the biological methylation of mercury suggested from methylation studies *in vivo* with *Neurospora crassa*. Nature. 230:452-454.

Latinwo, L., L. Overbye, J. Caguiat, R. Blake, N. Revis, and J. H. Jackson. 1990. Reduction of selenium, mercury, and lead salts encoded by a single plasmid in a *Pseudomonas* strain from a toxic waste site. Abstract. 90th Annual Meeting of the American Society for Microbiology. Anaheim, CA., May, 1990. Q-244. p. 329.

Losi, M.E., C. Amrhein, and W.T. Frankenberger, Jr. 1994a. Bioremediation of chromate contaminated groundwater by reduction and precipitation in surface soils. J. Environ. Qual. 23:1141-1150.

Losi, M.E., C. Amrhein, and W.T. Frankenberger, Jr. 1994b. Factors affecting chemical and biological reduction of hexavalent chromium in soil. Environ. Toxicol. and Chem. 13:1727-1735.

Mackenzie, F.T., R.J. Lantzy, and V. Paterson. 1979. Global trace metal cycles and predictions. J. Int. Assoc. Math. Geol. 11:99-142.

Macy, J.M. 1994. Biochemistry of selenium metabolism by *Thauera selenatis* gen. nov. sp. nov. and use of the organism for bioremediation of selenium oxyanions in San Joaquin drainage waters. p. 421-444. *In* W. T. Frankenberger Jr. and S. Benson (eds.) Selenium in the Environment. Marcel Dekker. New York, NY.

Macy, J. M., S. Lawson, and H. DeMoll-Decker. 1993a. Bioremediation of selenium oxyanions in San Joaquin drainage water using *Thauera selenatis* in a biological reactor system. Appl. Microbiol. Technol. 40:594-598.

Macy, J.M., T.A. Michel, and D.G. Kirsch. 1989. Selenium reduction by a *Pseudomonas* species: A new mode of anaerobic respiration. FEMS Microbiol. Letts. 61:195-198.

Macy, J.M., S. Rech, G. Auling, M. Dorsch, E. Stackbrant, and L. Sly. 1993b. *Thauera selenatis* gen. nov. sp. nov., a member of the beta-subclass *Proteobacteria* with a novel type of anaerobic respiration. Int. J. Syst. Bacteriol. 43:135-142.

Marquez, A.M., M.J. Espuny, F. Congregado, and M.D. Simon-Pujol. 1982. Accumulation of chromium by *Pseudomonas aeruginosa*. Microbiol. Letters 21:143-147.

Mattison, P.L. 1992a. Bioremediation of metals--putting it to work. Ch. 3, Chromium. Cognis, Inc., Santa Rosa, CA.

Mattison, P.L. 1992b. Bioremediation of metals--putting it to work. Ch. 4, Mercury. Cognis, Inc., Santa Rosa, CA.

Mattison, P.L. 1992c. Bioremediation of metals--putting it to work. Ch. 6, Redox treatment of other metals. Cognis, Inc., Santa Rosa, CA.

Mattison, P.L. 1992d. Bioremediation of metals--putting it to work. Ch. 5, Selenium. Cognis, Inc., Santa Rosa, CA.

Mayland, H. F., L. F. James, K. E. Panter, and J. L. Sondregger. 1989. Selenium in seleniferous environments. pp. 15-50. *In* L. W. Jacobs (ed.) Selenium in

Agriculture and the Environment. SSSA Special Publ. No. 23. ASA. Madison, WI.

McBride, B.C., and R.S. Wolfe. 1971. Biosynthesis of dimethylarsine by *Methanobacterium*. Biochem. 10:4312-4317.

McBride, B.C., H. Merilees, W.R. Cullen, and W. Pickett. 1978. Anaerobic and aerobic alkylation of arsenic. *In* F. E. Brickman, and J. M. Bellama (eds.) Organometals and Organometalloids Occurrence and Fate in the Environment. Am. Chem. Soc. Symp. Ser. 82:94-115.

McCready, R.G., L., J.N. Campbell, and J.I. Payne. 1966. Selenite reduction by *Salmonella heidelberg*. Can. J. Microbiol. 1:703-714.

McGrath, S.P., and S. Smith. 1990. Chromium and nickel. p. 137. *In* B. J. Alloway (ed.) Heavy Metals in Soils. John Wiley and Sons, New York.

McNeal, J.M., and L.S. Balistrieri. 1989. Geochemistry and occurrance of selenium: An overview. p 1-15. *In* L. W. Jacobs (ed.) Selenium in Agriculture and the Environment. SSSA Special Publ. 23. ASA. Madison, WI.

Mosher, B.W., and R. A. Duce. 1989. The Atmosphere. pp. 295-325. *In*: M. Ihnat (ed.) Occurrance and Distribution of Selenium. CRC Press, Boca Raton, FL.

Ohtake, H., E. Fujii, and K. Toda. 1990. Reduction of toxic chromate in an industrial effluent by use of a chromate-reducing strain of *Enterobacter cloacae*. Environ. Technol. 11:663-668.

Oremland, R. S., J. T. Hollibaugh, A. S. Maest, T. S. Presser, L. G. Miller, and C. W. Culbertson. 1989. Selenate reduction to elemental selenium by anaerobic bacteria in sediments and culture: biogeochemical significance of a novel, sulfate-independent respiration. Appl. Environ. Microbiol. 55:2333-2343.

Oremland, R.S., N.A. Steinberg, T.S. Presser, and L.G. Miller. 1991. *In situ* bacterial selenate reduction in the agricultural drainage systems of western Nevada. Appl. Environ. Microbiol. 57:615-617.

Osborne, F.H. and H.L. Ehrlich. 1976. Oxidation of arsenite by a soil isolate of *Alcaligenes*. J. Appl. Bacteriol. 41:295-305.

Palmer, C.D., W. Fish, and J.F. Kelly. 1988. Inorganic contaminants: Recognizing the problem. p. 555-579. Proceedings of the Second National Outdoor Action Conference on Aquifer Restoration, Groundwater Monitoring, and Geophysical Methods.

Phillips, S.E. and M.L. Taylor. 1976. Oxidation of arsenite to arsenate by *Alcaligenes faecalis*. Appl. Environ. Microbiol. 32:392-399.

Reamer, D.C. and W.H. Zoller. 1980. Selenium biomethylation products from soil and sewage sludge. Science 208:500-502.

Roda, I. G., and G.F. Smirnova. 1989. Biochemical treatment of chromium-containing wastewater. Khimiya i Tekhnologiya Vody. 11:169-172.

Romanenko, V.I., S.I. Kusnetsov, and V. N. Koren'kov. 1976. Koren'kov method for biological purification of wastewater. USSR Patent SU 521,234.

Romanenko, V.I. and V.N. Koren'kov. 1977. A pure culture of bacteria utilizing chromates and bichromates as hydrogen acceptors in growth under anaerobic conditions. Institute of Biology of Inland Water, Academy of Sciences of the USSR. Translated from Mikrobiologiya. 46:414-417.

Rosenfeld, I. and O.A. Beath. 1964. Selenium. Geobotany, Biochemistry, Toxicity, and Nutrition. Academic Press, New York.

Salisbury, F.B. and C. Ross. 1969. Plant Physiology. p. 164. Wadsworth Publ. Co . Belmont, CA.

Shrift, A. 1973. Metabolism of selenium by plants and microorganisms. pp. 763-814. *In*: D.L. Klayman and W.H.H. Gunther, (eds.) Organic Selenium Compounds: Their Chemistry and Biology. John Wiley & Sons. New York.

Shamberger, R.J. 1983. Biochemistry of Selenium. Plenum Press. New York.

Shimada, K. and K. Matsushima. 1983. Isolation of potassium chromate-resistant bacterium and reduction of hexavalent chromium by the bacterium. Bull. Fac. Agric. Mie. Univ. 67:101-106.

Smillie, R.H., K. Hunter, and M. Loutit. 1980. Reduction of chromium(VI) by bacterially produced hydrogen sulfides. Water Res. 15:1351-1354.

Thompson-Eagle, E.T., W.T. Frankenberger, Jr., and U. Karlson. 1989. Volatilization of selenium by *Alternaria alternata*. Appl. Environ. Microbiol. 55:1406-1413.

Thompson-Eagle, E.T. and W.T. Frankenberger, Jr. 1991. Selenium biomethylation in an alkaline, saline environment. Water Res. 25:231-240.

Tilton, R. C., H.B. Gunner, and W. Litsky. 1967. Physiology of selenite reduction by enterococci. I. Influence of environmental variables. Can. J. Microbiol. 13:1175-1185.

Trelease, S. F. 1945. Selenium in soils, plants, and animals. Soil Sci. 60:125-130.

U.S. Environmental Protection Agency. 1975. Federal Register. vol. 40, No 248, pp. 59566-59588. Dec. 24. National Interim Primary Drinking Water Standards Administrator: Russel E. Train.

Wakao, N., H. Koyatsu, Y. Komai, H. Shimokawara, Y. Sakurai, and H. Shiota. 1988. Microbial oxidation of arsenite and occurrance of arsenite-oxidizing bacteria in acid mine water from a sulfur-pyrite mine. Geomicrobiol. J. 6:11-24.

Wilkinson, S.C., K.H. Goulding, P.K. Robinson. 1989. Mercury accumulation and volatilization in immobilized algal cell systems. Biotechnol. Lett. 11:861-864.

Wood, J.M., F.S. Kennedy, and C.G. Rosen. 1968. Synthesis of methyl-mercury compounds by extracts of a methanogenic bacterium. Nature. 220:173-174.

Wood, J.M. and H-K. Wang. 1983. Microbial resistance to heavy metals. Environ. Sci. Technol. 17:582A-590A.

Woolfolk, C.A., and H.R. Whiteley. 1962. Reduction of inorganic compounds with molecular hydrogen by *Micrococcus lactilyticus*. J. Bacteriol. 84:647-658.

Woolson, E.A. 1977. Generation of alkylarsines from soil. Weed Sci. 25:412-416.

Woolson, E. A. and P. C. Kearney. 1973. Persistence and reactions of [^{14}C]-cacodylic acid in soils. Environ. Sci. Technol. 7:47-50.

Wu, L. 1994. Selenium accumulation and colonization of plants in soils with elevated concentrations of selenium and salinity. p. 279-325. *In* W. T. Frankenberger Jr. and S. Benson (eds.) Selenium in the Environment. Marcel Dekker, New York, NY.

Yamada, M., and K. Tonomura. 1972a. Formation of methylmercury compounds from inorganic mercury by *Clostridium cochlearium*. J. Ferment. Technol. 50:159-166.

Yamada, M., and K. Tonomura. 1972b. Further studies on formation of methylmercury compounds from inorganic mercury by *Clostridium cochlearium* T-2. J. Ferment. Technol. 50:893-900.

Zehr, J.P., and R.S. Oremland. 1987. Reduction of selenate to selenide by sulfate-respiring bacteria: experiments with cell suspensions and estuarine sediments. Appl. Environ. Microbiol. 53:1365-1369.

Zumriye, A., Y. Sag, and T. Kutsal. 1990. A comparative study of the adsorption of Cr(VI) ions to *C. vulgaris* and *Z. ramigera*. Environ. Technol. 11:33-40.

12 Bioremediation of Valdez and Prall's Island Oil Spills

S.M. Hinton
Exxon Corporate Research Center
Annandale, New Jersey

This is a brief account of the development of the strategies for the bioremediation of the Valdez and Arthur Kill oil spills. I hope to make one important point that is the combination of variables, such as the environment and type of oil, creates a unique character for every oil spill. Therefore, experience on the implementation of bioremediation for one marine oil spill may not give the practitioner as much comfort as one may expect in addressing the next crisis. The magnitude of the Exxon Valdez crude oil spill, and the largest single application of bioremediation technology created an historical event. On the other hand, the Arthur Kill home fuel oil spill was by comparison small, but small petroleum distillate spills may represent the greatest challenge in the future because they typify the vast majority of oil spills. As a microbiologist the most intriguing aspect of bioremediation of oil spills is the process in determining the treatment. Some of the considerations that will be discussed in the development of a bioremediation (nutrient application) strategy for marine oil spills are logistics of widescale application, commercial availability of the fertilizer, method of fertilizer application, and fertilizer performance in the delivery of nutrients. Other important considerations that will not be addressed are toxicity of the fertilizer, eutrophication, and monitoring.

THE EXXON VALDEZ OIL SPILL

At 12:04 a.m. on March 24, 1989, the Exxon Valdez oil tanker went aground on Bligh Reef in Prince William Sound (PWS), Alaska (approximately 25 miles from the port of Valdez). The grounding caused ruptures in eight of the eleven cargo tanks (258,000 barrels or 10.8 million gallons) of which approximately one-fifth or 1.2 million barrels of crude oil was released in less than five hours. Early the next day the released crude oil covered an area of about 6 square miles, and by mid-day on March 27, gale force winds had driven the oil slick 37 miles to the shorelines

 Bioremediation: Science and Applications. SSSA Special Publication 43.

of Knight, Naked and Smith Islands. Despite attempts to contain the spill, tidal currents and winds directed a significant portion of the oil to float ashore. Approximately 350 miles of the more than 3,000 miles of shoreline in PWS, and 750 miles of more than 6,000 miles of shoreline in the Gulf of Alaska, in other terms, approximately 15% of the total shoreline, was impacted by oil to some degree (see Fig. 12-1).

Oil spill response technology is divided into three broad areas; oil containment and recovery, shoreline treatment, and marine operations and logistics. Free-oil containment and recovery operations began within hours after the Valdez oil spill. The first priority was to protect shorelines and key environmentally sensitive areas (i.e., fish hatcheries). using booms and skimmers. Until June 40 to 50 skimmers were deployed, after June there was little free oil remaining on the open water. Another top priority is tanker lightering which began on the day of the accident and ended after over 1 million barrels of the Exxon Valdez oil cargo was off loaded onto three separate vessels.

With an oil spill of this magnitude a large effort is devoted to assessing what portion of the shoreline was impacted and to what extent. In 1989 assessments determined that 357 miles of shorelines within PWS and 732 miles in the Gulf of Alaska needed cleanup treatment. In general, the shoreline in PWS was more heavily impacted than the shoreline in the Gulf. All cleanup techniques were thoroughly reviewed by committees from Exxon, landowners, federal, and state agencies which provided recommendations and technical support to the Federal On Scene Coordinator, the Commander of the Coast Guard, who directed the overall cleanup response. The removal of the heaviest concentrations of oil was approved to minimize contact with wildlife and to prevent the oil from weathering to a more viscous state that is resistant to natural degradation processes. In PWS alone the 1989 treatment operations were conducted by six Task Forces, each consisting of 500-700 personnel and 100 marine vessels. The primary shoreline treatment was manual removal of oiled debris and sediment, and the mechanical treatment with warm or cold water washing. The cleanup generated large quantities of solid and liquid oil waste. More than 25,000 tons of solid waste was transported to Oregon in 1989 for disposal in a toxic landfill, while the recovered oil was sent to an oil refinery.

The oil recovery and shoreline treatment of 1989 were dependent upon marine and logistics support because the cleanup operations were conducted from vessels due to the lack of land access. As many as 1400 vessels, 84 aircraft and more than 11,000 persons were involved in the 1989 cleanup operations. To give you an appreciation of the extent of logistic support required for the cleanup operation almost all equipment, food, and personnel had to be brought in from outside Alaska and delivered offshore over distances of hundreds of miles.

THE 1989 BIOREMEDIATION VALDEZ OIL SPILL RESPONSE

The bioremediation response to the Valdez Oil Spill had its public beginning at a conference hosted by the EPA's Office of Research and Development on April 17-18, 1989. The invitees were international scientific experts on hydrocarbon microbiology. The conclusion of the meeting was to recommend a field demonstration to evaluate the use of nutrients containing nitrogen and phosphorous

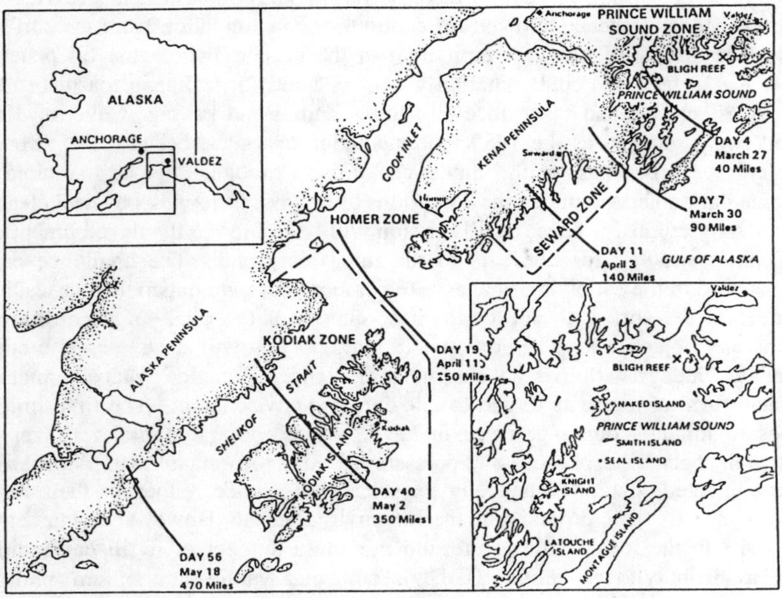

Fig. 12-1. Chronology of oil impact from the Exxon Valdez.

to accelerate biodegradation. The EPA and Exxon on June 2, 1989, signed a cooperative agreement under the Federal Technology Transfer Act of 1986 to conduct field and laboratory testing of bioremediation for its application in PWS.

The major objectives of the Exxon/EPA field project was to determine the extent of natural biodegradation of oil, to determine if the rates of oil biodegradation could be enhanced by the addition of nutrients, to develop an application strategy for nutrients to an inter-tidal zone, to establish methods to monitor ecological effects of nutrient addition, and to develop a model of the movement of nutrients in beach sediment. The urgency or crisis of an oil spill response meant that these objectives were being developed simultaneously under many constraints; the main one being time.

In 1989, bioremediation was not intended as a primary treatment, but would be used as a secondary or tertiary treatment to remove any residual oil remaining after physical removal of gross oil. Therefore, any bioremediation application had to be integrated into the overall cleanup operation. Other general restrictions on the use of bioremediation were that it was not to be used on vertical surfaces, mud flat nor marshes, and additional restrictions were placed on application near wildlife habitat and anadromous streams. The general environment in PWS is a forge type estuary in which the vast majority of the high-wave energy shoreline is vertical rock (approximately 72%) interspersed with bays and beaches of boulder and cobble (approximately 24%), gravel (approximately 3.5%), and fine grain sand and mud (approximately 0.5%). Therefore, approximately 27% of the PWS beaches with small boulders (defined in the regulation guidelines as less than the size of a chair), cobble and gravel were deemed potential targets for the application of bioremediation. Upon inspection of these flat pebble/cobble beaches revealed that the porosity of the beach allowed for continuous contamination from the surface to subsurface depths of two feet or more. In the beaches where the oil penetrated below the surface, the contamination was considered for technical reasons to be in two zones, surface and subsurface oil. Snug Harbor and Passage Cove, on Knight Island, were selected as the 1989 bioremediation test sites, because they were flat cobble/gravel beaches with moderate oiling thought to be suitable for bioremediation and simulated the conditions of beaches following physical cleanup.

As previously stated, the discussion will be limited to the development of an application strategy for nutrients to the inter-tidal zone. The challenge can be summarized in the following images. Remember your own experience in wading or standing in the surf at the beach, and the energy of the tidal surges pushing and pulling any object in it. The hope to add nutrients that will survive long enough to stimulate biodegradation is an enormous challenge. Imagine what chemical and physical form of nutrients to add to this dynamic environment. The first thing that comes to mind is our experience in lawn, garden, or agricultural fertilizers that contain nitrogen, phosphorous and potassium. Numerous manufacturers of fertilizer were contacted, but none had any experience or advice, either in their fate, or performance of their products in marine environments. However; manufacturers, and visits to the local garden store, did provide a number of fertilizer candidates, such as isobutylidene diurea (IBDU), and magnesium ammonium phosphate (MagAmp). Several of these fertilizers were tested in the EPA Cincinnati laboratory under simulated tidal conditions to assess their nutrient release rates (US EPA, 1990). Generally, the rate of release was dependent upon surface area with the exception of MagAmp. MagAmps slow linear rate of release and physical character, which allowed it to be compressed like concrete into any size or shape, made it a leading candidate. But, in the context of a large-scale application MagAmp did not meet other requirements. The estimated amount of MagAmp required to treat the shoreline in PWS was greater than the world supply. The scale of application was one of the main constraints in developing a fertilizer application strategy for the Valdez oil spill.

Another factor that had a major impact on the considerations for large-scale application of bioremediation was that the chemical, and physical form of the fertilizer, dictated the application strategy. As an example, a liquid fertilizer would be sprayed onto or into the beach sediment while fertilizers in solid form would have to be deposited on the surface. This may seem obvious but let's examine how it impacted the implementation of large-scale bioremediation of PWS. One of the Exxon engineers suggested that a liquid fertilizer could be applied to the surface of the beach at low tide using a "sprinkler system." It was proposed that sea water could be pumped from the Sound, mixed with water soluble fertilizer, and sprayed over the contaminated area during low tide. This idea has merit in that the rate and frequency of nutrient application could be adjusted, and the amount of water delivered would determine the depth of nutrient penetration into the sediment. This strategy was tested in the 1989 Exxon/EPA bioremediation test plots in Passage Cove and showed promise. However, the flat sand/gravel shorelines deemed appropriate for bioremediation were discontinuous segments of beach stretching over hundreds of miles located in very remote areas. Therefore, accessibility, distances, cost of materials, daily maintenance, danger to workers (i.e., bears), and the lack of time for implementation made this strategy for large-scale application in 1989 impractical.

The magnitude of the problem and the time constraints were the main driving forces that dictated testing the feasibility of a single application strategy. A single application of nutrients meant that the fertilizer and the method of application be designed to survive tidal surges and slowly over weeks or months release forms of nitrogen and phosphorous. There were two candidates that meet these requirements, one, slow-release solid fertilizers, and the second, a microemulsion formulation, Inipol EAP22, manufactured by a subsidiary of Elf Aquitaine (CECA S. A., 92062 Paris La Defense, France).

The time-release solid fertilizers are developed as residential and agricultural fertilizers. It was thought for survival in the tidal zone that the physical form of the fertilizer have the approximate density of the beach gravel (specific gravity of 1.8). The first solid fertilizer with the desired physical, time-release character, and commercial availability that was field tested is called Woodace briquettes. The briquettes have a N:P:K ratio of 14:3:3 which refers to the weight percent nitrogen, ortho-phosphate, and potassium oxide, respectively. Field studies showed that briquettes broadcast did not retain their position on the beach surface therefore, the briquettes were bagged in a fish net which was secured to the beach. The application strategy for the briquettes was as point sources of fertilizer which depended upon the tidal action to distribute the nutrients. Unfortunately, the nutrients were not distributed laterally across the beach and instead streamed directly toward the water line with each ebb tide. The problem with this fertilizer was the physical form dictated an application strategy that did not uniformly nutriate the beach. The application problem was solved by broadcasting granules of fertilizer over the entire surface of the beach. The granular fertilizer consists of inorganic nutrients (N:P:K ratio of 28:8:0) that was a custom blend specific for the Valdez bioremediation application, hence the name "Customblen." Commercially the fertilizer is called Osmocote, manufactured by Grace Sierra Chemicals (Milpitas, CA 95035), which describes the slow-release mechanism. The inorganic nutrients are coated with a derivative of vegetable oil. When the oil coat is penetrated by water, the inorganic salts are solubilized creating the osmotic pressure required to slowly release the nutrients.

The other fertilizer tested in Snug Harbor in 1989, and applied wide-scale was the formulation called Inipol EAP22. A little history of the development of this material may be appropriate because of the misconception many have about the formulation and the proposed mechanism. In 1978, the Amoco Cadez oil spill impacted the coast of France. Subsequently, the French oil company, Elf Aquitaine, developed several dispersants and a biodegradation accelerating agent as tools for future oil spills (Ladousse & Tramier, 1991). Inipol EAP22 (Elf Aquitaine Product 22), has been referred to as an oleophilic or oil-adhering fertilizer which is a misnomer. Inipol EAP22 is a microemulsion, a stable water-in-oil formulation, which is important for effective application to an oil spill. The oil on the shoreline is approximately a 50:50 mixture of oil and water, called mousse. For chemical/physical reasons a microemulsion mixes with mousse more effectively than a comparable oleophilic or hydrophobic substance. In addition, the continuous or oil phase of the formulation, oleic acid containing trilaureth(4)-phosphate (the source of phosphate), is an interfacially active material. This is distinct from oil-adhering. Oleophilic or oil-adhering substances "stick" or mix throughout the oil. Interfacially active material functions at an interface. In this case, the oil-water interface is the site of microbial oil-degrading activity. The nitrogen source is urea dissolved in the aqueous phase which yields a formulation with a N:P:K ratio of 7.3:0.8:0. The ratio and amount of nitrogen and phosphorous nutrients are constrained by the stability of the microemulsion, and it is not an optimization for biodegradation. In addition, as stated in the patent (US 4,460,692), the formulation should contain "a directly-assimilable source of carbon", such as a fatty acid. The purpose of the organic carbon source was to "permit rapid development of the microorganisms necessary for the degradation .... crude petroleum." In other words, it is proposed that an easily-degradable organic carbon source is required to accelerate the biodegradation of hydrocarbons. The proposed mechanism of Inipol EAP22 is that the microemulsion readily mixes with weathered crude oil which destabilizes the microemulsion releasing the urea nitrogen source. The surface-active organic material is a readily degradable carbon and energy source serving to increase the number of hydrcarbon-degrading bacteria. When the organic carbon is exhausted, the increased number of hydrocarbon-degrading bacteria support an increased rate of petroleum biodegradation. This biological agent was originally tested and developed for treatment of oil spills at sea, but the application to oil on open water is no longer recommended. The reason, in part, is if the treatment is applied to an oil slick at sea, and the oil impacts a shoreline due to the time requirements for biodegradation, it may be misjudged by the public as ineffective.

The Inipol formulation has the consistency and appearance of honey at room temperature. The physical character of the material requires it to be heated to a temperature of 90-100°F to reduce viscosity for surface application with airless sprayers. The application rate of Inipol was not dictated by the amount of oiling as one might expect. Instead, the manufacturer recommends the microemulsion be applied as a thin coat (an application of 0.75 gallons per 100 square feet which is approximately the thickness of four coats of paint) to surface oil. The reason for this application strategy is governed by the chemical and physical form of the formulation. The continuous phase of the microemulsion is oleic acid which is a liquid when applied, but changes phase to a waxy substance at 15°C, the average water (and beach) temperature during the summer. Therefore, the microemulsion would gel before penetrating into the cool subsurface. In addition, an excessive application would cause the material to pool which would create its own problem for biodegradation. Inipols limitation to surface oil application is one factor in the overall wide-scale bioremediation guidelines recommended by the EPA. The 1989 application guidelines stated that Inipol was to be applied to surface oil and Customblen was to be applied to beaches with subsurface oil (at an application rate of 0.0033 lbs ft^{-2}). In the event there was surface and subsurface oil, both treatments were to be applied. Overall, in 1989, approximately 68,000 gallons of Inipol EAP 22 and 20,000 pounds of Customblen were applied to 73 miles of shoreline in PWS.

ARTHUR KILL HOME HEATING OIL SPILL

The inter-refinery pipeline that connects the Exxon Bayway refinery with the Bayonne terminal was damaged on or about January 2, 1990. As a result of the damage to the pipeline, 567,000 gallons of home heating oil No. 2 was discharged

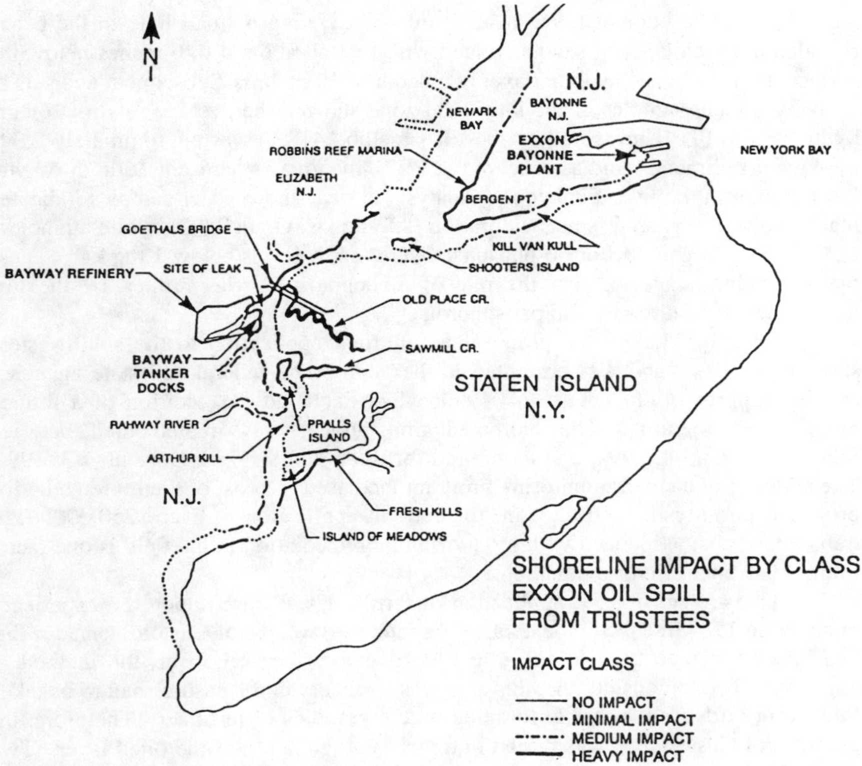

Fig. 12-2. The Arthur Kill shoreline impacted by the oil spill.

into the Arthur Kill estuary. Wind and tides spread the oil over 15 miles of the waterway impacting the New York and New Jersey shorelines (Fig 14-2). The primary cleanup response was containment and mechanical cleanup techniques to recover surface oil. Subsurface oil was recovered by trenching the shoreline in which the oil (and water) collected was recovered by using sorbents and portable vacuum equipment. However, the mechanical subsurface cleanup activities were suspended in early March 1990, in areas designated as bird rookeries. It was at this point bioremediation was proposed as a nonevasive cleanup technique for the bird sanctuary on Prall's Island, New York. Prall's Island is a low-lying island about a mile long created in the early part of the century from deposits of dredge spoils from the Arthur Kill. The southwestern shore of the island about 3/8 of a mile long composed of sand/gravel beach over a base of clay and peat was the target for bioremediation.

THE 1990 BIOREMEDIATION PRALL'S ISLAND RESPONSE

On June 26, 1990, the southwestern shoreline of Prall's Island, was surveyed for the feasibility of bioremediation as a nonevasive tool to remove residual fuel oil. In contrast to the contaminated beaches in PWS, there was no visible surface oil. The first evidence or impression of contamination detected by the survey team was the characteristic odor of home heating oil. Excavating a small hole in the beach revealed a 3 inch layer of sand saturated with fuel oil about 4 to 6 inches below the surface, trapped by a clay underlayer at a depth of 12 inches. Subsequent analysis of "core" sediment samples of the inter-tidal zone showed the average total petroleum hydrocarbons (TPH measured by the EPA method 418.1) was approximately 2,000 mg kg^{-1} of sediment. And as observed in PWS, the greatest concentration of oil was located high in the inter-tidal zone. Analysis of near shore water samples indicated that dissolved oxygen was near saturation (8.5 mg kg^{-1} O_2 at 24°C), while ammonia- and nitrate-nitrogen, and ortho-phosphate were at low levels below 1 mg kg^{-1}. These measurements suggested that the rate of biodegradation was limited by the low nutrient levels of nitrogen and phosphorous.

On July 31, 1990, a proposal for nutrient application to the southwestern shore of Prall's Island was presented to the appropriate federal and state agencies, and was approved on August 24, 1990, which included the addition of a limited monitoring program. The bioremediation project was implemented between September 1 and 3, 1990, and monitored for three months until December 3, 1990. The addition of a small monitoring program increased the cost of the bioremediation project from an estimated $25,000 for nutrient application to over $250,000. The tremendous cost associated with monitoring bioremediation in the field is one factor limiting advances in its application.

The strategy for the application of fertilizer was based upon the experience obtained in 1989 in PWS. The application rate was 0.02 lbs of Customblen/sq ft that had been shown to be safe and effective for subsurface oil. But, the method of application had to consider that this area was a rookery and measures had to be taken to prevent birds from possible foraging on the granules of fertilizer. Therefore, the granules of Customblen was buried in trenches at the depth of the oiled layer. Two parallel trenches positioned at mid- and mid-high tide were each filled with 0.4 lbs linear ft^{-1} based upon an application to an estimated oil zone width of 40 ft. The midhigh tide trench was just below the low-high tide line of the beach, and approximately 12 feet below that the mid- tide trench was installed. The trenches were not straight, but curved with the elevation of the beach which can be seen by the tide line. In contrast to the bags of briquettes tested in PWS, which were point sources of nutrients, the trench provided a continuous line of fertilizer. The concept was the tide would solubilize the fertilizer and the ebb tide would "draw a continuous curtain" of nutrients through the sediment. Two trenches of fertilizer were installed to maintain a sufficient concentration of nutrients in the entire 40 ft oil zone.

The efficacy of this application strategy was verified by monitoring the nutrient levels in the interstitial waters and the reduction in TPH. The first week after application of fertilizer, the average dissolved oxygen concentration declined while an increase in the levels of ammonia-, nitrate-nitrogen and phosphate were detectable in the interstitial waters. The greatest concentration of nitrogen and phosphorous were detected near the trench and gradually decreased with distance toward the low tide line. Increased nutrient levels were not detected above the fertilizer trench suggesting the fertilizer moves as predicted with the ebb tide. On average the ammonia and phosphate concentrations increased to 5-10 mg kg^{-1} in the first week of application and gradually decreased over a two month period to background levels (approximately 1 mg kg^{-1}). On average the TPH concentration measured in core samples of beach sediment was reduced from 2,000 mg kg^{-1} to less than 400 mg kg^{-1} in three weeks.

As recommended, for the monitoring program, in-situ containers of "homogenized" oiled sediment were buried at various locations in the beach. This is a technique used to standardize the oil concentration in the contaminated beach in an attempt to reduce the variability in measuring TPH reduction. This approach is not appropriate for mid-distillate petroleum spills because the oil is mobile. The in-situ container technique requires the oil to adhere to the sediment placed in the container. However, the fuel oil migrated out of the containers into the surrounding beach. This can be easily tested in the field by simply digging a hole in the contaminated area of the beach and replacing the oil-sand with "clean" sand. Depending on the relative mobility of the oil and time, the contamination in the surrounding area will move into the clean test area.

One recommendation for future nutrient applications to small distillate marine oil spills is to use a modification of the "sprinkler system" of nutrient application that was successfully tested in Alaska. Instead of using a permanently based sprinkler system use a "wand" with a sprayer manifold delivered from a shallow boat or land vehicle. This would allow the nutrient to be sprayed as soluble nutrient on the beach just below the high tide line. The soluble nutrients in the high tide region of the beach (usually the area with the heaviest contamination) would be drawn into the lower regions of the beach with the next outgoing tide. Again, the advantages are ease of application, and control of the application rate and area over the course of the treatment.

In conclusion, the Valdez and Prall's Island bioremediation projects both demonstrated that nutrient application can be safe environmentally and effective in accelerating the rate of biodegradation. There are numerous questions and technical hurdles that need to be addressed for general acceptance of bioremediation. One area that limits advances in field application, in the authors opinion, is in the area of monitoring the ecology of the biodegradation process.

REFERENCES

Ladousse, A. and B. Tramier. 1991. Results of 12 years of Research in Spilled Oil Bioremediation: Inipol EAP22. p. 577-581. *In* Proceedings of the 1991 International Oil Spill Conference (Prevention, Behavior, Control, Cleanup). US Coast Guard, American Petroleum Institute, U.S. Environmental Protection Agency. March 4-7, 1991. San Diego, CA

U.S. Environmental Protection Agency. 1990. Alaskan Oil Spill Bioremediation Project. Office of Research and Development, United States Environmental Protection Agency. U.S. Gov. Print. Office, Washington D.C EPA/600/8-89/073.

Pritchard, P.H., Coasta, C.F., and L. Suit. 1991. Alaska Oil Spill Bioremediation Project, Science Advisory Board Draft, U.S. Environmental Protection Agency, Office of Research and Development. U.S. Gov. Print. Office, Washington D.C. EPA/600/9/91/046a.

13 Accelerated Bioremediation of Triazine Contaminated Soils: A Practical Case Study[1]

H.C. Finklea and M.F. Fontenot Jr.
Ciba-Geigy Corporation
St. Gabriel, Louisiana

As the costs of cleaning up chemical contamination continue to rise, the increased attention given bioremediation comes as no surprise. Narrow-interest groups are stepping up their attacks on virtually all chemical and physical treatment technologies. Biotransformation frequently presents a cost advantage. These trends, and the inherent appeal of things "natural" to a public that is becoming increasingly chemophobic, seem to augur well for this oldest of cleanup methods. Soils contaminated with pesticides, typically caused by chronic, inadequate past practices or episodes such as spills and warehouse fires, pose remediation problems that have few easy or inexpensive solutions.

The case presented here, in which soil bioremediation was taken through to a successful outcome, provides a number of pointers for those considering similar ventures. Ciba-Geigy Corporation at its St. Gabriel, Louisiana plant site carried out an on-site remediation of 19,000 m^3 (25,000 yd^3) of atrazine-contaminated soils using land-farming techniques (Felsot et al., 1994 See Chapter 14) in a 8-hectare (20-acre) "Biological Cleanup Unit" (BCU). Treatment time was about six months with degradation of total s-triazines from an initial range of 100 to 300 mg kg^{-1} in some soil parcels to values at or below 2 mg kg^{-1}.

[1]Dr. Ralph J. Portier, Institute for Environmental Studies, 18 Atkinson Hall, Louisiana State University, Baton Rouge, LA 70803 provided early encouragement and guidance at the inception of the project. The prime contractor was Dames & Moore, Suite 900, 4949 Essen Lane, Baton Rouge, LA 70809. Microbiology consultant was Compliance Consultants, Inc., 11745 Bricksome Avenue, Suite A-4, Baton Rouge, LA 70816. The Ciba-Geigy project manager for the BCU project was Mr. C. M. Hansen. Technical assistance was provided by Tom S. Isacks, Geraghty & Miller, Inc., and administrative assistance by Kay Kellen, Ciba-Geigy.

BACKGROUND

Site Description

The St. Gabriel Plant is located on a 600 hectare (1500 acre) site about 32 kilometers (20 miles) south of Baton Rouge on the east bank of the Mississippi River near River Mile 200 in Iberville Parish. Immediately adjacent are the facilities of four other chemical manufacturers. The communities of St. Gabriel, Sunshine and Carville lie within eight kilometers (5 miles). Much of the site is wooded and a portion, at the turn of the century, was a sugar plantation. About 60 hectares (150 acres) of the site are developed and are devoted to chemical synthesis, formulation and packaging. Hydrogen cyanide, triazine herbicides, other crop-protection chemicals, specialty chemicals and dyestuffs are produced at this site.

Ciba is located on the erosional or cut bank side of the Mississippi River. The sedimentary deposits associated with this depositional setting are largely backswamp clays with thin, interbedded silty clays and silts which are associated with overbank deposition. These Holocene backswamp clays and overbank silts and silty clays extend in depth to approximately 30 m (100 ft) below land surface and are underlain by the Plaquemine aquifer of Pleistocene age.

Remediation Area

The area requiring remediation, designated Block F-5, was the site of a former temporary surface impoundment used during the 1970s to decant process solids from a raw wastewater effluent stream. The solids consisted of calcium carbonate, calcium sulfate, substituted *s*-triazine compounds which are process impurities from the synthesis of atrazine, as well as significant amounts of the herbicide active ingredient. The impurities were substituted triazines of low solubility and soil mobility (typically, trisalkylamino-, dealkylamino- and hydroxyl triazine compounds). After dewatering, the solids were periodically removed and landfilled off-site in permitted facilities.

Geotechnical testing of soils typical of the site indicate a pH of 7.9, a cation exchange capacity of 40.4 meq 100 g^{-1}, and an organic matter content of 9%. The soils are typically comprised of a fine-grained silty clay (5% sand, 37% salt and 58% clay). X-ray diffraction analysis of the clay fraction indicates that the clays are composed of illite (48%), smectite (38%), kaolinite (11%), chlorite (2%) and quartz (1%).

Meaningful waste management regulations first appeared in 1980, with the State of Louisiana requiring permitting ahead of federal regulations. As a result of these regulations, the impoundment was closed during the period 1981-83. The last of the solids and contiguous soil had been removed, and the excavation back-filled with clean adjacent soil.

By 1985 Ciba had plans to construct a multi-purpose rotary kiln incinerator and chose the site of the former pond. In that year the Ciba unilaterally conducted a rigorous site assessment: A total of 61 borings covering the area were taken at a minimum of 0.6-m (2-ft) intervals and analyzed for selected aromatics, halocarbons and triazine herbicides. The average results for these constituents were 0.187, 0.018 and 12 mg kg^{-1}, respectively. The triazine levels were highly variable, with one parcel at 500 mg kg^{-1}. Ciba deemed these levels to be of such significance that thorough site preparation would be sufficient remedy. Consequently, this information was not made known to the Louisiana Department of Environmental Quality until 1989 when the pace of the permitting of the incinerator quickened. This delay in reporting the soils data proved to be a significant deterrent to progress. The regulatory climate had shifted rapidly so that virtually any level of residual contamination had become unacceptable.

Regulatory Climate

Site assessment, waste characterization, technology selection and process development are the usual, expected concerns of the remediation manager, but it is a serious mistake to neglect the regulatory arena. "Arena" is an appropriate term here, for the specifics of the pertinent regulations and the underlying law comprise only one aspect of an unfolding drama. A number of forces interact in this arena. There is the federal agency that oversees the state agency that has jurisdiction, providing funding and critiquing results. There are local special-interest groups that are as persistent as they are vociferous, and increasingly competent in seizing (or creating) issues that effectively pursue their chosen agendas. There are the news media, drawn to controversy, quick to report conflicting opinion without challenge, and seldom willing to seek out and report facts that could promote rather than hinder consensus. There is the local community, living at close quarters with large industrial facilities, who often feel they bear a disproportionate exposure to chemical releases and receive little benefit, especially jobs. There are the large corporations with distant headquarters, some revitalizing their organizational cultures and living down past practices, all are straining under the economic pressures of global competition and the soaring costs of meeting social demands in addition to the demands of their shareholders. Finally, there are the politicians.

In 1987 the village of St. Gabriel, Louisiana appeared on the national scene with stories of high miscarriage rates (Burnett, 1987; Donavan, 1987). The issue grew out of efforts of a local citizen who was ignored by the state agency to whom she originally reported her findings. The *Washington Post* (Maraniss & Weisskopf, 1987) gave significant space and pictorial coverage. Oprah Winfrey came to Iberville Parish, more to champion a cause than cover an issue. A dark-horse gubernatorial candidate sympathized publicly with the activists, and it was well after his term in office that an independent epidemiological study essentially laid the issue to rest as invalid (White et al., 1989).

The new governor's administration however changed the regulatory landscape, particularly in regard to groundwater protection. An assistant secretary of the Louisiana Department of Environmental Quality (LDEQ) was installed that imposed arbitrarily stringent groundwater protection standards, effectively slowing the pace of permitting. Her controversy with a single permittee, another large multinational chemical manufacturer located in the New Orleans - Baton Rouge corridor, centered on a herbicide-contaminated area on which the company had built a new facility. As a result, all responsible parties in the state with remediation applications were suddenly faced with stern new permitting requirements. A new acronym appeared: NIMTOO - not in my term of office.

THE PROBLEM

Nature, Extent and Degree of Contamination

The extent of soil contamination beneath Block F-5 (Fig. 13-1) was determined by a site investigation conducted in 1985-86. The investigation involved the drilling of a series of soil borings to a uniform depth below grade over a predetermined grid pattern. The investigation indicated that approximately 19,000 m^3 (25,000 yd^3) of material exceeded the established cleanup levels of 0.1 mg kg^{-1} aromatics, 0.1 mg kg^{-1} halocarbons, and 10 mg kg^{-1} triazines. The laboratory analyses indicated that the measured total aromatics (chiefly toluene but also traces of benzene and chlorinated benzenes) ranged from 0.1 to 3 mg kg^{-1} and averaged 0.2 mg kg^{-1}. The concentrations of total halocarbons (chiefly carbon tetrachloride but also traces of dichloromethane and the trihalomethanes) ranged from 0.01 to 0.4 mg kg^{-1} and averaged 0.2 mg kg^{-1}; the concentrations of total triazines (chiefly atrazine, but also propazine and simazine) ranged from 0.01 to 258 mg kg^{-1} and averaged 12 mg kg^{-1}.

Five additional borings were made in 1990 to determine if the soil contamination had migrated beyond the limits delineated in the earlier study. As expected the results showed the concentrations to be somewhat less than the concentrations obtained during the earlier investigation. The data also confirmed the 1985-1986 data, which showed the herbicide to be confined to the upper 3 m (10 ft) of soil.

Remediation Options

Ciba faced the classical choices in selecting a remediation technology.

1) Leave undisturbed; extended groundwater monitoring.
2) Excavation and off-site land-fill disposal.
3) Excavation and incineration.
4) Excavation and on-site storage or disposal.
5) *In-situ* treatment.
6) Excavation and on-site treatment.

Others will obviously have to deal with site-specific and situational factors that differ, but the selection process is similar. A complex mix of factors are involved: technical feasibility, economics, company culture, regulatory climate, interest group activism, political influences, community acceptance and broader public relations. Ciba screened each of the six remediation approaches and formed the following conclusions regarding each:

1) Not acceptable; land needed for immediate use.
2) Usual company option, especially for relatively small quantities.
3) Excessive cost; excessive response to a low risk situation.
4) Not a final solution; unpalatable to company and agency.
5) Not acceptable; area needed immediately for construction.
6) Acceptable; reasonable cost and potentially quick results.

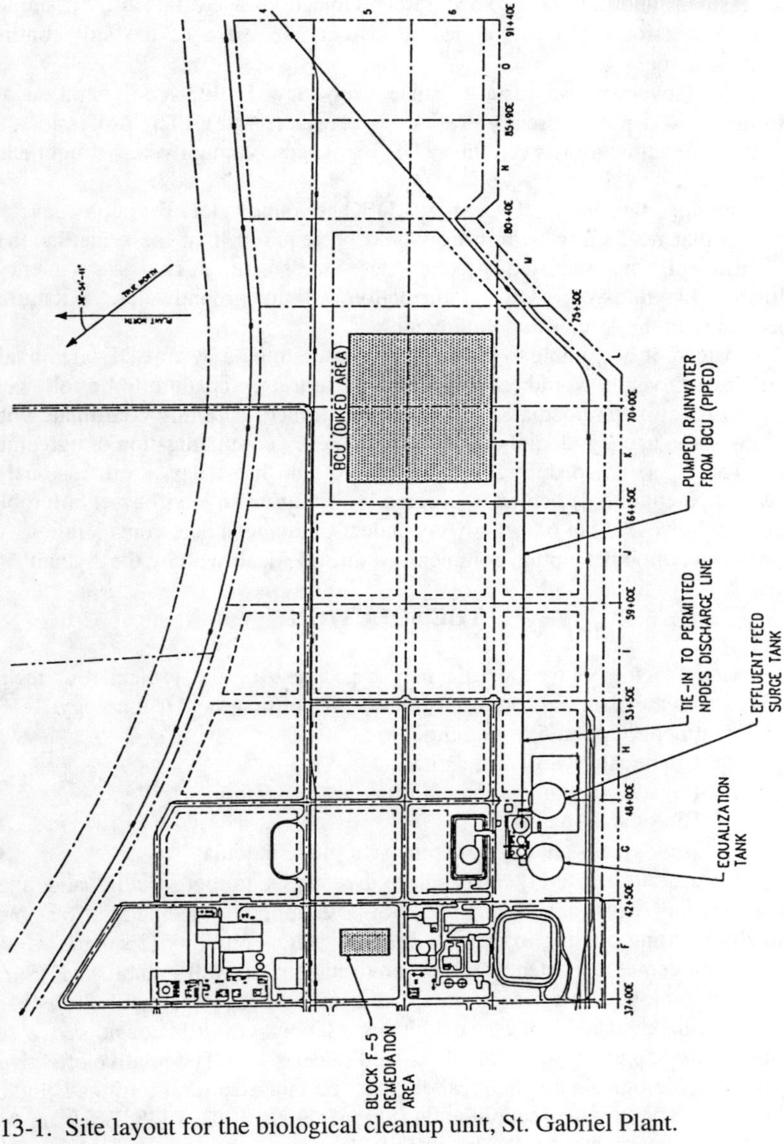

Fig. 13-1. Site layout for the biological cleanup unit, St. Gabriel Plant.

Regulatory Agency Role

Ciba had first mentioned Block F-5 as the site of a proposed multipurpose, rotary kiln incinerator in 1986, with the application to the LDEQ for the site's hazardous waste permit revision. It was not until late 1989 that the company reported on its earlier assessment, when it proposed remediation of the area.

In early 1990 the required remediation plan was submitted and sent to public notice. A public meeting was held in which Ciba presented detailed information on the assessment and plans for remediation. Objections were raised by a single but well organized local group that had formed in the wake of the still continuing miscarriage issue.

The Governor had installed an entirely new LDEQ administration and a confrontational deputy secretary presided over water issues. The professional staff was having to adjust to new realities. The regulated community was caught entirely by surprise.

Initially, in early 1990, the LDEQ Groundwater Division gave Ciba assurance that no facility permitting would be required, that the remediation plan was sufficient, but required groundwater assessment. This assessment was completed by mid-year with confirmation of no groundwater contamination associated with the former impoundment.

Instead of being able to proceed, Ciba was stunned by an LDEQ reversal; the Solid Waste Division would require its full permitting procedure to be followed for the proposed on-site remediation. The lesson learned: carefully coordinate with all divisions in the agency; do not count on intra-agency communication or agreement.

The Agency had throughout this period, and has today, a professional staff that is competent and dedicated to progressive solutions to environmental problems. Outcomes, however, can be heavily dependent on nontechnical considerations when political and public perception complexities must be dealt with by the Agency.

THE SOLUTION

Many factors played a role in the progress of the project, but technical feasibility was the essential factor in Ciba's choice of treatment technology:
On-site treatment (Option 6) possibilities were:

Chemical (hydrolysis, oxidation, etc.)
Physical (solubilization, thermal methods, etc.)
Biological - natural flora
Biological - natural flora plus xenobiotic inocula.

Atrazine is relatively resistant to hydrolysis in both acidic and alkaline aqueous media (Armstrong et al., 1967). Soil washing was deemed not feasible due to atrazine's strong binding to clays and its low water solubility (33 mg L^{-1}). While atrazine's moderate persistence in the environment is well established (Sumner, 1987), and biological treatment of atrazine effluent has met with little success because of inadequate kinetics, soil biodegradation was still considered a strong option. Atrazine's environmental fate is well understood. The relative effectiveness of combined biological and chemical degradation in agricultural settings is indicated by the absence of atrazine accumulation (Skipper et al., 1967, 1978; Erickson & Lee, 1989; Mandelbaum et al., 1993). While treatment time was not a significant consideration, Ciba elected to use techniques to accelerate the biotreatment.

Microbial Culture Development

The development of an adapted, acclimated microbial culture capable of degrading triazines was an essential primary step in developing predictions of field degradation rates and target degradation times for the soils. This initial task, which was assigned to Compliance Consultants, Inc. (Baton Rouge), included the evaluation of serial microbial cultures, wherein both soil samples from Block F-5 and triazines were used to begin the acclimation process. The addition of proprietary stock microbial cultures developed by the firm was necessary in order to facilitate observable growth of the heterogenous cultures. A microbial consortium was used to treat the soils. The consortium consisted primarily of motile and nonmotile rod-shaped bacteria, tentatively identified as pseudomonads. Other forms, such as ovoid motile, were also present but in lesser numbers. Actinomycetes and fungal forms were not observed. Growth rate determination and substrate reduction were two of the techniques used to identify increasingly acclimated cultures. The final selected culture which exhibited the greatest growth rate was used as the seed to prepare large quantities of inoculum (approximately 2,000 L) for application to the soils placed in the Biological Cleanup Unit (BCU). The inoculum was in liquid form, with all densities in the range of 10^9 to 10^{12} cells per milliliter.

BCU Design

The BCU has a surface area of approximately 81,000 m^2 (20 acres) to facilitate a one-lift application of 0.3 m (12 in.), assuming a 20% soil expansion factor. This surface area was selected to allow for the placement of roughly 23,000 m^3 (30,000 yd^3) of soil. The interior of the BCU was sloped to the grades as shown on Figures 13-2 and 13-3. The dikes are approximately one meter high, with a top width of 0.2 m and side sloped at 2.5:1. After the application of contaminated soil, these levees provide a minimum freeboard of approximately 30 cm (1 ft) which would contain the design storm rainfall amount (25-year, 24-hour rainfall is approximately 25 cm). The interior ditches vary in depth from approximately 0.3 to 1 m at the deepest point, with a bottom width of 1 m and side slopes of 1:1. The interior ditches flow to the center of the BCU where a collection point for rainwater is located. Cross sections of the dike and ditch system are presented in Fig. 13-3.

Native soil in the area proposed for dike construction was stripped to a sufficient depth to remove all unsuitable materials (e.g., topsoil, vegetation, roots, and stumps). The material chosen for dike construction was taken from the soils excavated during the construction of the internal BCU drainage ditches. The excavated soil was classified and visually inspected to ensure that all deleterious material had been removed. Construction of the dike was on a dry, prepared surface and consisted of placing and compacting 23 cm (9-in) loose lifts until the final lines and grades were achieved. Each lift of soil placed during the construction of the dikes was compacted to at least 90 percent of the maximum dry density as determined by the Standard Proctor Test (ASTM D698).

The 52 m by 52 m area in the center of the BCU serves as a collection point for the rainwater which falls into the BCU. Water which falls into this area is pumped to the plant's effluent discharge control system (Fig. 13-1).

Permitting

As noted earlier, Ciba intended to operate the BCU on a short-term basis to remediate the soils in Block F-5. The LDEQ abandoned their initial stance and

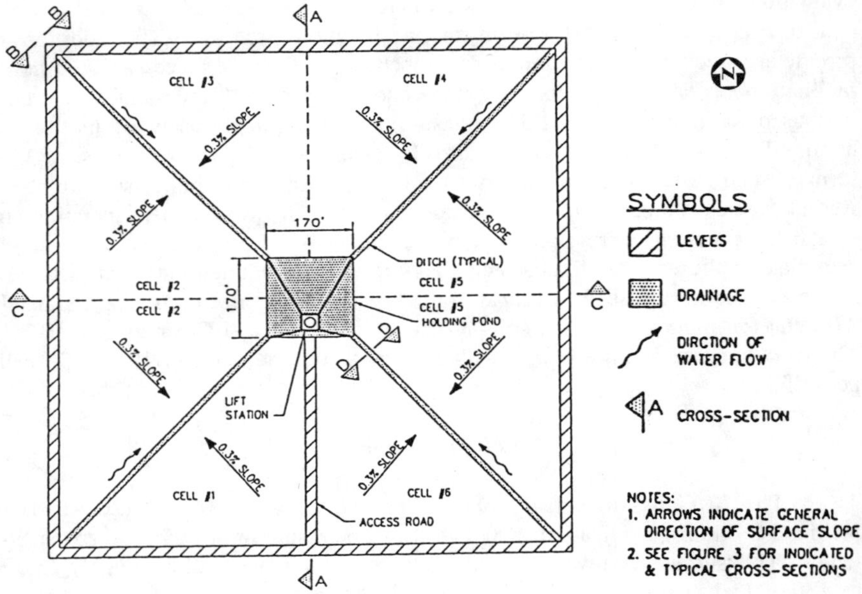

Fig. 13-2. Aerial view of the BCU treatment cell, St. Gabriel Plant.

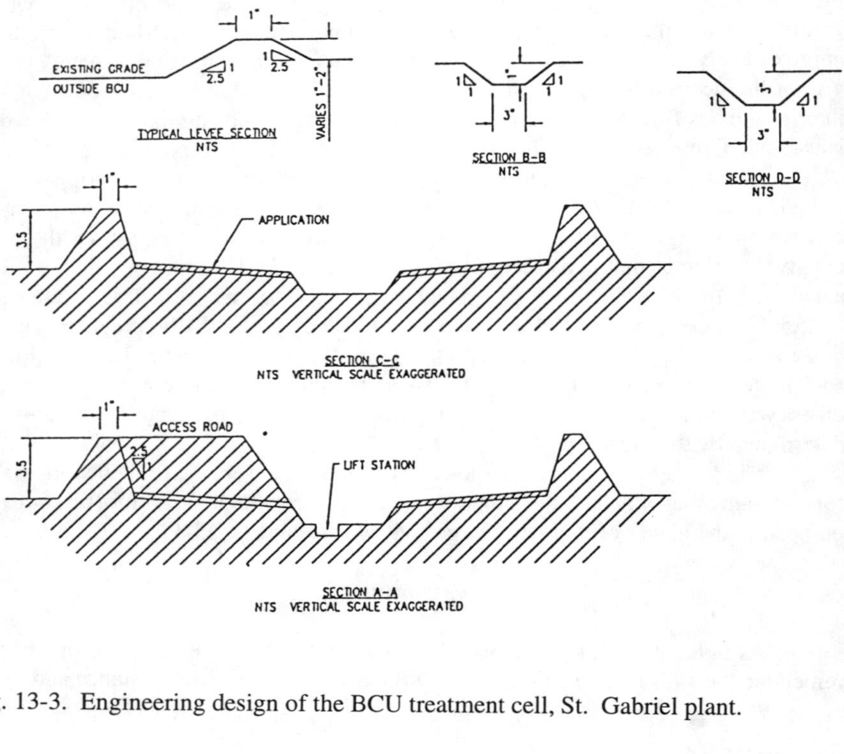

Fig. 13-3. Engineering design of the BCU treatment cell, St. Gabriel plant.

deemed the BCU to be a Solid Waste Disposal Facility, issued Ciba an "Order to upgrade" and required the submittal of a solid waste permit application. In spite of the fact that only a year of operation was envisioned, the full rigor of a permanent facility, 10-year permitting procedure was imposed. This was a major setback and inevitably led to significant delay and increased expenditures. As is often true in situations like this, the environment was not affected one way or another. The order and subsequent requests specified that the permit application include the following elements:

Public Notice Requirement
Environmental Impact Questionnaire
Solid Waste Annual Report
Demonstration of Financial Responsibility
Standard Permit Application

The requirements specified in the Standard Permit Application, include preparation of the following plans:

Area Master Plan
Facility Plans and Specifications
Operational Plan
Implementation Plan
Monitoring Plan
Closure and Post Closure Plans
Special Requirements

The process of obtaining a Solid Waste Permit is iterative, usually requiring multiple submittals of the requested information and plans for review over an extended time period to fully address the concerns of the regulatory agency. This process requires close coordination and interaction with the agency in order to clear up issues which may arise during the review process and also to ensure that the permit is granted in a timely fashion. Lesson learned: it is essential that inappropriate, environmentally unnecessary regulatory permitting be avoided.

Operations

The bioremediation of Block F-5 soils began on June 5, 1991 with placement of soils in the BCU proceeding simultaneously with the excavation of Block F-5. The placement of soil in the BCU was complete on July 2, and bioremediation of the soil continued through the end of the year.

The physical manipulation of the treatment surface was important from several aspects: distribution of the inoculum, moisture control, and aeration. Because the surface was divided into separate cells, it was possible to cultivate certain sections with greater frequency where the concentration of triazines was elevated. Initial deep disking to a depth of approximately 60 cm (24 in) was performed in order to remove construction debris (wood, concrete), the presence of which would have interfered with the efficient operation of the system. Specialized equipment including plows designed to break hardpan soil (chisel plows) were used to assure complete turnover of the treated soils to a depth of 30 to 40 cm (12 to 16 in). Surface disking was performed to break up large clods which formed. On average, each cell was disked or plowed eight times throughout the treatment period.

Initial disking of the BCU was conducted on July 16 and 17 during which time two passes were made on all cells using a D-8 dozer and 60 cm (24-in) blades. After the initial disking, a four wheel drive tractor and 36 cm (14-in) disk blades were used for weekly management practices.

On July 25 approximately 90 g m^{-2} (800 lb per acre) of fertilizer (13-13-13) were spread throughout the BCU. The operational plan called for disking and/or plowing the unit four days per week followed by samplings on Fridays. The inoculation process began on August 30 when Cell #1 was completely inoculated with microbial cultures developed and adapted for triazine degradation. The remaining cells were inoculated during the period of September 16 through September 19. The cells were periodically disked and plowed at various times from September to December 1991. BCU management activities were interrupted several times by the passage of storm fronts and heavy rains which are prevalent at this time of year. The BCU was plowed for the last time in May 1992 in preparation for a later confirmatory sampling event.

Ciba collected weekly soil samples from the BCU to verify satisfactory treatment of compounds of concern. All sampling procedures were conducted in accordance with regulatory requirements specified in the solid waste permit issued by LDEQ. For sampling purposes, the BCU was sectioned into six cells (Fig.13-2). Four grab samples were collected from each cell and composited into one sample for each cell. Each grab sample was collected by retrieving soils from the surface through the complete depth of the target soil lift (30 cm). Because of the project sensitivity and the regulatory framework of the BCU within the solid waste program, Ciba employed strict decontamination procedures of sampling equipment between sample collection points and chain-of-custody protocol to ensure sample integrity.

The composite soil samples were analyzed for the target compounds of concern: total triazines, chlordimeform, chloroaminotoluene, o-toluidine, total halogenated hydrocarbons, and total aromatic hydrocarbons. All analyses used in the project conformed with standard methods in SW-846 (USEPA, 1989). Volatile halogenated organics and volatile aromatic compounds in the soil were determined by use of the purge and trap procedure (Method 5030) with gas chromatographic analysis by Method 8021. The triazines were analyzed using a Ciba-Geigy procedure which uses a gas chromatography equipped with a nitrogen-phosphorus detector. In addition, the samples were analyzed for the biotreatment operational parameters: ammonia, nitrates, orthophosphate, pH, percent moisture and ATP. These measurements were to serve as guidance for the project microbiologist. The permit work plan specified no target values or operating ranges. The intervention of repeated rainstorms early in the operation period of the project limited the usefulness of these measurements.

RESULTS

Triazine Degradation

Weekly sampling of the BCU was conducted over the operational period of the project (19 weeks). The sample results for triazines are given in Fig. 13-4. The total triazine levels in all six cells were below the target level of 10.0 mg kg^{-1} by December, five months after inoculation.

Graphical plots of atrazine concentration for each of the six cells show the gradual reduction in triazine concentration (e.g., Fig.13-5a & 13-5b) The plots also show high variability. The EXTEME example is shown in Fig. 13-5b. This variability is most likely due to heterogeneity of the treated soils. The 4-point sampling grids for each cell were changed on a weekly basis. This procedure was intended to ensure that all portions of each cell were represented in the sequential sampling.

This concentration variability is likely due to the fine-grained nature of Mississippi River back-swamp clays known locally as "blue-gumbo." Although the BCU soils were plowed several times during the life of the project, the mixing and breakdown of clay agglomerates may not have been complete. Agglomerates not sufficiently broken down to allow proper contact time with microbes, nutrients, and oxygen may show higher levels of triazines.

The laboratory results for the total halogenated hydrocarbons and the total aromatic hydrocarbons of all six BCU cell composite samples were below the target cleanup levels of 0.1 mg kg^{-1}, and are in fact, below quantification limits. As noted earlier, samples were also analyzed for biotreatment operations parameters. Soil ammonium concentrations prior to fertilization were approximately 36 mg kg^{-1} within the BCU. After fertilization, ammonium concentrations for the next three weeks averaged approximately 63 mg kg^{-1}, and then declined to around 10 mg kg^{-1} for the duration of the BCU operation. Initial nitrate concentration, 23 mg kg^{-1}, increased to greater than 90 mg kg^{-1} after fertilization and then gradually declined to approximately 40 mg kg^{-1} through October. November analyses indicated nitrate levels decreased to a consistent 10 mg kg^{-1} through the remainder of the operation.

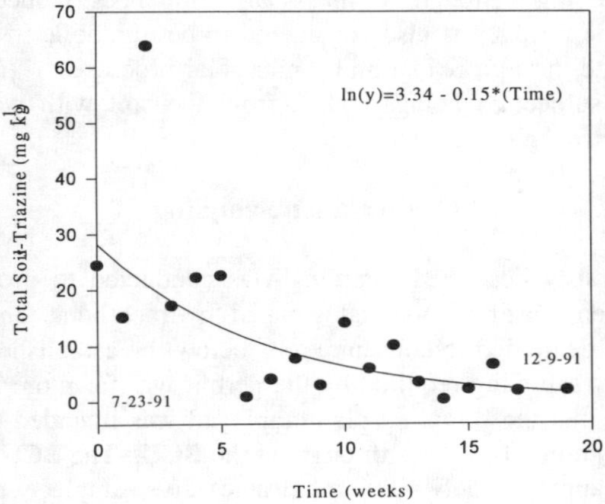

Fig. 13-4. Averaged total *s*-Triazine concentration in BCU soils from site in Louisiana.

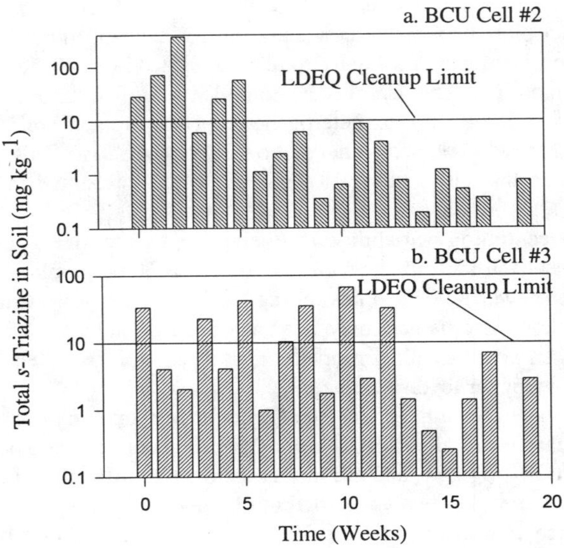

Fig 13-5. Total *s*-Triazine concentration for BCU Cells #2 and #3 over the 19 week incubation.

Orthophosphate concentrations reflected near constant concentrations of approximately 250 mg kg^{-1} throughout the BCU operation.

Orthophosphate is generally expected to be immobile in soils, and concentrations are high enough to mask any influences induced by biological activity. Ammonia nitrogen is also considered to be immobile in soils, but can be converted to nitrate through oxidation by microbial processes. Nitrates are mobile within soils and subject to being flushed from the soil with water, resulting in eventual loss of nitrogen from the system.

Confirmation Sampling

The previously described sampling was conducted to provide preliminary indication of when triazine, total halogenated hydrocarbons, and total aromatic hydrocarbons had decreased to concentrations below the established cleanup levels. Final confirmatory sampling required by the permit was far more rigorous than the weekly sampling, and the sampling plan employed was intended to ensure a more widespread distribution of samples throughout the BCU. The BCU was divided into 20 sections, each approximately 4,000 m^2 (1 acre); five samples were obtained from each section. The confirmation sampling techniques were the same as those utilized in collecting the weekly samples from the six plots. In the laboratory, a single composite sample was formed from each set of five samples from a section. The laboratory analyzed the 20 samples for the target compounds. The volatile organic compounds were analyzed utilizing Method 8021. The triazines were analyzed utilizing Method 8141, while o-toluidine and chloroaminotoluene were analyzed by Method 8270.

For the confirmation sampling event conducted in March 1992, the laboratory reported concentrations of total triazines ranging from 0.034 mg kg^{-1} (Section 20) to 48.2 mg kg^{-1} (Section 5). The triazine levels from five of the twenty sections (5, 9, 10, 14 and 15) exceeded the cleanup level of 10 mg kg^{-1}. The other target compound concentrations were reported at levels below the cleanup levels for the 20 sampled sections.

Ciba and the Solid Waste Division agreed that the five sections not meeting the cleanup criteria should be recultivated using a chisel-type plow to further expose clay agglomerates and enhance the further biodegradation of the triazine herbicides. The recultivation work was performed in May 1992, and the five sections were resampled in June 1992. The samples were obtained using the same procedures described previously for the first confirmatory sampling event. The samples were submitted to two laboratories for triazine analysis.

Results were well under the cleanup levels except for a single section. A third confirmatory sampling event was performed in September 1992 in section number 5 only to resolve a question of analytical validity. Five samples were obtained within the one-acre area utilizing the same methodology outlined previously and submitted to the two laboratories for compositing and subsequent triazine analysis. The analytical results obtained from this event indicate that section 5 had achieved the established cleanup levels.

Costs

Ciba has expended approximately $1,050,000 on the BCU project. This estimate includes costs associated with permitting, soils excavation and placement, analytical services, BCU operations and closure. This figure is a realistic estimate of the costs associated with the remediation of Block F-5 soils and excludes the initial contamination assessment costs.

These costs are far in excess of those expected in nonindustrial settings. Typical non-urban agricultural settings, free of concerns associated with industrial chemical operations, should encounter far less cost.

Ciba estimates that a total of approximately 19,000 m^3 (34,000 tons) of soils were treated in the BCU, at an approximate cost of $31 per ton. Ciba has a long-standing policy of disposing all chemical wastes at permitted hazardous waste disposal facilities; therefore the off-site transportation and disposal of soils at a non-hazardous disposal facility was not an option. In 1991, it would have cost approximately $5,300,000 to excavate and dispose of Block F-5 soils in accordance with standard policy. On this basis, the successful operation of the BCU resulted in a savings of approximately $4,250,000.

Closure and Post Closure

Ciba considered continuing interim maintenance activities with the objective of potential future use for bioremediation of additional triazine-contaminated soils at the BCU. Louisiana has recently promulgated new solid waste rules and regulations which are far more encompassing than the previous regulations. The costs associated with re-permitting and the requirements of operating the BCU under the new regulations were considered to be cost-prohibitive. Therefore, Ciba decided to close the BCU.

The final report for the bioremediation of Block F-5 soils in the BCU was submitted to the LDEQ in April 1993. Ciba subsequently received approval to close the BCU in accordance with the closure plan in the solid waste permit application. Closure of the unit, completed in April and May 1994, consisted of leveling the dikes and site grading to restore the site's topography and surface water drainage to original conditions. In July 1994 LDEQ accepted Ciba's closure certification report, agreeing that clean closure was achieved. There are no post closure requirements. In the future, Ciba may plant sweet corn or pasture at the BCU site.

CONCLUSIONS

Enhanced biodegradation can be an effective, economic technique for on-site remediation of triazine-contaminated soils.

By implication, many other herbicide, fungicide and insecticide residues can be likewise successfully treated.

The observed half-life of atrazine under the conditions of this project, approximately 30 days, confirmed predictions based on previously published values.

If permitting and regulatory agency interventions are encountered, they are likely to be more important (and controlling) than the technology employed.

Agency cooperation and timely achievement of bioremediation objectives are accomplished by early notification, complete disclosure, gaining agency agreement step-by-step and coordinating the project with other agencies which could conceivably have an interest.

The project must be thoroughly coordinated with the various divisions within the permitting agency; to rely on intra-agency communication is to risk costly delay.

REFERENCES

Armstrong, D.E., G. Chesters, and R.F. Harris. 1967. Atrazine Hydrolysis in Soil. Soil Sci. Soc. Am. Proc. 31:61-66.

Burnett, J. 1987. All Things Considered. (National Public Radio). Chemical Plant Emissions An Issue In Louisiana. August 28, 1987.

Donavan, S. 1987. USA Today. La. Women Fear Plants, Birth Problems Linked. August 17.

Erickson, L.E., and K.H. Lee. 1989. Degradation of atrazine and related *s*-triazines. Crit. Rev. Environ. Contr. 19:1-14.

Mandelbaum, R.T., L.P. Wackett, and D.L. Allen. 1993. Rapid Hydrolysis of Atrazine to Hydroxyatrazine by Soil Bacteria. Envir. Sci. Technol. 27; 1943-46.

Maraniss, D., and M. Weisskopf. 1987. *Washington Post*. Louisiana Contends With Pollution. December 22.

Skipper, H.D., C.M. Gilmour, and W.R. Furtick. 1967. Microbial versus chemical degradation of atrazine in soils. Soil Sc. Soc. Amer. Proc. 31:653-656.

Skipper, H.D., V.V. Volk, M.M. Mortland, and K.V. Raman. 1978. Hydrolysis of atrazine on soil colloids. Weed Sci. 26:46-51.

Sumner, D., M. Huber, and K. Balu. 1987. Summary of Environmental Fate Studies for the Reregistration of Atrazine. Report No. EIR-87014. Ciba Crop Protection Division, Ciba-Geigy Corporation, Greensboro, NC.

U.S. Environmental Protection Agency. 1989. Test Methods for evaluating Solid Waste, Physical/Chemical Methods, SW-846,3rd. ED.

White, L., F.J. Mather, and J.R. Clarkson. 1989. St. Gabriel Miscarriage Investigation, Eastbank, Iberville Parish, LA. Tulane School of Public Health, and Tropical Medicine.

14 Remediation of Herbicide-Contaminated Soil by Combinations of Landfarming and Biostimulation[1]

A.S. Felsot
Washington State University
Food & Environmental Quality Lab
Richland, Washington

J.K. Mitchell
University of Illinois
Department of Agricultural Engineering
Urbana, Illinois

E.K. Dzantor
Tennessee Valley Authority
Muscle Shoals, Alabama

Since the passage of the Resource Conservation and Recovery Act (1976), public attention has been focused on problems of waste disposal. The Superfund program has directed attention to chemical contaminants listed as hazardous pollutants, but many other chemicals may be regulated by the states as special waste. For example, pesticide waste generated at the thousands of agrochemical retail facilities has been particularly problematic. These facilities provide farmers with a variety of services including the custom application of fertilizers and pesticides. Chemicals are often mixed and loaded at one location where frequent spillage may result in the accumulation of high concentrations. Rinsing of equipment and empty containers also produces contaminated discharges that can move off-site as runoff if not handled properly. Similar conditions of spillage and rinse out may occur on individual farms, especially when the same site is used repeatedly for loading and cleanup. Another waste-generating problem common to commercial operations and

[1]This review is a contribution from the Washington State University Agricultural Research Center; field studies on landfarming were conducted while A. S. Felsot was stationed at the Illinois Natural History Survey and Illinois Agricultural Experiment Station, University of Illinois at Urbana-Champaign. Funding was received from the Illinois Hazardous Waste Research and Information Center, Project No. HWR 91-084, and the Illinois Fertilizer and Chemical Association.

individual farms is the storage of fuels in leaky tanks either above or below ground. Fuel may also be spilled during loading or discharged during maintenance of equipment.

OVERVIEW OF LANDFARMING AS A BIOREMEDIATION PRACTICE

When present in the soil at unusually high concentrations, biodegradable chemicals can be extremely persistent, which increases the risk of surface and groundwater contamination (Wolfe et al., 1973; Staiff et al., 1975; Davidson et al., 1980; Schoen & Winterlin, 1987; Felsot & Dzantor, 1990). In a survey of 56 agrochemical retail facilities, the Illinois Department of Public Health has documented that 61% operate water wells that are contaminated by one or more pesticides at levels 1-2 orders of magnitude above U.S. EPA Health Advisories (Long, 1989). The Illinois Environmental Protection Agency (IEPA) has investigated 16 agrochemical retail facilities where pesticides have contaminated well water (Taylor, 1989). In addition to pesticides, well water at commercial facilities has been contaminated with formulation solvents, fuels, and oils (Long, 1989). Problems with waste at commercial facilities has adversely affected soil and groundwater in nearby residential neighborhoods (Long, 1989).

Agricultural chemical retail facilities may be investigated by State and Federal Environmental Protection Agencies and subjected to specific enforcement of regulatory programs like the Federal Superfund List. Although new containment regulations mandated by states like Illinois and proposed by the U.S. EPA should prevent widespread contamination in the future, many sites have been contaminated for years as a result of standard operating procedures. The regulatory programs will require the facilities to undertake remedial action to remove and dispose of contaminated soil. An average cleanup at one site could involve several thousand cubic meters of soil. The contaminants may include insecticides, herbicides, and fungicides mixed in the same soil. Even after cleanup and the installation of containment facilities, however, contamination can still occur and may require further remediation (Habecker, 1989). Two of the most common contaminants found at agrochemical retail facilities are alachlor and atrazine, class B and C carcinogens, respectively.

In addition to remediation of pesticide waste, new federal and state regulations will require agrochemical facilities (as well as other commercial interests) to upgrade or remove underground fuel tanks. In Illinois alone, the program will require the upgrading of approximately 60,000 fuel tanks over the next 10 years. As many of the tanks are removed, soil contaminated with fuel will require removal and disposal in special waste landfills. The constituents of regulatory concern are benzene, ethylbenzene, toluene, and xylene (BETX).

When a business has a major chemical spill, state or federal regulatory agencies can order a cleanup, but such action is more difficult for private farms and residences. The nature of the cleanup is more problematic. Contaminated soil is excavated and removed to a "secure" landfill. The end result is perhaps a cleaner site, but the waste has not been detoxified. The most reliable technologies that have been proposed for on-site and/or off-site detoxification of pesticide waste have focused on liquids (Seiber, 1987). Remediation of wastes in soil is more difficult and expensive, especially for small businesses and individual homeowners.

The ubiquity of pesticide and fuel waste, the need for cost-control, and the lack of adequate landfill space has stimulated a prolific research interest in decontamination of soils by microbial systems. Several strategies, which are currently used or are being studied include (i) spreading of waste on agricultural or noncropped land to stimulate degradation, transformation, and/or immobilization of contaminants (variously known as landfarming, land treatment, land application, or land spreading), (ii) pretreatment of contaminants with various reagents to produce degradates more amenable to microbial mineralization; (iii) treatment of soil with microbial enzymes; (iv) amendment of soils and groundwater with nutrients and/or alteration of physical conditions to stimulate indigenous microbial metabolism (biostimulation); (v) inoculation of wastes with contaminant-adapted microorganisms (bioaugmentation). The latter four strategies, which have been reviewed from the perspective of pesticide detoxification (Felsot & Dzantor, 1990), are largely experimental or under development. Landfarming, on the other hand, has been used for several decades to dispose of municipal sludge and oily wastes (Torrey, 1979; Crites, 1984; Marshall & Devinny, 1986).

The objectives of this chapter are twofold. The literature is briefly assessed to determine the state-of-the-art regarding landfarming of waste fuels, petroleum sludges, and pesticide-contaminated soils. The potential of biostimulation to be used in conjunction with landfarming of pesticides is also reviewed. The second objective is to examine some results from a recently reported experiment that assessed the feasibility and safety of landfarming soil contaminated with the herbicide alachlor.

DISPOSAL OF CHEMICAL WASTES BY LANDFARMING

The appellation "landfarming" was probably introduced into the scientific literature in an article describing disposal by biodegradation of oily sludges in soil (Dibble & Bartha, 1979). From an engineering perspective, landfarming is a "managed treatment and ultimate disposal process that involves the controlled application of a waste to a soil or soil-vegetation system" (Loehr et al., 1985b). For landfarming soils contaminated by agricultural chemicals, the objective is placement of contaminated soil within the plow layer of uncontaminated soil where dilution lowers the pesticide concentration sufficiently to facilitate both chemical and aerobic microbial degradation. To have long-term utility, landfarming must not irreversibly affect any potential future uses of a site nor contaminate surface and groundwater. Of the various strategies for bioremediation, landfarming has been studied the longest and therefore is more amenable to quick implementation.

In successful landfarming systems, wastes are degraded, transformed, or immobilized; virtually any type of waste can be treated (Loehr et al., 1985b). In practice, however, most studies and currently operational systems have dealt with municipal wastewater and sludge. The constituents of concern in these wastes are excessive concentrations of nutrients, such as nitrate, and heavy metals. After many years of study, treatment guidelines were developed (Torrey, 1979); wastewater and sludge can now be considered recyclable for use in irrigation and as soil fertilizer, respectively.

Although few investigations have focused specifically on landfarming of fuels and pesticides, insights from studies of trace synthetic organics in municipal wastewater and sludge are useful for focusing research needs. With the advent of highly sensitive chromatographic and spectrometric instruments in the 1970s, new concerns like groundwater contamination developed about the behavior of trace organics during permitted land treatment operations. The importance of sorption (Novak et al., 1995) and biodegradation processes to the fate of trace organics during land application of municipal wastewater has been reviewed by Hutchins et al., (1985). Reviewed compounds included a detergent, polynuclear aromatics (PNAs), halogenated benzenes and aliphatics, and substituted benzenes.

FUEL COMPONENTS

Fate in Simulated Landfarming Studies and Remediation

Specific studies indicated that aromatic fuel components like toluene and ethylbenzene were volatilized from soils held in rapid infiltration microcosms (Piwoni et al., 1986). Volatilization was also an important route of benzene loss from two soils (Rogers et al., 1980). Organic carbon normalized sorption coefficients for benzene in soil were less than 150 (Garbarini & Lion, 1986; Uchrin & Mangels, 1987). According to the U.S. EPA guidelines on soil leaching potential (Cohen et al., 1984), these sorption data indicate that aromatic fuel components have a comparatively high potential for leaching in addition to volatilizing; therefore, landfarming of gasoline contaminated soils must be undertaken with appropriate regard for off-site movement.

Other studies with fuel components in soils show significant biodegradative potential. More toluene degraded when applied to sandy soil columns at a higher application rate than at a lower rate, and less was detected in the column effluent (Wilson et al., 1981). In river water to groundwater infiltration systems, dimethyl benzenes are degraded by denitrifying bacteria under anaerobic conditions and can be utilized as sole carbon and energy sources (Kuhn et al., 1985). In such systems, aromatic hydrocarbons were more readily degradable than chlorinated aromatics (Schwarzenbach et al., 1983). In microcosms containing methanogenic aquifer materials collected near municipal landfills, BETX were extensively degraded presumably by biological transformations following a long lag period (Wilson et al., 1986). Gasoline components are also known to be degraded in aerobic subsurface environments (Barker et al., 1987).

PNAs were very stable under field conditions in small plots to which excessively high rates (200 dry tons/acre) of municipal sludge was applied; lower rates (25 dry tons/acre) of application resulted in significant dissipation (Demirjian et al., 1987). One study has indicated that persistent organics like PCBs may be significantly accumulated by grazing animals when sludge is surface-applied (Crites, 1984). Sewage sludges and receiving soils have been found to contain the persistent organochlorine insecticides DDT and dieldrin; therefore use of sludge to biostimulate degradation must consider the possibility of introducing additional contaminants.

Bioremediation of fuel components seems to be proceeding along two fronts, *in situ* biodegradation and landfarming. An aquifer contaminated with alkylbenzenes from leaky underground storage tanks has been experimentally remediated *in situ* using injection of mineral nutrients and oxygen or hydrogen peroxide (Wilson et al., 1989). In another study, groundwater was pumped into sawdust/sand pits that were amended with acetate and nitrate to stimulate benzene biodegradation (Harrison & Barker, 1987).

Biodegradation and leaching of landfarmed fuel hydrocarbons (nC_{12}-nC_{21}) were studied in a field plot and in a lysimeter (Oudot et al., 1989). After 3.5 yr, 94% and 83% of the hydrocarbons were removed from the field plot and lysimeters, respectively. Fuel components were found at 60 to 80 cm; leachate contained no unmodified hydrocarbons but did contain suspected oxidative metabolites. Jet fuel, heating oil, and diesel oil were remediated in landfarming lysimeters by treatment with liming, fertilization, and weekly tilling (Wang & Bartha, 1990). Both hydrocarbons and polyaromatic hydrocarbons (Wang et al., 1990) dissipated within 20 weeks after bioremediation.

The Iowa Department of Natural Resources (IDNR) has issued guidelines for landfarming soils contaminated by leakage from underground storage tanks (IDNR Underground Storage Tank Program--Notification for Land Farming Petroleum Contaminated Soil). Parameters for landfarming of fuel and oil wastes have been developed in the Netherlands (Soczo & Staps, 1988).

LANDFARMING OF PETROLEUM SLUDGES

Petroleum wastes in the form of oily sludges have routinely been landfarmed at oil refineries and petrochemical plants. Until recently, however, little attention, had been given to the fate of priority pollutants like aromatic hydrocarbons and 4-5 membered-ring PNAs. During land treatment of an oily waste from refinery lagoons on experimental plots, naphthalenes, alkanes, and certain aromatics were rapidly lost from soil, especially during warmer months ($T_{1/2}$<30 d; Loehr et al., 1985a). Earthworms did not bioaccumulate any of the measured components during the study. Mutagenic activity and microbial assays like Microtox have been used to monitor degradation of land-applied oily waste constituents (Sims et al., 1987) and contamination in runoff and leachate (Brown & Donnelly, 1984). In small plots receiving oil sludges, greater amounts of mutagenic activity were detected in runoff water than in leachate. The activity, however, decreased with time due to degradation and infiltration (Brown & Donnelly, 1984). In an earlier study, oil was not detected in runoff water and leachate, but the concentration of residual oils or oxidation products in treated plots were large enough after nine months to cause significant inhibition of plant growth (Raymond et al., 1976).

Several laboratory studies have delineated the environmental and operational factors (e.g., soil moisture, pH, mineral nutrients, micronutrients, organic supplements, loading rate, treatment frequency, and temperature) that affect the biodegradation of oil sludge during landfarming (Dibble & Bartha, 1979; Bossert et al., 1984). Micronutrients and organic amendments did not stimulate biodegradation of refinery sludge, and sewage sludge interfered with biodegradation. Biodegradation was most rapid when C:N ratios in soil were 60:1 (Dibble & Bartha, 1979). In other studies, biodegradation of refinery sludge was optimal at a C:N ratio of 9:1, but petrochemical sludge degraded fastest at a C:N ratio of 124:1 (Brown et al., 1983). Several studies agreed that repeated applications of small amounts of sludge resulted in faster biodegradation rates than single large applications (Dibble & Bartha, 1979, Brown et al., 1983). The fastest degradation rates of aromatics in refinery sludge were associated with higher rather than lower application rates (Dibble & Bartha, 1979). Pertinently, xylene at the highest rate of application from a simulated spill degraded more quickly than at the lowest rate of application (Aurelius & Brown, 1987).

REMEDIATION OF PESTICIDES BY LANDFARMING AND POTENTIAL ROLE OF BIOSTIMULATION

Compared to studies of organics in municipal wastewater and petroleum sludges, land application of pesticide-contaminated soils has been given little attention. Most research efforts have focused on bioremediation of chlorophenolic wastes in soil that arise from wood preserving operations (Crawford & Mohn, 1985; McGinnis et al., 1989). Soils previously exposed to pentachlorophenol exhibited enhanced rates of biodegradation when retreated, suggesting microbial adaptation and utilization of the pesticide as an energy source. A combination of bioaugmentation and biostimulation has been used in combination with composting in windrows to treat chlorophenol-contaminated soils in Finland (Valo & Salkinoja-Salonen, 1986). The effects of tilling and fertilization on biodegradation of PCP were studied in "landfarming chambers" (Mueller et al., 1991); degradation rates of PCP were too slow to meet acceptable treatment standards after 90 days of incubation.

Liquid pesticide wastes have been successfully detoxified by pretreatment prior to soil disposal. Chlorinated hydrocarbon pesticides biodegraded during sewage treatment, and diazinon, parathion, and dieldrin degraded rapidly when composted with cannery wastes (Wilson et al., 1983). Liquid wastes of alachlor and atrazine have been exposed to UV light and ozonation prior to disposal in soil (Kearney et al., 1988; Somich et al., 1988). These pesticides are cometabolically degraded, and the pretreatment degrades the compounds into mineralizable products. Land application of composted cotton gin wastes containing pesticides resulted in very low to undetected residues after incorporation of the waste into the soil (Winterlin et al., 1986). Solid-state fermentation bioreactors have been tested to detoxify liquid wastes (Berry et al., 1993); the resulting spent lignocellulosic matrix upon which the pesticides are initially sorbed was found to be safe for subsequent land application.

The feasibility of remediating pesticide-contaminated soils by combining landfarming with biostimulation has been suggested by the results of several studies. Pesticide waste taken from a highly contaminated soil evaporation pit in California was partially detoxified by amendment with a variety of inorganic and organic nutrients and by incubating under aerobic or anaerobic conditions (Winterlin et al., 1989). Corn plant residue, corn meal, and municipal sewage sludge effectively stimulated degradation of high concentrations of alachlor in soil (Dzantor & Felsot, 1992). A 10 fold dilution of alachlor and metolachlor-contaminated soil with uncontaminated soil to simulate landfarming and an altered physical environment also stimulated pesticide degradation (Felsot, 1991). Corn meal has been used in solid state fermentation bioreactors as a source of C to increase microbial activity for pesticide degradation (Berry et al., 1993). Peat increased the degradation of several pesticides that had been applied to soil at high concentrations (Schoen & Winterlin, 1987). Toxaphene degradation was enhanced in contaminated soil kept under anaerobic conditions and amended with cotton gin waste (Mirsatari et al., 1987). A combination of landfarming, bioaugmentation and biostimulation were used successfully to bioremediate an atrazine contaminated site (Finklea and Fontenot, 1995).

Soil contaminated with the herbicides alachlor, metolachlor, atrazine, and trifluralin were excavated and land applied to small field plots of corn and soybean fields (Felsot et al., 1988). These herbicides are cometabolized and microbial adaptations for enhanced biodegradation do not occur (Felsot & Dzantor, 1990). Alachlor and metolachlor were more persistent when applied in the contaminated soil than when applied as fresh herbicide sprays (Felsot et al., 1990a). Minor phytotoxicity to soybeans treated with the highest rates of waste soil occurred in the field and in greenhouse assays. The mixture of corn and soybean herbicides was seen as a potentially limiting factor in attempting to landfarm the pesticide wastes. Unresolved questions included the translocation of the pesticides and the mechanism of prolonged persistence in the contaminated soils.

Trifluralin-contaminated soil excavated from the site of a pesticide warehouse fire was applied to farmland in late summer using a manure spreader (Bicki & Felsot, 1994). Initially, contaminated soil was left on the soil surface, but 20 days later, small plots were delineated to test the effects of rototilling and dissipation of trifluralin. Concentration of trifluralin residues recovered from rototilled plots (upper 5 cm) were not significantly different than residues from untilled plots beyond the day of tillage. Residues were still high enough, however, to prevent oat growth the following spring. These results indicated that landfarming may require at least two growing seasons to remediate residues to levels nontoxic to sensitive crops.

DESIGN AND REMEDIAL ACTION CONSIDERATIONS FOR LANDFARMING

Cleanup of a contaminated site is actually a multistage process, and landfarming is only one step that deals with the contaminated soil itself. Successful remediation of a site will first depend on a thorough assessment of the extent and depth of contamination. The decision of how much soil to remove for the landfarming phase will depend on the cleanup targets as dictated by the appropriate regulatory agency. Consideration of the extent of excavation will influence the decision about land requirements for landfarming. A guiding principle for efficacious and safe landfarming can best be described by the simple maxim "do no harm." To accomplish this goal and meet the objective of landfarming, remedial action strategies must be developed for pretreatment of the contaminated soil, choosing and assessing the candidate landfarming site, and preparation and management of the site during landfarming. Considerations in developing the plan of action have been discussed by Bicki and Felsot (1994) and are outlined below.

I. Pretreatment of Contaminated Soil
  - A. Excavation strategy and estimation of soil volume.
  - B. Provisions for storage of soil.
  - C. Grinding, blending, mixing of soil.

D. Determination of identity of contaminants and average concentration.
E. Decontamination of pretreatment site.

II. Assessment of Landfarming Site
A. Distance of site from source site.
B. Property ownership.
C. Contractual relationships with owner.
D. Vulnerability of ground and surface water resources.
E. Potential for off-target movement.
F. Identification and characterization of wells.
G. Proposed travel route for transporting contaminated soil.
H. Demarcation of site boundaries.
I. Installation of monitoring wells.
J. Landscape position.
K. Soil properties.
L. Background contaminant levels.
M. Tillage and cropping history.

III. Preparation and Management of Landfarming Site
A. Marking of application area.
B. Seeding and establishment of vegetative borders.
C. Soil fertility and pH adjustments.
D. Tillage treatments.
E. Field leveling.
F. Plant residue management.
G. Timing of application.
H. Potential phytotoxicity due to the mix of pesticide contaminants.
I. Pesticide registration label restrictions applicable to subsequent crops.
J. Calculation of application rate.
K. Calibration of application equipment.
L. Depth of incorporation.
M. Periodic monitoring of pesticide residue levels.
N. Closure plan.

TESTING THE FEASIBILITY AND SAFETY OF LANDFARMING HERBICIDE-CONTAMINATED SOILS

Some states have passed legislation enabling certain agencies to permit landfarming; the motivation for these actions are the urgent need to cleanup sites and the comparative low cost and ease of landfarming. The technique would seem particularly suitable for pesticide-contaminated soils because the practice involves a technology that has been used for many years to dispose of municipal wastewater and sludges (SSSA, 1986). Furthermore, at many sites, the pesticides in contaminated soils are usually registered by the U.S. EPA for application to soil.

In the Corn Belt, candidate land to receive contaminated soils will likely be fields that cannot be taken out of production. Unresolved questions from previous landfarming experiments included the translocation of pesticides from the receiving land, prolonged persistence of pesticides in contaminated soils, and potential for crop phytotoxicity (Felsot et al., 1988; 1990a). Because landfarming is a leading candidate for treatment of pesticide-contaminated soils, these questions must be accurately assessed to help define appropriate guidelines and regulations.

The remainder of this chapter describes some results from a field experiment that studied degradation, translocation, and phytotoxicity of pesticide contaminants in landfarmed soil. The following is a report of the comparative degradation of alachlor after contaminated soil or normal sprays were applied to small field plots. The degradation of landfarmed alachlor-contaminated soil was also studied following amendment of soil with corn meal or sewage sludge. Results of runoff and phytotoxicity assays will be published elsewhere.

OVERVIEW OF EXPERIMENTAL PROCEDURES

A thorough description of the experimental design, sampling procedures, initial soil residues, and runoff monitoring was recently published (Felsot et al., 1992). Preliminary results of the use of biostimulation in combination with landfarming have also been reported (Dzantor et al., 1993). Briefly, soil contaminated with alachlor and three other herbicides, trifluralin, atrazine, and metolachlor were applied at three different loading rates based on the legal application rate of alachlor (i.e., 1X, 4.5 kg ai ha^{-1}; 5X, 22.4 kg ai ha^{-1}; 10X , 44.8 kg ai ha^{-1}). To assess the efficiency and environmental safety of alachlor degradation during landfarming, similar rates of application were simulated by spraying a herbicide mixture containing Lasso 4E (480 g alachlor L^{-1}) on replicate field plots. It was hypothesized that if landfarming were efficacious and safe, then degradation rate of alachlor would be very similar in plots receiving contaminated soil or sprays.

Plots comprised an area 10-m long by 3-m wide and were outfitted for collection of runoff. The soil was a Catlin silt loam (Fine-silty, mixed, mesic Typic Argiudoll). Thirty such plots were constructed with 24 devoted to pesticide applications either by landfarming or spraying, and six plots devoted to no pesticide applications for determination of background pesticide residues.

Degradation of alachlor was monitored by collecting 5 cm diameter by 10 cm long soil cores at random within each plot on 0, 30, 96, 187, 335, and 425 days after application. To reduce variability known to occur when sampling field plots, six individual cores from each of four replicated experimental units were analyzed individually before averaging together the residue values (Felsot et al., 1992). To compare different pesticide loadings, all pesticide residue data was normalized to a percentage based on the initial residues recovered.

Plots were bermed along all sides so that all runoff and erosion would be directed across a trough into collection barrels. Water and sediment were sampled after every runoff-producing rainfall by stirring the contents and immersing a collection bottle. Sediment was separated from the water by filtration and analyzed separately for residues of alachlor (Felsot et al., 1992). Alachlor concentration was multiplied by runoff volume or eroded sediment weight to obtain total mass of pesticide per runoff event, which was then summed over 37 events from growing season 1991 to August 1992. Data were expressed as a percentage of the initial application mass.

In separate plots (1 m^2) adjacent to the runoff plots, contaminated soil and sprays were applied at 1X, 5X, and 10X equivalent alachlor rates, which were similar to rates applied in the larger landfarming plots. These plots were designated "biostimulation plots." Sewage sludge (25 g kg^{-1} soil based on a 7.5 cm depth) and corn meal (50 g kg^{-1} soil) were incorporated into the top 5 cm of selected plots to stimulate the degradation of herbicide residues. Unamended, pesticide-treated plots were used as positive controls, and amended, no-pesticide plots were used as controls for determination of analytical background and microbial activity. Three soil cores (7.5 cm deep x 5 cm diam.) were collected from each plot through 428 days after application of soil and sprays. Individual cores were sieved and extracted as described by Felsot et al. (1992). Also, microbial activity (soil dehydrogenase activity and microbial populations) was determined on triplicate subsamples of bulked cores from each plot (Dzantor & Felsot, 1991).

COMPARATIVE DEGRADATION OF ALACHLOR IN LANDFARMED AND SPRAYED PLOTS

The actual rates of alachlor applied were determined by analysis of spray and soil intercepted by pans placed in three locations in the plots and compared to the targeted rates of application (Felsot et al., 1992). Because average initial residues of the herbicides in sprayed plots were approximately 2-3 times lower than the average residues in the landfarmed plots, degradation was compared by normalizing herbicide residues to a percentage of actual amounts applied and plotting the percentages recovered relative to time after application. Degradation curves for alachlor at the 1X, 5X, and 10X application levels are discussed in this report (Fig. 14-1). All curves exhibited a fast degradation phase within the first 96 days after application, followed by a much slower degradation over the next year. Time to 50% and 90% loss (i.e., DT-50% and DT-90%, Hamaker, 1972) was estimated by visual inspection of each curve.

At the 1X rate of application, alachlor had an estimated DT-50% of 25 days in landfarmed and sprayed plots (Table 14-1); however, by the end of the experiment (425 days after application), recovery of alachlor residues in landfarmed plots were greater than in sprayed plots (0.47 mg kg^{-1} vs. 0.06 mg kg^{-1}). Alachlor also seemed to persist longer in landfarmed plots than in sprayed plots at the 5X and 10X application levels. Furthermore, initial rates of degradation in both landfarmed and sprayed plots were much slower at the higher loading rates than at the 1X rate. Based on the degradation curves for alachlor, 1X rates of application seemed safe for landfarming because compared to sprayed residues, there was little difference in proportion of residues remaining following two growing seasons. At higher rates of application, however, initial rates of degradation in landfarmed plots were slower. It is pertinent that plots at the higher rates of application had little vegetative cover, even after the second growing season, which suggested that herbicide residues resulting from landfarming at the higher loading rates were still phytotoxic after two growing seasons. In contrast, 1X plots had an abundant weed population, and the rhizosphere may have stimulated degradation of alachlor (See Anderson & Coats, Chapter 8).

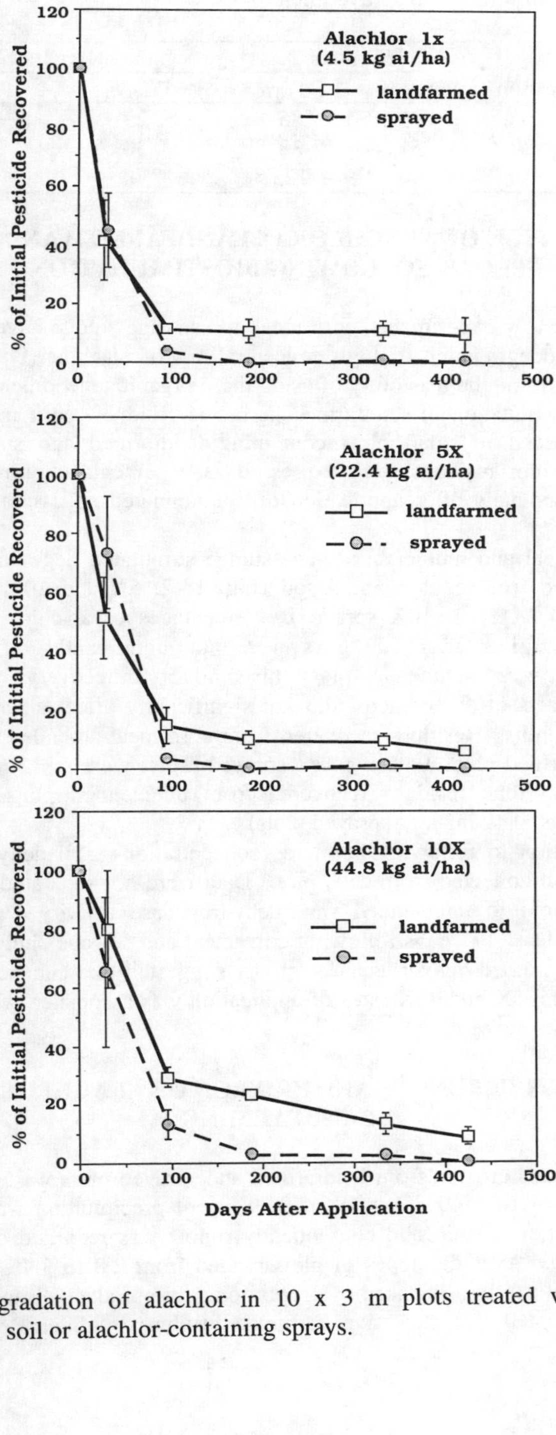

Fig. 14-1. Degradation of alachlor in 10 x 3 m plots treated with alachlor-contaminated soil or alachlor-containing sprays.

Table 14-1. Estimated times to 50% ($DT_{50}\%$) and 90% ($DT_{90}\%$) disappearance of alachlor in landfarmed and sprayed field plots.

	Sprayed		Landfarmed	
Rate of Application	$DT_{50}\%$	$DT_{90}\%$	$DT_{50}\%$	$DT_{90}\%$
1X	26	85	25	390
5X	51	88	30	180
10X	50	125	70	400

DEGRADATION OF ALACHLOR IN LANDFARMED AND SPRAYED PLOTS FOLLOWING BIOSTIMULATION

In laboratory experiments, corn meal and sewage sludge have been shown to enhance the biodegradation of high concentrations of alachlor (Dzantor & Felsot, 1992). To determine the feasibility of using these organic amendments in the field to augment the degradation of landfarmed herbicide residues, corn meal and sewage sludge were added to small plots containing landfarmed and sprayed herbicide residues. Alachlor residues were expressed as a percentage of residues initially recovered immediately after application of contaminated soil, sprays, and organic amendments.

Corn meal and municipal sewage sludge stimulated degradation of alachlor residues derived from contaminated soil (Fig. 14-2, 5X and 10X landfarmed) or sprays (Fig. 14-2, 5X & 10X sprayed). Percentages of alachlor recovered from unamended 1X plots after 100 days were not significantly different than the percentages recovered in amended plots; this similarity in recovery suggested that the stimulatory effects of amendments did not significantly affect alachlor degradation beyond two months. Residues recovered from corn meal-amended 10X plots (3.37 mg kg^{-1} landfarmed plots, 0.05 mg kg^{-1} sprayed plots) were still lower at 428 days post application than residues recovered from unamended plots (6.74 mg kg^{-1} landfarmed plots, 0.42 mg kg^{-1} sprayed plots).

In addition to following herbicide concentrations, soil dehydrogenase was measured in soil collected from each plot. Data were not separated by application level. Compared to unamended soil, dehydrogenase activity was significantly elevated for at least 177 days following corn meal and sewage sludge amendments (Fig. 14-3). Increased dehydrogenase activity suggested that enhanced degradation of alachlor at the 5X and 10X rates of application was associated with increases in general microbial activity.

CUMULATIVE RUNOFF AND EROSION OF ALACHLOR DURING LANDFARMING

Runoff and erosion from landfarmed and sprayed plots were measured from July 1991 to August 1992. A total of 82.9 cm of precipitation was recorded; the greatest proportion of rain and consequently runoff was recorded during the 1992 growing season. Average slopes of plots ranged from 2.8 to 3.4%. Plant residue cover was significantly influenced by proportional rate of alachlor application (Table 2), which suggested that excessive herbicide residues inhibited development of

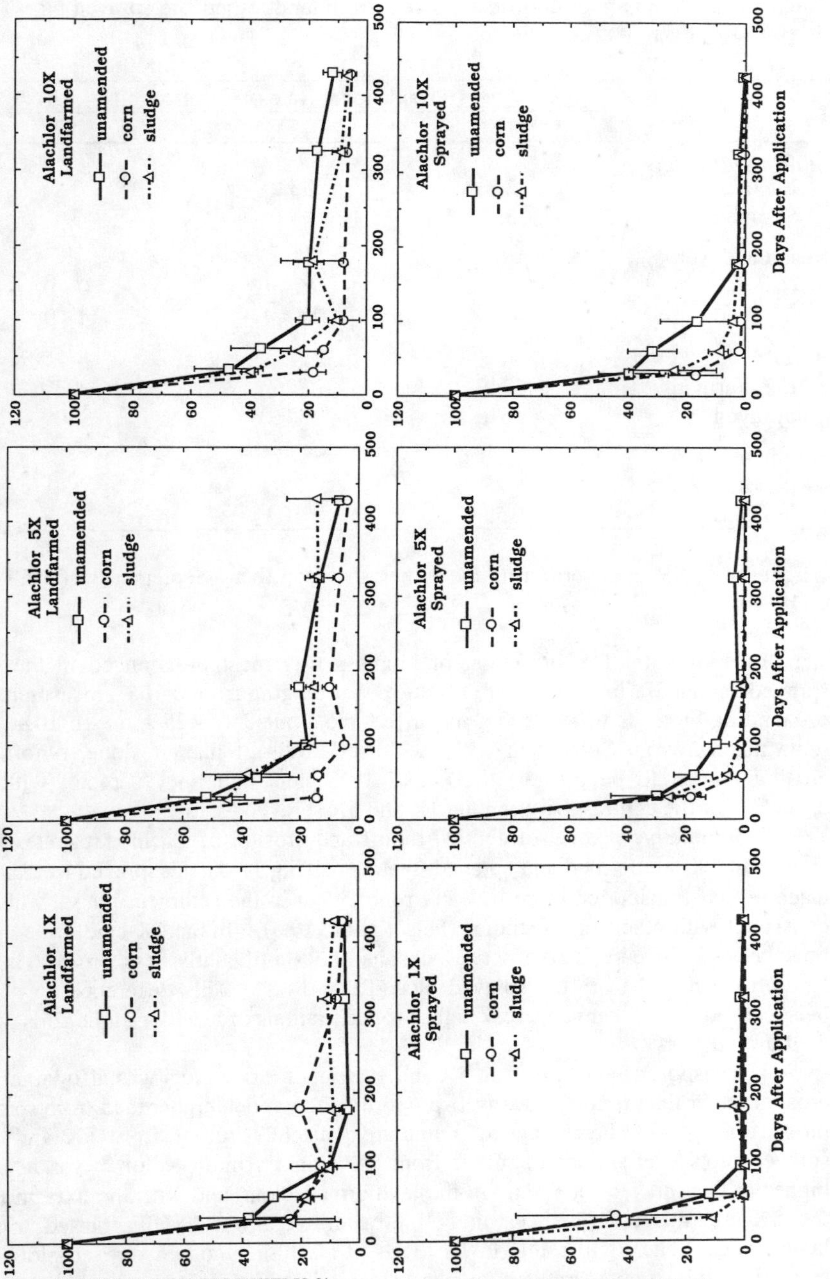

Fig. 14-2. Degradation of alachlor in 1 m^2 plots amended with corn meal or sewage sludge and treated with alachlor-contaminated soil or alachlor-containing sprays.

Table 14-2. Cumulative runoff and erosion from landfarmed and sprayed plots (July 1991 - August 1992).

	Proportional Rate of Application		
	1X	5X	10X
Slope (%)			
Landfarmed	2.8	3.2	3.1
Sprayed	3.4	3.0	3.1
Residue Cover (%)[†]			
Landfarmed	42 - 92	26 - 13	0 - 0
Sprayed	54 - 86	14 - 44	4 - 9
Runoff (L)			
Landfarmed	4249	7963	9030
Sprayed	7696	8460	9409
Sediment (kg)			
Landfarmed	25	236	339
Sprayed	153	104	294

[†]Highest % residue cover during the growing season (May-September) of 1991 and 1992.

standing biomass. The inhibition of biomass was most pronounced in the 10X sprayed and landfarmed plots with residue cover ranging from 0-10%; measurements of standing biomass were >200 g m^{-2} in 1X plots but only <25 g m^{-2} in 10X plots (data not shown). Coincident with the lower levels of plant residue, runoff and erosion tended to be greater in 5X and 10X plots than in 1X plots (with the exception of the contrast between the 1X and 5X sprayed treatment).

Cumulative percent-of-applied runoff and erosion of alachlor residues were similar in 1X landfarmed and sprayed treatments (Fig.14-4). In sprayed treatments, alachlor was transported more in water runoff than in the sediment phase, which is consistent with other runoff studies (Felsot et al., 1990). In the 1X landfarmed plot, however, the reverse was observed despite the significantly less erosion in the landfarmed plot than in the sprayed plot (Table 14-2). This difference may have been caused by the comparatively higher concentrations of alachlor in the soils of the landfarmed plots.

Transport of alachlor from 5X and 10X landfarmed plots was also largely by erosion rather than runoff and was disproportionately much higher than from sprayed plots (Fig. 14-4). Percentage of cumulative alachlor runoff from 5X and 10X sprayed plots was similar to runoff from 1X plots. Runoff volume was actually higher in the sprayed plots than in the landfarmed plots, and with the exception of the 5X treatment, differences in soil erosion were not heavily biased toward landfarming. The significantly lower levels of alachlor runoff in sprayed plots may be explained by a comparatively quicker degradation of alachlor in the sprayed plots resulting in lower residues than in the landfarmed plots.

The extremely high levels of alachlor in sediment compared to water could reflect "residue aging," where nonequilibrium sorption on intraparticle micropores

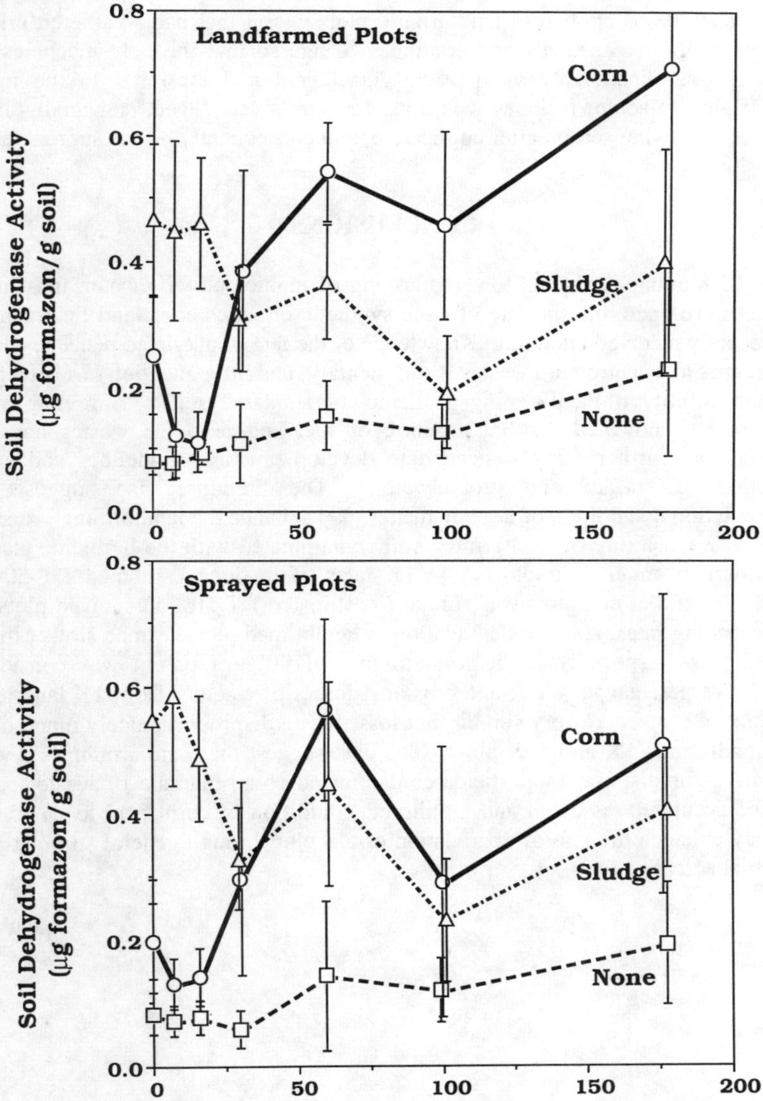

Fig. 14-3. Dehydrogenase activity in soil amended with corn meal or sewage sludge.

provides protection from microbial degradation and a very slow diffusion back into bulk solution (Steinberg et al. 1987; Pignatello & Huang 1991; Scribner et al., Thus, alachlor was applied to landfarming plots in soil that had weathered prior to the start of the experiments and continued to age so that the only mechanism of transport was functionally via particle detachment and erosion. In the freshly sprayed plots, alachlor residues would have progressively "aged," especially during the second growing season after application, and consequently be transported largely by erosion.

CONCLUSIONS

In summary, systems for landfarming of municipal wastewater and sludges are well developed, but the fate of trace synthetic organics after land treatment has only recently received attention. Knowledge of the fate of alkylated benzenes during land treatment of petroleum wastes is rudimentary, and little attention has been given to the feasibility of landfarming pesticide-contaminated soils. A review of the literature has indicated that landfarming of fuel and pesticide wastes has great potential, but further study is needed to develop practical guidelines and ensure protection of surface and groundwater. The literature also suggests that biostimulation by addition of organic matter may enhance the landfarming process.

The feasibility of landfarming soil contaminated with the herbicide alachlor was tested in small field plots. At 1X rates of loading, estimated DT-50% of alachlor in landfarmed plots was similar to estimated DT-50% in sprayed plots. At higher loading rates, alachlor degradation in landfarmed plots seemed slower than in sprayed plots, especially if the time to loss of 90% of parent was considered. Cumulative percentage runoff and erosional losses of alachlor from 1X landfarmed and sprayed plots were very similar, but losses were disproportionately much higher from landfarmed 5X and 10X plots. The data suggest that landfarming is a viable alternative for disposal of pesticide-contaminated soils, but rate of waste loading must be accurately assessed and monitored. Addition of corn meal to landfarmed soil may enhance the rate of degradation of alachlor through general stimulation of microbial activity.

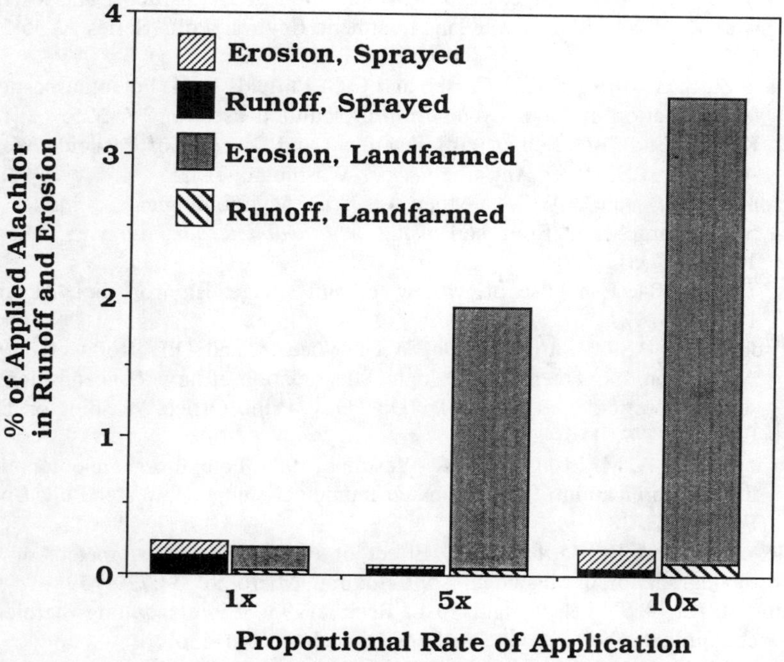

Fig. 14-4. Cumulative runoff and erosion of alachlor from landfarmed and sprayed plots.

REFERENCES

Aurelius, M.W., and K.W. Brown. 1987. Fate of spilled xylene as influenced by soil moisture content. Water Air Soil Poll. 36:23-31.

Barker, J.F., G.C. Patrick, and D. Major. 1987. Natural attenuation of aromatic hydrocarbons in a shallow sand aquifer. Ground Water Monit. Rev. 7:64-71.

Berry, D.F., R.A. Tomkinson, G.H. Hetzel, D.E. Mullins, and R.W. Young. 1993. Evaluation of solid-state fermentation techniques to dispose of atrazine and carbofuran. J. Environ. Qual. 22:366-374.

Bicki, T.J., and A.S. Felsot. 1994. Remediation of pesticide-contaminated soil at agrichemical facilities. p. 81-99. *In* R.C. Honeycutt and D. Schabacker (ed.) Advances in Understanding Mechanisms of Movement of Pesticides Into Groundwater. Lewis Publishers, Chelsea, MI.

Bossert, I., W.M. Kachel, and R. Bartha. 1984. Fate of hydrocarbons during oily sludge disposal in soil. Appl. Environ. Microbiol. 47:763-767.

Brown, D.W., D.C. Donnelly, and L.E. Deuel, Jr. 1983. Effects of mineral nutrients, sludge application rate, and application frequency on biodegradation of two oily sludges. Microb. Ecol. 9:363-373.

Brown, K.W., and K.C. Donnelly. 1984. Mutagenic activity of runoff and leachate water from hazardous waste land treatment. Environ. Poll. (Series A) 35:229-246.

Brown, K.W., and K.C. Donnelly. 1984. Mutagenic activity of runoff and leachate water from hazardous waste land treatment. Environ. Poll. (Series A) 35:229-246.

Cohen, S.Z., S.M. Greeger, R.F. Carsel, and C.G. Enfield. 1984. Potential pesticide contamination of groundwater from agricultural uses. p. 297-325. *In* R. F. Krueger and J. N. Seiber (ed.) Treatment and Disposal of Pesticide Waste. ACS Symp. Ser. 259, Am. Chem. Soc., Washington, DC.

Crawford, R.L., and W.W. Mohn. 1985. Microbiological removal of pentachlorophenol from soil using a *Flavobacterium*. Enzyme Microb. Technol. 7:617-620.

Crites, R.W. 1984. Land use of wastewater and sludge. Environ. Sci. Technol. 18:140A-147A.

Davidson, J.M., P.S.C. Rao, L.T. Ou, W.B. Wheeler, and D.F. Rothwell. 1980. Adsorption, movement and biological degradation of large concentrations of selected pesticides in soil. 111p. U.S. Gov. Print. Office, Washington D.C. EPA 600/2-80-124

Demirjian, Y A., A. M. Joshi, and T. R. Westman. 1987. Fate of organic compounds in land application of contaminated municipal sludge. J. Water Poll. Contr. Fed. 59: 32-38.

Dibble, J. T., and R. Bartha. 1979. Effect of environmental parameters on the biodegradation of oil sludge. Appl. Environ. Microbiol. 37:729-739.

Dzantor, E.K., A.S. Felsot, and M.J. Beck. 1993. Bioremediating herbicide-contaminated soils. Appl. Biochem. Biotech. 40:621-630.

Dzantor, E.K., and A.S. Felsot. 1991. Microbial responses to large concentrations of herbicides in soil. Environ. Tox. Chem. 10:649-655.

Dzantor, E.K., and A.S. Felsot. 1992. Landfarming and biostimulation for decontaminating herbicide wastes in soil. p. 46-67. *In* Proc. International Workshop on Research in Pesticide Treatment/Disposal/Waste Minimization. T. D. Ferguson (ed.) U.S. Gov. Print. Office, Washington D.C. EPA 600/S9-91/047.

Felsot, A., E.K. Dzantor, L. Case, and R. Liebl. 1990a. Assessment of problems associated with landfilling or land application of pesticide waste and feasibility of cleanup by microbiologial degradation. 68p. Final Project Report , IL Hazardous Waste Research & Information Center, HWRIC RR-053, Champaign, IL..

Felsot, A.S. 1991. Land application of contaminated soils. p. 45-59. *In* 1991 Illinois Fertilizer Conference Proceedings, R. G. Hoeft (ed.) Coop. Ext. Serv., University of Illinois, Urbana-Champaign.

Felsot, A.S., and E.K. Dzantor. 1990. Enhancing biodegradation for detoxification of herbicide waste in soil. p. 249-268. *In* K. D. Racke, and J. R. Coats, (ed.) Enhanced Biodegradation of Pesticides in the Environment. Am. Chem. Soc. Symp. Ser. No. 426, Am. Chem. Soc., Washington, D.C.

Felsot, A.S., J.K. Mitchell, and A.L. Kenimer. 1990. Assessment of management practices for reducing pesticide runoff from sloping cropland. J. Environ. Qual. 19:539-545.

Felsot, A.S., J.K. Mitchell, T.J. Bicki, and J.F Frank. 1992. Experimental design for testing landfarming of pesticide-contaminated soil excavated from agrochemical facilities. p. 244-261. *In* J. Bourke, A. Felsot, T. Gilding, J.

Jensen, and J. Seiber, (ed.) Pesticide Waste Management: Technology & Regulation. Am. Chem. Soc. Symp. Ser. 510.

Felsot, A.S., R. Liebl and T. Bicki. 1988. Feasibility of land application of soils contaminated with pesticide waste as a remediation practice p. 55. Final Project Report (HWRIC RR 021), Hazardous Waste Research and Information Center, Illinois State Water Survey Division, Savoy, IL.

Finklea, H.C. and M.F. Fontenot, Jr. 1995. Accelerated bioremediation of triazine contaminated soils: a practical case study. p. 217-231. *In* H.D. Skipper and R.F. Turco (ed.) Bioremediation: Science and applications. SSSA Spec. Publ. 43. ASA,CSSA and SSSA, Madison WI.

Garbarini, D.R., and L.W. Lion. 1986. Influence of the nature of soil organics on the sorption of toluene and trichlorethylene. Environ. Sci. Technol. 20:1263-1269.

Habecker, M.A. 1989. Environmental contamination at Wisconsin pesticide mixing/loading facilities: case study, investigation and remedial action evaluation. 80p. Wisconsin Department of Agriculture, Trade, and Consumer Protection, Agricultural Resource Division, Madison, WI.

Hamaker, J.W. 1972. Decomposition: quantitative aspects. p. 253-340. *In* C.A.I. Goring, and J. W. Hamaker (eds.), Organic chemicals in the soil environment. Vol. 1. Marcel Dekker, Inc. N.Y.

Harrison, E.M., and J.F. Barker. 1987. Sorption and enhanced biodegradation of trace organics in a groundwater reclamation scheme--Gloucester Site, Ottawa, Canada. J. Contaminant Hydrology 1:349-373.

Hutchins, S.R., M.B. Tomson, P.B. Bedient, and C.H. Ward. 1985. Fate of trace organics during land application of municipal wastewater. CRC Crit. Rev. Environ. Control 15:355-416.

Kearney, P.C., M.T. Muldoon, C.J. Somich, J.M. Ruth, and D.J. Voaden. 1988. Biodegradation of ozonated atrazine as a wastewater disposal system. J. Agric. Food Chem. 36:1301-1306.

Kuhn, E.P., P.J. Colberg, J.L. Schnoor, O. Wanner, A.J.B. Zehnder, and R.P. Schwarzenbach. 1985. Microbial transformations of substituted benzenes during infiltration of river water to groundwater: laboratory column studies. Environ. Sci. Technol. 19:961-968.

Loehr, R.C., J.H. Martin, E.F. Neuhauser, R.A. Norton, and M.R. Malecki. 1985a. Land treatment of an oily waste-degradation, immobilization, and bioaccumulation. U.S. Gov. Print. Office, Washington D.C. EPA 600/S2-85/009.

Loehr, R., M. Asce, and M.R. Overcash. 1985b. Land treatment of wastes: concepts and general design. J. Environ. Engineering 111:141-159.

Long, T. 1989. Groundwater contamination in the vicinity of agrichemical mixing and loading facilities p. 139-149. *In* Proc. Illinois Agricultural Pesticides Conf. '89. Coop. Ext. Serv. Univ. of Ill., Urbana-Champaign,.

Marshall, T.R., and J.S. Devinny. 1986. Methods for enumeration of microorganisms in petroleum land treatment soils. Hazar. Waste Hazar. Mater. 3:175-181.

McGinnis, G.D., H. Borazjani, L.K. McFarland, D.F. Pope, and D.A. Strobel. 1989. Characterization and laboratory soil treatability studies for creosote and pentachlorophenol sludges and contaminated soil. U.S. Gov. Print. Office, Washington D.C. 600/S2-88/055.

Mirsatari, S. G., M.M. McChesney, A.C. Craigmill, W.L. Winterlin, and J. N. Seiber. 1987. Anaerobic microbial dechlorination: an approach to on-site treatment of toxaphene-contaminated soil. J. Environ. Sci. Health B22:663-690.

Mueller, J.G., S.E. Lantz, B.O. Blattmann, and P.J. Chapman. 1991. Bench-scale evaluation of alternative biological treatment processes for the remediation of pentachlorophenol- and creosote-contaminated materials: solid-phase bioremediation. Environ. Sci. Technol. 25:1045-1055.

Oudot, J., A. Ambles, S. Bourgeois, C. Gatellier, and N. Sebyera. 1989. Hydrocarbon infiltration and biodegradation in a landfarming experiment. Environ. Pollution 59:17-40.

Pignatello, J.J., and L.Q. Huang. 1991. Sorptive reversibility of atrazine and metolachlor residues in field soil samples. J. Environ. Qual. 20:222-228.

Piwoni, M.D., J.T. Wilson, D.M. Walters, B.H. Wilson, and C.G. Enfield. 1986. Behavior of organic pollutants during rapid-infiltration of wastewater into soil: I. Processes, definition, and characterization using a microcosm. Hazardous Waste Hazardous Materials 3:43-55.

Raymond, R.L., J.O. Hudson, and V.W. Jamison. 1976. Oil degradation in soil. Appl. Environ. Microbiol. 31:522-535.

Rogers, R.D., J.C. McFarlane, and A.J. Cross. 1980. Adsorption and desorption of benzene in two soils and montmorillonite clay. Environ. Sci. Technol. 14:457-460.

Schoen, S. R., and W.L. Winterlin. 1987. The effects of various soil factors and amendments on the degradation of pesticide mixtures. J. Environ. Sci. Health B22:347-377.

Schwarzenbach, R.P., W. Giger, E. Hoehn, and J.K. Schneider. 1983. Behavior of organic compounds during infiltration of river water to groundwater. Field studies. Environ. Sci. Technol. 17:472-479.

Scribner, S.L., T.R. Benzing, S. Sun, and S.A. Boyd. 1992. Desorption and bioavailability of aged simazine residues in soil from a continuous corn field. J. Environ. Qual. 21:1150-120.

Seiber, J.N. 1987. Pesticide waste disposal: review and 1985 National Workshop summary. *In* National Workshop on Pesticide Waste Disposal; JACA Corp., p. 11-19. U.S. Gov. Print. Office, Washington D.C. EPA 600/9-87/001.

Sims, R.C., J.L. Sims, D.L. Sorensen, W.J. Doucette, and L.L. Hastings. 1987. Waste-soil treatability studies for four complex industrial wastes: methodologies and results, volumes 1 and 2. U.S. Gov. Print. Office, Washington D.C. EPA 600/S6-86/003.

Soczo, E.R., and J.J.M. Staps. 1988. Review of biological soil treatment techniques in The Netherlands p. 663-670. *In* K. Wolf, W. J. Vanden Brink, F. J. Colon (eds.) Contaminated Soil '88 Kluwer Academic Publishers.

Soil Science Society of America. 1986. Utilization, treatment, and disposal of waste on land. 318p. Soil Science Society of America, Inc., Madison, WI.

Somich, C.J., P.C. Kearney, M.T. Muldoon, and S. Elsasser. 1988. Enhanced soil degradation of alachlor by treatment with ultraviolet light and ozone. J. Agric. Food Chem. 36:1322-1326.

Novak, J. M., K. Jayachandran, T.B. Moorman and J.B.Weber. 1995. Sorption and Binding of Organic Compounds in Soils and Their Relation to Bioavailability. 1995 p. 13-31. *In* H.D. Skipper and R.F. Turco (ed.) Bioremediation: Science and applications. SSSA Spec. Publ. 43. ASA,CSSA and SSSA, Madison WI.

Staiff, D.C., S.W. Comer, J.F. Armstrong, and H.R. Wolfe. 1975. Persistence of azinphosmethyl in soil. Bull. Environ. Contam. Toxicol. 13:362-368.

Steinberg, S.M., J.J. Pignatelo, and B.L. Sawhney. 1987. Persistence of 1,2-dibromoethane in soils: entrapment in intraparticle micropores. Environ. Sci. Technol. 21:1201-1208.

Taylor, A.G. 1989. Illinois water quality sampling update-pesticides. Ill. Environmental Protection Agency Memorandum. Springfield, IL.

Torrey, S. (ed.) 1979. Sludge disposal by landspreading techniques. Noyes Data Corporation, Park Ridge, New Jersey. 372 p.

Uchrin, G.C., and G. Mangels. 1987. Sorption equilibria of benzene and toluene on two New Jersey coastal plain groundwater aquifer solids. J. Environ. Sci. Health A22:743-758.

Valo, R., and M. Salkinoja-Salonen. 1986. Bioreclamation of chlorophenol-contaminated soil by composting. Appl. Microbiol. Biotechnol. 25:68-75.

Wang, X., and R. Bartha. 1990. Effects of bioremediation on residues, activity and toxicity in soil contaminated by fuel spills. Soil Biol. Biochem. 22:501-505.

Wang, X., X. Yu, and R. Bartha. 1990. Effect of bioremediation of polycyclic aromatic hydrocarbon residues in soil. Environ. Sci. Technol. 24:1086-1089.

Wilson, B.H., G.B. Smith, and J.F. Rees. 1986. Biotransformations of selected alkylbenzenes and halogenated aliphatic hydrocarbons in methanogenic aquifer material: a microcosm study. Environ. Sci. Technol. 20:997-1002.

Wilson, G.B., L.J. Sikora, and J.F. Parr. 1983. Composting of chemical industrial wastes prior to land application. p. 263-273. *In* J. F. Parr, P. B. Marsh, and J. M. Kla (eds.) Land treatment of hazardous wastes. Noyes Data Corp., Park Ridge, New Jersey.

Wilson, J.T., C.G. Enfield, W.J. Dunlap, R.L. Cosby, D.A. Foster, and L.B. Baskin. 1981. Transport and fate of selected organic pollutants in a sandy soil. J. Environ. Qual. 10:501-506.

Wilson, J.T., L.E. Leach, J. Michalowskik, S. Vendegrift, and R. Callaway. 1989. *In situ* bioremediation of spills from underground storage tanks: new approaches for site characterization project design, and evaluation of performance. U.S. Gov. Print. Office, Washington D.C. EPA 600/S2-89/042.

Winterlin, W., J.N. Seiber, A. Craigmill, T. Baier, J. Woodrow, and G. Walker. 1989. Degradation of pesticide waste taken from a highly contaminated soil evaporation pit in California. Arch. Environ. Contam. Toxicol. 18:734-747.

Winterlin, W.L., M.M. McChesney, S.R. Schoen, and J.N. Seiber. 1986. Chemical residues during screening, composting, and soil incorporation of cotton gin waste. J. Environ. Sci. Health B21:507-528.

Wolfe, H.R., D.C. Staiff, J.F. Armstrong, and S.W. Comer. 1973. Persistence of parathion in soil. Bull Environ. Contamin. Toxicol. 10:1-9.

15 Bioremediation Engineering

A.W. Bourquin and T.A. Pedersen
Camp Dresser & McKee
Cambridge, Massachusetts

Biotechnologies are being used increasingly for the remediation of solid, aqueous and gaseous phase media at sites contaminated by petroleum products, hazardous materials and industrial wastes. Successful implementation of bioremediation is contingent upon appropriate site and contaminant characterization coupled with an understanding of regulatory requirements and the site owner's needs. Design and performance of site assessments, engineering studies, and laboratory and field treatability studies are critical to the development of sound cleanup approaches and realistic cost estimates. Remediation goal definition and attainment verification approaches must be established early in project planning to ensure bioremediation technologies are appropriately evaluated during the feasibility study and to allow for the post treatment demonstration of system efficacy. Approaches that maximize the potential for regional, state, local regulatory agency approval, and public acceptance are necessary components of bioremediation projects.

The professional and public attention being given to bioremediation reflects a growing public awareness of environmental issues and an increased interest in natural biological processes. Industry and government are becoming increasingly accepting of biological systems used to minimize pollution and to cleanup contaminated environmental media. Bioremediation is the use of biological agents and processes to reduce contamination, or cleanup soil and water polluted by substances hazardous to human health and/or the environment. Compared to other processes, such as incineration or even landfill disposal, bioremediation is relatively inexpensive due in part to the technology's low energy requirements and the ability for on site implementation. Bioremediation often minimizes equipment mobilization and materials costs, and, in many instances, requires only simple and readily available equipment. These factors make bioremediation very attractive for reclamation of polluted environments.

Since the beginning of time, naturally occurring organic compounds have been recycled by biological processes. Scientists study these processes, including the carbon and nitrogen cycles, mediated by microorganisms. Bioremediation technologies control and enhance these and other natural biological processes to treat toxic and hazardous substances in the environment. The biological agents are usually bacteria and occasionally fungi. These microorganisms most often are indigenous to the contaminated area. Although they may be isolated from elsewhere and introduced to the contaminated site, little evidence of increased *in situ* treatment success through inoculation has yet been presented in the literature. The wide array of microbial metabolic capabilities enables microorganisms to remove or reduce pollutant concentrations to levels that can reduce the potential risk to human health or the environment posed by environmental contaminants. The natural capability of microorganisms to degrade even some complex chemical compounds, such as some synthetic organic pollutants, on a large scale makes bioremediation possible. Environmental remediation contractors and consultants have only recently begun to apply the full spectrum of known bioremediation capabilities identified by microbial physiologists in laboratory settings.

Bioremediation is an extension of the biological treatment processes traditionally used by engineers to treat wastewaters. Composting, sewage treatment, and landfills are biological treatment processes which rely upon the natural biologic activities of microorganisms to degrade organic waste products. These treatment processes serve to reduce contamination of surface and groundwater resources and to deal with the increasing quantities of solid waste produced by society.

The efficiency of many of these established technologies demonstrate how natural biological processes can be used to minimize environmental contamination. Although still practiced today, burying or burning wastes is becoming a less desirable option for many hazardous waste generators. Energy intensive incineration terminally destroys the contaminants, whereas the "dig and haul" approach merely moves the waste to another location which in many cases can lead to future problems. Since the passage of Superfund Amendments and Reauthorization Act (SARA), and the Resource Recovery and Conservation Act (RCRA) "land ban," regulatory agencies and responsible parties have focused attention on terminally destructive techniques, including bioremediation, to solve environmental problems.

ATTITUDES AND APPLICATIONS

Just a few years ago bioremediation was considered inappropriate for the treatment of contaminated media encountered at Superfund sites. The U.S. Environmental Protection Agency (U.S. EPA) did not advocate biological treatment of toxic and hazardous waste until late 1986, when it funded the "Microbial Systems for Hazardous Waste Treatment" research and development initiative. Even then U.S. EPA considered the technology to be in the research and exploratory stages. Following the Valdez incident, U.S. EPA enthusiastically endorsed bioremediation and since then has recommended its consideration as a potentially applicable cleanup technology for contaminated sites. Unfortunately even now, five years after the Valdez spill, certain segments of federal and state regulatory agencies and the remediation industry still have misconceptions of biological treatment capabilities and potentials. Some of these misconceptions include: the amount of time needed for bioremediation, the types and concentration of compounds that are amenable to biodegradation, the necessity for specialty cultures, the extent of remediation possible and achievable cleanup levels, and the level of understanding of science necessary to implement biological treatment in the field.

Application of bioremediation is being accomplished with varying degrees of success. Since municipal wastewater treatment has been the primary area of application of biological treatment, the engineering community has naturally attempted to direct this experience towards bioremediation of soils and groundwater. Initial reactions to bioremediation within the engineering community was one of "What's new, we have been doing this for many years"? However, it was soon evident that biological treatment of contaminated soils and groundwater was technically different from the treatment of municipal wastewater. The lack of process understanding resulted in a number project failures. The engineering community at times reacted by considering bioremediation as technically impractical or a technological "flash in the pan" which would not last long or have any broad application range. This attitude may have been an impediment to the acceptance of bioremediation and the resulting low numbers of field applications that occurred in the early and mid 1980s.

Remedial investigation and feasibility studies (RI/FS) often could not justify use of bioremediation because of a lack of request supporting demonstrating its technical feasibility data. Nevertheless many successful projects were undertaken to address petroleum product release even in the absence of scientific documentation. As specialty bioremediation firms, sometimes referred to as "bio-boutiques," came on the scene in the late 1980s, a renewed emphasis was placed on the application of the technology to hazardous waste site remediation. Many "bio-boutiques" were able to successfully apply the technology to affected environmental sites of relatively easy to degrade contaminants such as petroleum products. More importantly, in many cases they demonstrated good scientific data for biodegradation of the constituents. Successful applications of biotreatment lead to interest in bioremediation as a potentially cost saving and effective technique, by the news media, the general public, responsible parties, and regulatory agency staff. The engineering community reacted by offering this technology to clients and, in many cases, approached it as a technology that could be "bought" as needed from vendors (bioboutiques) with little internal expertise necessary to understand the technologies, intricacies, and potential implementation problems. Unfortunately, this was the same approach taken just a decade earlier leading to misapplications and generalities that did not support biotreatment. In many cases, bioremediation was perceived as a technology that differed little from conventional wastewater treatment processes and little understanding of biochemical processes were needed in order to implement. However, unlike municipal or industrial wastewater treatment processes, bioremediation of complex wastes, in many cases, requires a thorough understanding of the biochemistry to successfully implement the biodegradation of specific chemicals in the field.

As the hazardous waste site remediation industry has matured, an understanding of the differences between biotreatment of hazardous waste and of wastewater containing naturally occurring organic compounds has resulted. The engineering community has recognized the need for a multidisciplinary approach to solving complex environmental problems and has endeavored to access the expertise of the appropriate engineering and scientific disciplines. As such, the involvement of microbiologists and biochemists in remediation projects is becoming more commonplace.

CRITICAL SUCCESS FACTORS

Bioremediation can be effective only where environmental conditions permit microbial growth and activity. In some cases, the environment can be modified to support or accelerate microbial growth, for instance, by nutrient addition or pH adjustment to facilitate growth of indigenous microorganisms. In other cases, contaminated soils may be excavated and treated in on site bioreactors, or contaminated waters may be passed through biological treatment facilities that allow more control of the environmental conditions.

The most widely publicized application of bioremediation to date has been the treatment of the oil-contaminated shorelines of Prince William Sound. Recognizing that natural washing processes could not remove all the spilled oil that contaminated the rocky shores, the U.S. EPA conducted tests of bioremediation that demonstrated dramatic removal over control areas. These findings formed the basis for treatment of 70 miles of shoreline. Initially, the high degree of variability of pollutant concentrations in a natural setting made it extremely difficult to demonstrate statistically the benefits of bioremediation. Fortunately, the visible evidence of cleansing of the rocks due to bioremediation was enough to convince those responsible to permit nutrient additions to other areas of the Sound. Subsequent statistical analyses of treated areas showed that nutrient addition produced at least a doubling of the rate of removal over untreated areas (Atlas, 1990).

Although the Alaskan oil spill cleanup represents the most visible use of bioremediation to the general public, there have been many other successful field applications. Nutrients and oxygen have been added to stimulate microbial biodegradation for the remediation of groundwater contaminated by gasoline since the 1970s, (Raymond, Biosystems). The addition of oxygen from air or hydrogen peroxide for treating fuel-contaminated aquifers, has become an accepted treatment. In other cases, polluted soils have been remediated through stimulation of the natural microbiota to degrade contaminates such as oil, phenol, creosote, pentachlorophenol, and some pesticides. Hundreds, perhaps thousands, of sites have been cleaned through the use of biological treatment (U.S. EPA, 1993b). Unfortunately, few of these restorations have been extensively documented. This, as well as other technical drawbacks, has contributed to the slow acceptance of bioremediation in certain sectors. William K. Reilly, former administrator of the U.S. EPA, stated, "Having been to Alaska to check on the progress of the Prince William Sound cleanup, I've seen what bioremediation can do to minimize the effects of a massive crude oil spill. It is virtually the only good news to come out of that tragic situation" (Atlas, 1990). Despite this endorsement and the successful applications to date, the implementation of bioremediation has not advanced at the rate which would be expected for a technology with such widescale applicability. In the past five years, many factors have influenced the selection and application of bioremediation by engineering companies; many of the decision points for consideration of bioremediation are reached in the early phases of engineering projects.

PROJECT PHASES

In undertaking the remediation of contaminated sites, a series of investigations and studies are undertaken. The studies include site investigations, engineering studies, risk assessments, and treatability studies which are followed by design, corrective action implementation and, finally, site closure.

Site Characterization

Before any bioremediation project is initiated, an assessment of the site conditions and the nature and extent of contamination is required. Site investigations are referred to by various terms depending on the regulatory program under which it is implemented. For example, at the Federal level, the study phase for sites listed on the National Priority List (NPL) and being undertaken pursuant to the Comprehensive Environmental Response Compensation and Liability Act (CERCLA) and Superfund Amendments and Reauthorization Act (SARA), are termed remedial investigations (RI). Whereas an investigation pursuant to the Resource Recovery and Conservation Act (RCRA) is termed a RCRA Facility Investigation (RFI). States refer to these investigations by a variety of terms ranging from Site Assessments (SAs) to Corrective Action Studies. Although the degree of sampling and rigor imposed on each of these studies varies considerably, all are directed at gathering information on the nature and extent of contamination at the site. What often is missing in these early studies is the collection of data useful for the evaluation of bioremediation applicability. By identifying potential remedial actions early in the investigative stage, the data required for the evaluation of bioremediation applicability can be obtained. The process of identifying what data is needed as part of an investigation to ultimately select a remedy has been outlined in the Data Quality Objectives (DQOs) development process guidance document (U.S. EPA, 1987). Application of this process in investigative phase of any project should provide for increased consideration of data types needed for the evaluation of bioremediation options.

Engineering Studies

In most instances, the selection of a remedy is based on information gathered during the Feasibility Study (FS) for Superfund sites or Corrective Measures Studies (CMS) for RCRA facilities. These reports provide a comparative evaluation of a variety of technologies from the standpoint of implementability, reliability, cost and other factors to support an informed risk management decision regarding what remedial actions appear to be most appropriate given what is known about the site. In the past, comparative information on bioremediation has not been as readily available as that for the more established technologies such as incineration. As bioremediation processes have become more common, the database on effectiveness and costs have grown to be useful in conducting a more balanced assessment of the appropriateness of bioremediation during feasibility studies.

Treatability Studies

The dynamic nature of biological processes renders bioremediation more prone to site specific influences than many other remedial technologies. One way of overcoming the difficulty in extrapolating data from site to site is by the performance of biotreatability studies on the specific soils or groundwater found at the site in question. These studies are directed towards development of site-specific performance data for biotreatment.

Before implementation of any significant environmental remediation activity, treatability studies are undertaken to assess the contaminant and media characteristics related to treatment effectiveness. In many cases, it is preferable to conduct the treatability studies during the investigative or feasibility study phase of the project, rather than at the predesign phase if there is a reasonable probability of success or determined during the FS.

Treatability studies are generally conducted in a phased manner which progresses from simple laboratory studies, to bench scale systems, and finally to field pilot scale systems prior to design of a full scale system. Treatability studies are conducted to develop design criteria and to generate scaling factors for the full scale system and often include optimization components. Initial laboratory studies may be simple flask studies which give basic information on the applicability of the technology to the given site and the probability of success. Bench scale studies are usually more elaborate in nature and include information on optimization, extent and rate of bioremediation, and important scale up factors for the full scale design. Pilot demonstrations are conducted to develop full scale design criteria and cost estimates including verifications of laboratory data. The level and complexity of treatability study will depend much on the complexity of the site and the level of data needs for implementation of the technology.

Since many treatability studies are not conducted in a rigorous research mode, the data generated is often not reported in the scientific literature. In some cases, the data is of limited value because of the lack of attention given to experimental design particularly relative to mass balance or abiotic controls. U.S. EPA has developed a series of guidance documents prepared to assist in undertaking treatability studies for Superfund sites (U.S. EPA, 1991, 1992a,b). The guidance document developed for bioremediation treatability studies provides a good basis for designing studies. However, additional efforts are needed in standardizing the critical data needs for treatability study approaches.

Treatability study information available for bioremediation has been limited in the past. However, the VISITT database system (U.S. EPA, 1993b) provides a vehicle for the exchange of treatability data in a time efficient manner. Although many vendors have provided data to U.S. EPA for inclusion in the VISITT database, results for most studies do not find their way to this repository. The VISITT database as of April 1993 contained information on 231 remedial technologies of which 36 percent are bioremediation technologies. Information is available on 1887 completed or planned bioremediation projects. Although a significant improvement in available information, this is probably a low number of bioremediation projects completed to date. The database development is limited by the companies performing the work being able to document and submit the data to VISITT. They are limited in time dedicated to nonrevenue generating activities and in some cases by the client's reluctance to allow publication of information on their site.

Engineering Design

Once a remediation approach is selected, design criteria are developed which ultimately lead to the preparation of engineering contract documents. These contract documents include the contractual obligations of the site owner, the engineer and/or constructor and set out the plans and specifications for the system. Designers rely on the engineers/scientists that have been involved in the early phases of the project to develop site specific plans and specifications for bioremediation or other selected remedial action.

It is important for scientists to provide direction and review during the development of the remedial design criteria and through the contract document preparation phase to ensure that the systems constructed will provide for the control and monitoring ultimately required for bioremediation systems. Many field projects have suffered, or appeared to have failed, due to insufficient or inappropriate data acquisition during the monitoring schedule. It is important that the appropriate types and quality of data be acquired during the operational phase of a project. Otherwise, the design may not adequately demonstrate performance of the technology or even worse the insufficient data acquisition may not exercise adequate control over the process.

The incremental costs of environmental monitoring during remediation are often not included as part of a final design. In effect, this leads to the implementation of bioremediation, or other technologies, without the collection of the data to adequately review the process performance. Planning for the monitoring phase during the feasibility study and demonstrating to site owners the necessity for monitoring is an important aspect of project planning.

Remedial Action

Remedial action implementation extends from the initial construction activities through the operation and maintenance period for the remedy. It is during this stage of the program that a tremendous amount of data can be obtained regarding the process effectiveness. Interestingly enough, it is at this stage that major data gaps often result. The effectiveness of a remedial action should be monitored over the course of its implementation. In order for the funds to be available for monitoring, these activities must be anticipated during the feasibility study and treatability study phases.

Post Closure

Following closure of a remedial project, groundwater and, in some cases, soil gas or soil samples, are collected to verify that remediation has been complete. During this phase of a project, useful information can be obtained about the effectiveness of a bioremediation system.

Post closure monitoring data for a variety of technologies such as pump and treat and soil vapor extraction (SVE) have demonstrated that site remediation has in fact not been achieved. This is sometimes due to a lack of understanding of the mechanisms of the remediation technology or sometimes due to inappropriate design which results in untreated areas. The latter is a common occurrence where *in situ* treatment projects are implemented.

TREATMENT TECHNOLOGIES

The engineering community has historically taken the lead in designing site remedial action plans. Foremost, is the fact that, engineers have been providing potential clients with quality engineering services for many years. In the areas of hazardous waste, engineers have been relied upon to conduct site assessments and develop remedial designs. The engineering community is now in the position to influence site owners and regulatory agencies on the feasibility of remedial techniques for site cleanup, including using bioremediation. Engineering companies have a record of accomplishments and current site owner contacts and therefore must be looked to for advancing the application of bioremediation.

Engineers are traditionally the implementors of processes. Bioremediation connotes a general process rather than a technology, which must be understood in order to successfully implement. Since biodegradation processes (biochemistry) differ depending on the particular compound to be degraded, the approach to remediation is quite different from that taken to design treatment system for domestic sewage, where many different organisms are able to degrade the naturally occurring compounds. Bioremediation is not a technology that, "once built or bottled" for one site, can necessarily be applied to another site, regardless of how similar in nature the conditions appear. Biological processes are dynamic and, therefore change, with the environmental conditions of the natural system. In order to achieve degradation of specific problem contaminants in environmental systems, it is important to manage the specific process and not just the general process of biodegradation of organics. In addition, treated effluent streams may reflect combined fate processes and disappearance of contaminants can be mediated by volatilization or some unknown mechanism in addition to the affects of microbial agents. Hence, recent approaches have been to integrate engineering and scientific disciplines to achieve a higher quality product and fewer operational problems in the field. Generally, scientists identify the control parameters needed to maintain the desired reactions and engineers assure that these processes are monitored and maintained in the field.

Biological treatment technologies for contaminated soils and groundwater can be categorized in four primary areas: landfarming or solid phase bioremediation, slurry reactors, *in situ* biotreatment, and combined vapor-phase technologies. Choosing the best process is a function of the physical and chemical nature of the contaminant and the matrix in which the contamination is found.

Solid-Phase Bioremediation

Landfarming relies on principles applied in agriculture in the biocycling of natural compounds. The conditions for biodegradation by the natural indigenous microbial populations of soil are optimized by regular tilling of the soil and by the addition of nutrients and water. Landfarming has been successfully used for years in the managed disposal of oily sludge and other petroleum refinery wastes. Solid-phase biotreatment of petroleum- and creosote-contaminated soils is probably the most widely used and cost effective biotreatment technology applied today. However, federal regulations (Resource Conservation and Recovery Act, "land ban") now prohibit most of these types of land based treatment techniques without provisions to control fugitive emissions and leaching of organics and metals (no migration). Unfortunately, these restrictions may have been brought about by the mismanagement of landfarms in the past rather than nonperformance of the systems. Solid-phase biotreatments systems are designed on the same principals of landfarming but no sludge is "disposed on land," rather, the contaminated soils are treated in place to facilitate biodegradation of constituents. A modified solid-phase bioremediation system which controls volatiles can be completely covered by a modified plastic film commonly used in greenhouse (Bourquin, 1989).

Biofilters

Gaseous volatile organic compounds extracted from soils using soil vapor extraction, from groundwater by air sparging or from soil treatment processes, may be treated by biofilters. Biodegradable volatile organic compounds can be treated in a vapor-phase bioreactor; nonbiodegradable volatile organic compounds can be removed by adsorption on activated carbon. The nature of contaminant, concentration, and volume of the air emissions; regulatory controls; and cost effectiveness all influence the choice of design.

Bio-Oxidation Piles/Biopiles/Soil Piles

Biopiles or soil pile biotreatment can be considered a modification of solid-phase biotreatment or a modification of soil vapor extraction that is usually used where space is limited and/or the contaminants are volatile. In soil pile bioreclamation, contaminated soil is excavated and piled on a lined area. The pile is covered and an air emissions recovery system installed. An internal piping system blows air through the soil and accelerates biodegradation through the addition of oxygen or vacuums air and volatile contaminants out of the soils while pulling oxygen into the pile. Nutrients are applied to the surface of the stockpile and allowed to percolate down through the soil. A leachate collection system allows the fluid to be recycled. During operation, pH and moisture content are maintained within ranges conducive to microbial activity. This technique combines soil vapor extraction or vapor-phase bioreactors with biological treatment systems to effectively degrade soil contaminants (Bourquin, 1989).

Bioreactors

The biotreatment of organic waste in bioreactors is effective for control of manufacturing effluents, groundwater cleanup, and other tasks. Several commercially available bioreactor treatment systems have been used for hazardous wastewater treatment (Bourquin, 1989). The advantages of using a bioreactor include:

- greater process management and control,

- increased contact between microorganisms and contaminants (less heterogeneity),
- use of specific cultures or inoculum, and
- decreased acclimation times or faster biodegradation rates.

Slurry Phase Bioremediation

In slurry-phase bioremediation contaminated soils are treated as an aqueous slurry in a bioreactor. This system maintains intimate mixing and contact of microorganisms with the hazardous compounds and creates the ideal environmental conditions for microbial biodegradation of the target contaminants. The decontamination process, however requires excavation and handling of the contaminated material; the required slurry retention time may vary. The bioreactor has the potential to operate in batch or continuous modes as well as sequential batch reactors. Treated soils pass through a water separator and the water, which then contains high populations of acclimated microorganisms, is recycled. This process greatly reduces the acclimation and treatment times needed for subsequent batches (Bourquin, 1989).

In Situ Bioremediation

In situ bioremediation is the biological treatment of contaminated soils and groundwater without excavation, usually where contamination is deep in the subsurface or under buildings, roadways, etc. *In situ* treatment involves the controlled management and manipulation of microbial processes in the subsurface. This requires an understanding of the microbiological processes needed to degrade the target contaminants, and the soils physical and chemical environmental effects on the microbial processes. Typically these systems use aerobic processes and involve the addition of oxygen as air, oxygen gas, or hydrogen peroxide, and small amounts of inorganic nitrogen and phosphorous. Recent evidence indicates that anaerobic denitrification processes may be effective in the biotreatment of BTEX-type (gasoline) contamination (Hutchins, et al., 1991).

The engineering design for any site is highly dependent on types of contaminants present, permeability of the soils, regulatory constraints, contamination of vadose zone, and other characteristics. A remedial design is highly dependent upon the contaminated site and costs may vary greatly (Bourquin, 1989).

Bioventing

Bioventing is the latest in biological treatment application technologies which incorporates both *in situ* bioremediation and soil vapor extraction. It is the fastest growing technology application in the US today and is highly effective on volatile compounds such as gasoline and light fuel oils. *In situ* bioremediation, historically, has relied on recirculation of water through the contamination zone to carry oxygen to the contaminated area. The water is usually saturated with air, pure oxygen, or hydrogen peroxide. These efforts have occasionally met with only limited success, however, due to the physical limitations of transferring water through the soils, resulting in inadequate oxygen supplies to the subsurface contamination. The inherent disadvantage of utilizing water as the carrier medium for the transfer of oxygen to the subsurface is illustrated by the fact that over 100,000 lbs of air saturated water is required to deliver 1 lb. of oxygen to the subsurface, whereas only 4.5 lbs. of air is required to deliver 1 lb. of oxygen to the subsurface.

Bioventing describes a process in which air is utilized to deliver oxygen to the subsurface to stimulate *in situ* biodegradation of organic contaminants. Bioventing represents a hybrid physical/biological process utilizing soil vapor extraction (SVE) techniques for oxygen transfer into the vadose zone for *in situ* aerobic contaminant biodegradation rather than contaminant air stripping.

Conventional SVE systems are designed to optimize the maximum volatilization rate of compounds having high vapor pressures from contaminated soil. Bioventing systems are designed and configured to optimize the oxygen transfer and oxygen utilization efficiency to biodegrade organic constituents in the unsaturated contaminated soils. Bioventing uses the hardware for an SVE system for the transport of oxygen to the subsurface, where indigenous organisms are stimulated to aerobically metabolize fuel components. Bioventing systems operate at lower flow rates and with different well configurations than those of conventional SVE systems. In general, bioventing systems are required to operate at pore volume exchanges of approximately one pore volume every 2 to 10 days (Hinchee, 1989).

IMPLEMENTATION CHALLENGES

Two examples where a lack of process knowledge has resulted in problem of field implementation are biodegradation of polynuclear aromatic hydrocarbons (PAHs) and trichloroethylene (TCE). Attempts at bioremediation of creosote contaminated soils have often resulted in high residues of PAHs. Field implementors thought the lack of biodegradation at some sites was due solely to bioavailability factors. Knowledge of PAH biochemistry indicates that although bioavailability plays a significant role in biodegradation of the contaminants, the presence of inducer, or energy substrate, is equally important. Microbiologists have known for years that, in most cases, the four to six ring PAHs are degraded only when significant quantities of two to four ring compounds are present as energy substrates and inducers (Bourquin, 1989). However, this level of sophistication had never been examined in the field until the late 1980s. Pilot studies have (Bourquin, 1989) shown the same level of inducer substrate was needed to achieve degradation of the higher ring compounds as previously seen in the laboratory. To date, the bioremediation of creosote contaminated soils is still problematic; however, the uncertainty associated with cleanup attainment is being further elucidated.

Another example involves the biodegradation of halogenated aliphatic compounds such as trichloroethylene (TCE). Attempts to aerobically biodegrade these contaminants in groundwater systems have generally failed. Treatment strategies that work for gasoline contamination or other easily-degraded contaminants have been applied. The addition of nutrients and oxygen to stimulate microbial degradation of these contaminants, which generally results in reduction of petroleum products in groundwater, does not effectively decrease chlorinated solvent concentrations of impacted groundwaters. Early attempts at bioremediation of TCE contaminated groundwaters using pump and treat technologies in open bioreactors wrongly attributed treatment to microbes when, in fact, removal was due to volatilization (HazMat, 1987). Other early investigations reported biodegradation, *in situ*, when TCE was a co-contaminant with benzene, toluene, ethylbenzene, xylene (BTEX) (personal communication, Litchfield, 1987). Unfortunately, the latter case was not reproducible at other sites because no TCE degrading bacteria could be found. It was not until 1986, when it was first reported that toluene or phenol grown bacteria were capable of degrading TCE (Nelson, et al., 1990). This understanding opened the way, for realistic remedial technologies for these contaminated aquifers (Nelson, et al. 1990). This discovery also helped to clarify reasons that biodegradation occurred at some sites but not at others.

The difficulties in engineering field operations are due to the necessity for inducting a specific reaction (enzyme) in a controlled manner. The organisms that are responsible for TCE bioremediation do not utilize TCE as a carbon and energy source, but rather cometabolize it. Therefore, the process cannot be managed solely by the addition of oxygen and a nitrogen source. The specific production of a single enzyme is necessary to carry out the process and must be managed continuously in the field. Many operational and biological problems can develop in such a process. The degradation of TCE is both an energy drain for the microorganism and a competitive inhibitor for the energy substrate, phenol. Therefore, the organism loses any competitive advantage it may have for energy substrates (Nelson, et al., 1990) and washes out of the system or is replaced by another non-TCE-degrading phenol-degrader.

To date, no fully successful demonstration of this TCE bioremediation has been conducted in the field. However, we believe it is a matter of integrating expertise at the scientific and engineering levels to finally achieve a successful move of this science into the field in a controlled and demonstrable manner. The combined expertise and understanding of engineers and scientists will be need to finally manage the uncertainty which is now hampering the remediation of chlorinated solvent contaminated aquifers. Extensive knowledge of the biochemistry of TCE biodegradation is available. Unfortunately, much of this information is not developed with a focus on implementation. Integrated science and engineering will help to bring this technology to the field.

As more bioremediation projects are implemented, the database of information will expand allowing greater acceptance of the technology. The major impediments to implementation of bioremediation projects are:

- Lack of demonstrated ability to achieve cleanup goals
- Uncertainty in cleanup time estimates
- Cost data

In some cases, vendors anxious to implement remedial actions will undertake projects that may not be appropriate for bioremediation. "Snake oil sales" occur in all aspects of life and bioremediation vendors are not immune to the temptation to oversell capabilities. Such tactics have caused a great deal of harm to the bioremediation industry and hindered its acceptance. However, site owners and responsible parties are becoming more aware of what constitutes good science and are not as likely to accept the claims of overzealous marketers.

The major driving force for advancing the application of bioremediation is the availability of credible data on system performance. There exists a considerable scientific database regarding bioremediation in controlled laboratory or field studies, appropriately designed to evaluate the effectiveness of a system. Unfortunately, most of these studies have been conducted on similar compounds, mainly the petroleum hydrocarbons. The U.S. EPA is currently conducting field demonstrations at various sites across the country (under the Bioremediation Field Initiative), to further develop the database of useful and appropriate information on the technology's performance in the field. This work includes some innovative as well as new science applications for problem contaminants as well as the more common petroleum compounds.

The academic and engineering community must endeavor to collect meaningful data that can be used to support the appropriate application of bioremediation. No better means exist to stem the tide of misinformation than flooding the market with good science.

The engineering community is not involved sufficiently in the development of new technologies. However, the U.S. EPA and other government organizations are currently putting a significant sum of money into new research in an effort to further develop the technology and save considerable dollars in remediation costs for both industry and government. Engineering companies can assist in this effort through cooperative demonstration in the field and continual feedback on system performance from the field.

CONCLUSION

Environmental remediation projects require the application of a variety of engineering and scientific disciplines. The interdisciplinary nature of these projects is further complicated by the onerous regulatory setting and litigious nature of environmental remediation activities. These factors combined with the constraints placed on project managers with respect to schedule and budget will influence the final selection of a remedy. In the past, a strong bias existed for selecting proven technologies. Substantial documentation would be required to select any innovative technology. The U.S. EPA has fostered a more careful consideration of innovative remedial technologies by making technical information needed for decisions more readily available through its Technology Innovation Office. The U.S. EPA Superfund Innovative Technologies Evaluate (SITE) program and the Vendor Information System for Innovative Treatment Technologies (VISITT) are two programs that have served to encourage technology transfer in the environmental engineering community.

The U.S. EPA has boldly encouraged the use of bioremediation technologies specifically through the bioremediation forum, a consortium of academicians, industry scientists and engineers, and regulators formed to discuss and remove impediments to bioremediation implementation. In addition, U.S. EPA has provided funds to implement the Bioremediation Field Initiative to expand the nation's field experience in bioremediation technologies and to fully document performance of full-scale applications. These efforts have contributed greatly to the continuing increase in numbers of applications in bioremediation.

The multidisciplinary nature of bioremediation necessitates that those scientists who understand microbial processes be involved in the engineering of the system to be applied in the field. As more complex technological principals are applied, the more important it is for the scientific expertise needed to address the issues be brought together with the available engineering expertise. Otherwise, bioremediation projects will fail in the field and the risk is that bioremediation will again become known as an unreliable technology.

Bioremediation is a feasible and practiced technology for many pollutants. Many contaminants are effectively treated to regulatory standards with minimal involvement of scientists. However, most improvements to the field applications can be attributed to an integration of scientific input with field engineering expertise. New technologies will require the systematic integration of science with engineering to implement effective bioremediation.

REFERENCES

Atlas, R.M. 1990. Focus on Bioremediation, ASM News, 56:7. Amer. Soc. Microb.

Bourquin, A.W. 1989. Bioremediation of Hazardous Waste. Hazard. Mater. Cont. 2:16-59.

HazMat. 1987. Annual Meeting, Washington, DC.

Hinchee, R.E., D.D. Downey, J.K. Slaughter, and M. Westray. 1989. Enhanced Biorestoration of Jet Fuels: A Full Scale Test at Eglin Air Force Base, Florida. Air Force Engineering and Services Center Report ESL/TR/88-78.

Hutchins, S.R., W.C. Downs, G.B. Smith, J.T. Wilson, D.J. Hendrid, D.D. Fine, D.A. Kovacs, R.H. Douglas, and F.A. Blaha. 1991. Nitrate for Biostoration of an Aquifer Contaminated with Jet Fuel. U.S. Gov. Print. Office, Washington D.C. EPA/600/S2-91/009.

Litchfield, C.D. 1987. Personal Communication with A. Bourquin.

Nelson, M.J.K., P.H. Pritchard, and A.W. Bourquin. 1990. Biodegradation of Halogenated Aliphatic Hydrocarbons, U.S. Patent No. 4,959,315.

U.S. Environmental Protection Agency. 1987. Data Quality Objectives for Remedial Response Activities: Development Process. U.S. Gov. Print. Office, Washington D.C. EPA/540/G87/003.

U.S. Environmental Protection Agency. 1991. Guide for Conducting Treatability Studies Under CERCLA: Aerobic Biodegradation Remedy Screening. Interim Guidance. U.S. Gov. Print. Office, Washington D.C. EPA/540/2-91/013.

U.S. Environmental Protection Agency. 1992a. Bioremediation of Hazardous Waste. U.S. Gov. Print. Office, Washington D.C. EPA/600/R-92/126.

U.S. Environmental Protection Agency. 1992b. Guide for Conducting Treatability Studies Under CERCLA - Final. U.S. Gov. Print. Office, Washington D.C. EPA/540/R-92/071a.

U.S. Environmental Protection Agency. 1993a. Bioremediation Research Guide. U.S. Gov. Print. Office, Washington D.C. EPA/542-B-93-004.

U.S. Environmental Protection Agency. 1993b. VISITT: Vendor Information System for Innovative Treatment Technologies User Manual Version 2.0. U.S. Gov. Print. Office, Washington D.C. EPA-542-R-93-001.

16 Applying Soil Science Skills in the Remediation Business

S.D. Cunningham
DuPont
Central Research and Development
Newark, Delaware

M.S. Mather
Petro Environmental Technology
Cincinnati, Ohio

By 1995, spending on hazardous waste management should reach 17 billion dollars per year. Annual growth rates in certain sectors of the industry are currently in excess of 10% (Staff, 1993). By all measures, the environmental remediation industry is large and growing. This industry represents a significant new market for technology, research, and employment opportunities that has largely been untapped by soil scientists. Certain scientific disciplines, in particular geology and hydrogeology, have responded well to this opportunity. In these and certain other disciplines, academic curricula have been adjusted, students are actively recruited, and research results are actively utilized in the remediation marketplace. Unfortunately this has not been the case with soil science.

It would seem obvious that a background in soil science would be an asset in the remediation field. The vast majority of all pollutants are either heavy elements or organic compounds that have strong physical, chemical, and/ or environmental similarities to plant nutrients and agricultural pesticides. Remediating contaminated soils involves knowledge of chemistry, microbiology, and the physical movement of the material in the matrix in which it is found. Despite these similarities however, few soil scientists are actually employed here. This chapter seeks to explore: the potential market for soil science in the remediation arena, some of the barriers to their entering the field, and some areas where their input is needed in particular. Our intended audience are soil scientists interested in environmental remediation issues. The authors bring differing perspectives to the topic of soil remediation, the perspective of both consumers and providers of technology as well as the perspective of a large corporation and a private consulting firm.

SOIL SCIENCE-A MARKET ANALYSIS OF ITS OFFERINGS

Soil Science departments, like most university departments, have three major products: research, graduating students, and extension services. An active department produces research and extension services that are useful to a particular community and graduates who find jobs in their fields. As in any industry, if the market changes, production must change. These changes can be made in either product composition, quality, quantity or shifts in timing and geographic location. Without market feedback, departments, like industries, begin to suffer from a lack of funding as well as a lack of individuals willing to choose that department/profession as a career.

Demand for the three products of soil science departments is affected by the total size of the market, market growth rate, perception of the product (often referred to as "positioning"), and cost of production. To underline the significance of the remediation opportunity to the soil science community, we offer the following comparisons.

The U.S. hazardous waste remediation market is growing rapidly and is predicted to consume \$750 billion dollars over the next 30 years (Russell, et al. 1991). This \$25 billion/year average (in the US alone) compares almost exactly to the 1993 global pesticide market (Fig. 16-1). The US "bio"remediation market, (perhaps of most concern to soil microbiologists), is expected to increase at an annual rate of 15-16% through the next decade (Staff, 1993c). Additionally, as the market matures, growth will incorporate increases in efficiency (on a constant dollar basis). In contrast, the pesticide industry is predicted to grow only slowly (<2%) and has actually shrunk by nearly \$1 billion over the last year (due largely to changes in European farm policies). This near parity of our domestic hazardous waste industry with the global pesticide industry and this large differential in growth rates has gone unrecognized as an opportunity in many soil science departments and laboratories.

The marketability of three products of soil science departments are also affected by "cost of production." In general, research and education costs are field specific. We would argue, however, that the "cost" of producing research and graduates would seem to be similar whether the specific expertise is: a) the mobility, fate, and degradation of a pesticide in an agricultural soil, or b) the mobility, fate, and degradation of a priority pollutant in an industrial soil. For this reason, and a general belief that soil science departments can "retool" at relatively low expense, we have eliminated this consideration from our arguments.

Another way of accessing demand for research and people is to compare and contrast job opportunities for graduates in the two disciplines. Comparing the back of the *Agronomy News* letter (National in scope) to a large city newspaper should convince one of the uneven job prospects. If salaries are an additional reflection of supply and demand (and hence the ability of a field to support more practitioners), then Fig. 16-2. (Bishop, 1992) may also be enlightening. Similar comparisons may be available for "agronomic" scientists, but could not be located by these authors.

Our perception, however, is that soil scientist when employed in green-plant based industries earn significantly less than equivalently trained people associated with the remediation field.

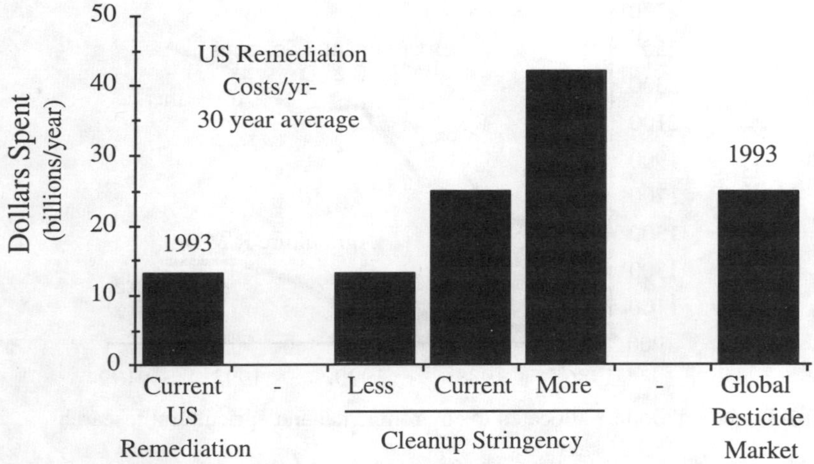

Fig. 16-1. Market estimates of the size of the U.S. remediation market relative to the global pesticide market (Russell et al., Staff 1993c, and DuPont internal documentation).

Another variable that is often important to recruiting students into a field and funding research is perception of "worthiness of the cause." This is a nebulous and highly personal factor and reflects the general public perception of the field. In one author's experience (who has worked in the soil science related subjects of nitrogen fixation, pesticide development, biological control of crop pests, and hazardous site remediation research), introducing himself to an unaffiliated group as a "practitioner of these arts" receives widely differing group responses. The interest and/or positive feed back expressed by the general public also seems to be directly correlated with the ability to attract funds at governmental, institutional, and commercial levels.

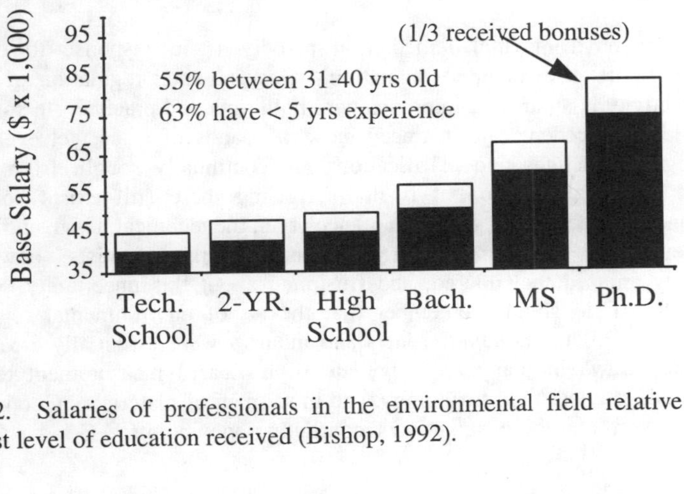

Fig. 16-2. Salaries of professionals in the environmental field relative to their highest level of education received (Bishop, 1992).

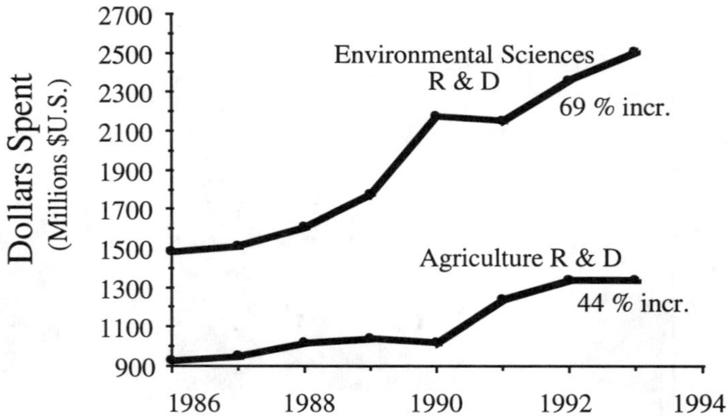

Fig. 16-3. Federal dollars allocated to environmental and agricultural research.

To some degree, funding of the sciences is "faddish" and cyclical. Research in nitrogen fixation and biological pest control had peaks of "fashion" in the last decade. "Environmental" concerns by the public, as well as the search for new missions for DOD and DOE laboratory facilities, is currently "in fashion." "Fashion" acts as a strong driving force in increased governmental and industrial research in the area. Fig. 16-3 plots the trends in federal spending on environmental and agricultural sciences over the past nine years.

Evidence points to an oversupply of soil science research and students in the agronomic field and an undersupply in hazardous waste remediation. Soil science departments that do not at least consider urban, industrial, and aberrant soils in their research and teaching efforts do damage on two fronts. They deprive their students of an opportunity to join an expanding industry and they deprive the industry of needed technology and innovation.

THE GROWTH OF THE REMEDIATION FIELD

Most environmental remediation activity is in response to current or perceived future governmental regulation. Governmental regulation of any legal societal activity is an art that improves with both time and practice. In some bodies of law, legal precedent can be traced back thousands of years, yet even in these areas, legal challenges and clarifications are continually sought from the court system. To legal scholars, it is not surprising there still exists considerable confusion, misunderstanding, and ignorance about the technical, legal, and economic consequences of environmental regulations. Many scientists, however, are continually amazed, bewildered, and frustrated about this uncertainty. Scientists entering this field should remember that the art of environmental regulation is relatively new. The remediation laws and industry were essentially created in the early 70s. Laws in many areas predate sophisticated measurement techniques, understandings about risk, toxicology, and movement of materials. Good science is needed to sort out these sometimes conflicting viewpoints. Good science will eventually prevail in setting appropriate regulations, however scientists should remember that, as is the case in many events where there are large financial stakes, lawyers are often employed on all sides. The American Bar association currently indicates that environmental law is the fastest growing sector of the legal profession. There are now 200,000 lawyers in the US who claim primary emphasis in Environmental Law.

The environmental industry was essentially born with the passage of the Clean Air Act in 1970. It received another big boost with the passage of the Clean Water Act in 1972 and RCRA (the Resource Conservation and Recovery Act) in 1976 and CERCLA (Comprehensive Environmental Response, Compensation and Liability Act) in 1980. These laws, amendments, and rulings set the stage for companies to begin to hire or train technical and legal expertise to comply with the requirements. By and large, the early staff was drawn from their own research and production staffs. These were largely people with engineering and chemistry backgrounds who were familiar with the chemistry of the solvents, waste streams, and escaped materials. Most remediation technologies and staff still reflects this initial engineering origin, a factor and a bias which must be considered in attempting to break into the remediation field.

Almost all of the initial regulatory activity centered on protecting the general public from mobile toxicants. This regulatory bias toward mobilility is clear in that here in the U.S., we have both the Clean Water and Clean Air Acts and yet there is no equivalent clean soils act. In part this reflects the legal precedent where individuals can have relatively free rights to manipulate soil, yet considerably less freedom in control over the water and air within their property boundries. From a toxicological perspective, however, this bias also is understandable in that water and air are nearly almost always the most likely routes of exposure to many of the original toxicants of concern. Soils are regulated under other laws, including RCRA and CERCLA, as well as by individual state; however, measurement devices, techniques and interpretations are often less clear than those set out under the clean air and water acts (Dzomback et al., 1993). With the exception of direct ingestion of soil by a child, which can often be prevented by containmnent techniques, regulatory and remediation efforts are often driven by the small percentage of the spilled material entering a mobile phase rather than the larger percentage absorbed into a soil matrix. Due to this focus on the mobile phases, it has been less apparent to the regulators and remediation managers concerned that soil science has a significant body of knowledge to offer. Soil scientists, therefore, are both late and intrinsically disadvantaged in establishing themselves in the field. All researchers who can claim primary expertise in water and air are at an advantage in funding and job hunting in the remediation field. This is reflected in many remediation and industrial firms which have numerous geologists, hydrogeologists and air modeling and sampling experts and few soil scientists. This disadvantage is, however, by no means insurmountable as evidenced by the growing number of soil scientists who have found work in the field.

THIS IS A GOOD TIME TO GET SOIL SCIENTISTS INVOLVED

Five years ago, most US companies did not have a good idea of the scope of their soil contamination problem. More recently, many companies have begun to designate potential Solid Waste Management Units (SWMU) under RCRA. To many the extent of their contaminated land has been a rude awakening. The result of this tabulation exercise has been that many companies are now acutely aware of their soil liabilities and are actively seeking technologies to address them. To a large degree the remediation community that responded to the water and air clean up challenge has begun to tackle soil issues as well.

Originally, most polluted soil was simply identified, excavated and moved into a secured landfill. This work involved primarily geological and civil engineering professionals. As the political, legal, and economic climate has shifted, moving contaminated soils into secure landfills has become more expensive and more difficult. The focus has begun to shift to removing the pollutant from the media rather than the media from the site.

There is now a window of opportunity for soil scientists to demonstrate their usefulness in this debate. Some universities as well as the Soil Science Society of America (SSSA) have begun to make efforts in this direction. This volume is an excellent step but more is needed. Surface soils are significantly different from aquifer materials that many in the remediation field are used to handling. Soil scientists are needed to help in the handling, risk analysis, sampling, analytical, microbial, physics, and kinetics of soil processes. The engineering and soil science communities are, however, generally unaware of each others activities and the synergy that could exist between them. This is particularly true as the industry turns to biologically based strategies.

Politics in the U.S. may also be on the side of introducing new techniques, analytical methods, and experience into the hazardous waste arena. On September 30th, 1993, President Clinton signed an executive order directing that future environmental regulations incorporate cost benefit analysis, risk prioritization, sound science, performance based standards, international competitiveness, and harmonization of standards. In a precedent setting move, he further ordered agencies to review existing regulations for consistency with these reform principals. Many in the agronomic community are familiar with pesticide issues, toxicology, and risk assessment. Numerous parallels exists in pollution remediation. Additional expertise, taken from the pesticide model and moved into the remediation field, is needed.

For soil microbiologists, microbial degradation is gaining a foothold as a respectable technology. Microbial degradation, although more than 20 years old in its use in gasoline clean up, is currently gaining ground in other areas as well. Increasingly, biological *ex-situ* and *in-situ* techniques are being studied and implemented (Staff, 1993a). Biological reactors are used to treat water and soil slurries. Composting and biopiles reduce contaminants in excavated soils. New *in-situ* technologies are changing underground conditions to support microbial growth and it's remediation effectiveness.

The remediation community, however, does not necessarily welcome soil microbiologists with open arms. There exists a heritage of "snake oil" salesman, with dubious microbiological heritage, in the remediation industry. Many relatively sophisticated remediation engineering audiences have extremely jaundiced views of microbiology in general and are specifically skeptical of bioaugmentation. This mistrust is born of a combination of past experience, bad microbiology, and slick marketing. "Wonder Microbe" & "Remarkable Enzymes" are available on the market. Their successes are often widely exaggerated for marketing purposes.

This barrier to soil microbiologists exists, but it is by no means insurmountable. Technical and commercial success has begun to whittle it away. The largest hazardous waste management firm in the US has recently made significant investments in microbial based technologies. Steve Hinton's chapter (See Chapter 12) in this volume provides additional support for the acceptance of microbial based cleanup. Microbiologists are gaining respect in the field especially as aerobic and now anaerobic remediation is rapidly developing and has been successful. The dubious heritage of microbiology in the industry may pose, however, particular problems to those interested in testing microbes or microbial consortia in the field. We are confident, however, that good science and favorable economics will win over the business community.

UNDERSTANDING HOW A REMEDIATION TREATMENT TECHNOLOGY OR TRAIN IS CHOSEN

On all sites, remediation activities begin with an initial site investigation often triggered by an "event." This "event" can be externally imposed (regulatory audit, pollutant detected in a nearby well, sale of the facility, etc.) or internally triggered (dying vegetation, concern over regulatory compliance, a spill with inadequate containment and cleanup, etc.).

The goal of the initial investigation phase is to obtain data which will delineate the extent of a problem and allow the formulation of a preliminary technical plan, budget, and timetable. In most cases, external consulting companies do this work. Several large companies, however, have internal resources which they employ. During the initial investigation past employees are often interviewed, water and soil samples taken, and geological, hydrological, and anthropogenic site features are mapped in general terms.

This data is usually collated in some fashion and passed out to a small group of "experts." These experts usually encompass the following expertise bases: chemistry, geology, engineering, health and safety, law, finance and occasionally biology. All of these individuals then prepare for a general meeting where remediation options are discussed by reviewing compliance issues, technical options, and collating relevant chemical and physical data (e.g., vapor pressures, toxicity, water solubility etc.) It is during this first meeting that an initial remediation plan is formulated. Later sampling and data collections will refine or alter this plan. It is often the case that the biological, microbial, and integrative skill based training of the members of this team is experienced based and may be extremely limited.

One particular result of this process is often a source of frustration to the R & D community. In general, this process produces a plan of action based on the "low risk" technology option. The R & D community often views the chemical (and petroleum) remediation industries as "risk adverse." In the abstract, this might be true, however this risk aversion grows directly from the consequence of failing to meet a regulatory imposed cleanup objective that many in the R & D community do not fully appreciate. In a remediation process, failure to meet specific cleanup goals or a regulatory imposed deadlines can have serious economic and legal ramifications. R & D scientists are often "process" oriented with goals couched in terms of increased "understanding." The engineering process, as a whole, tends to be more product oriented. "Engineering standards" and "best practices" are often more rigidly defined than parallel research goals and objectives. "Failure" has different thresholds in different communities. The research community, and especially the agronomic research community, accepts a general failure rate many orders of magnitude higher than the engineering community. A trivial, yet apt example, is that farmers expect a year of abnormal rains (drought or flood) and yet "on the average" manage to make a living. In contrast, to guard against a "100 year event" an engineering "best practice" would be to upgrade the thickness of a stainless steel casing. The two cultures are significantly different in their approaches, goals, and definitions of both success and failure. These differences can be a significant source of frustration on both sides. For soil scientists to make inroads into the remediation forum, they must learn to function well in this engineering dominated field. Without this capacity, they will tend to be relegated to the fringes of an area in which they have much to offer. Working between these two cultures requires good communication skills and a learned appreciation of each other's talents. The time is right for doing that now.

An important note in the business of remediation is that each site is owned, often reluctantly, by a person, corporation, or government entity with limited monetary and technical resources. Often with technically unsophisticated owners, the legal system will provide comfort in designating certain techniques as BDAT (Best Demonstrated Available Technology). This legal definition has benefits to many site owners, regulators, and consulting firms; however, it also may act to restrict the number of technologies considered or developed. Significant questions exist including: what if a site were to chose to use a technology that was not "Best"?; what if it failed? The concept of "Best" technology coupled with the urge to reduce both liability and risk can greatly limit research flexibility. At times, new technological solutions may not be considered for field testing without significant legal discussions. The owner of such a site may not wish to pursue these discussions due to a wide variety of reasons including cost, timing, or unwillingness to draw attention to the contaminated site. These are all issues that are often discussed in the remediation R & D field.

There are currently many techniques for remediating contaminated soils. As additional technologies are developed, one can expect the remediation manager to rely increasingly upon his technical staff to sort them out. Most technology decisions, however, are made on fairly clear criteria: cost and risk of failure. This is again a source of significant frustration to advocates of new and unproven technologies. Many times a site is cleaned up in a manner that is: a) at the lowest risk of failure, and b) in the most expedient manner. Biological solutions often don't fit this scenario. In many remediations, the time frame weighs heavily in the remediation decision. It is typically easier for the DOD or DOE to completely remove a contaminated soil rather than treat the contaminates within the soil.

The very lack of educated soil scientists and other scientific professionals has led to a mistrust of new technology. "Snake oil" salesmen have tarnished the reputation of remediation technology; and they have also played havoc with the decision makers and the regulatory authorities. Those organizations or industries in which there is a dramatic need for new technologies are the same ones where the probability of success needs to be shown for every project. An informed, well-trained, and "cross-cultural" soil scientist has a definite role to play in this arena.

BARRIERS FOR SOIL SCIENTISTS TO ENTER INTO THE FIELD

Legal Barriers

Most soil scientists are familiar with the economic, political, and environmental driving forces in agricultural production. Farmers make choices based on these issues on a daily basis. Federal policy also greatly influences the nature of agricultural production in the form of farm policy, subsidies, set-asides, conservation tillage, and other matters. To some degree, these same forces act in surprisingly similar ways in the remediation field. There is one additional driving force in the remediation field that usually surprises, annoys, and confounds the arriving soil scientist. This area is the legal aspects of the environmental remediation field. We have chosen to break the legal issues into five areas as they would most likely affect soil scientists entering the remediation field. These are:

Jargon. The environmental industry has developed a significant body of abbreviations, terms, and jargon that can seem daunting to the uninitiated. One popular summation runs more than 50 pages. In reality, however, there are perhaps 40 common terms that the new entrant needs. Without an understanding of certain terms, the new arrival is at a disadvantage in discussing their contributions to the field. A brilliant technical suggestion can be wiped from the discussion slate by the statement "That is not BDAT." Heads will nod around the table and the discussion will continue to another point. A reply like "No, but SITE is considering it" will save the concept and may allow a valuable technical contribution. Jargon is both an offensive and a defensive weapon and should be in every remediation soil scientist's vocabulary. To some degree this is not difficult, one author spent an afternoon with 50 flash cards. Universities, however, should take this responsibility and educate students that are aware of these terms and can use them coherently in a discussion. It lends credibility to research proposals and job candidates. This is particularly true when the target audience is the remediation community and less so when it is a "scientific" peer review.

Site access. OSHA requires that people that work on hazardous waste sites take a 40 hour course directed at the health and safety aspects of working there. There is also a requirement for an 8 hour annual update. One author had positions open for summer interns. A B.S. level soil scientist would seem like a likely candidate to take soil samples in a field, look at vegetation and put out field plots. In practice, without 40 hour training however, he can not legally work on the site. The 40 hour training can be and is offered by certain universities. It would save a potential employer approximately $1,000 in course fees, a week's pay, room and board at the training site and the aggravation of having to coordinate it. OSHA training will give soil scientists, both graduates and faculty, access to industrial soils. It provides a significant "leg-up" in a hiring procedure, especially for interns and graduate students, and will allow professors the legal right to access research materials without going through a sampling intermediary, which may or may not provide the quality of service and samples you might like.

Sample handling, manifesting, and analytical techniques. All remediation efforts have a legal component. The forms and record keeping can be cumbersome, but are workable. Some analytical techniques, if used for compliance issues need to be done under Good Laboratory Procedures (GLP). These are a set of protocols and record keeping requirements that many labs feel cumbersome, but: a) students leaving should be aware of their existence as GLP requirement are proliferating in many industries, and b) these protocols can be worked around if one is doing R & D work in a specific field. The results of non GLP work, may, however have to be repeated to get regulatory approval. GLP protocols significantly increase the cost of having outside analytical services performed. A complete analysis for priority pollutants and other contaminants of concern can run as high as $2,000 a sample. Soil scientists, which often work in labs heavily subsidized by government funding, can seem naive in suggesting extensive analytical work without concern for the costs. One proposal received contained a sampling plan for 200 analysis over the 6 month study. At the university's $8 price for nitrogen analysis this is easily acceptable. At $350 a sample (for PAH's in soil) the $70,000 analytical bill kills the project. Part of the art of remediation is learning to work on the minimum amount of information necessary, despite the incredible heterogeneous nature of the contaminated substrate.

Field actions. The legal ramification of certain behavior needs to be learned to prevent unintentional violation of the complex regulations. There are many examples of legal requirements on field behavior. As a general rule, they are easily learned despite some initial resistance on most people's part. One such example relevant to this group is soil sampling. The first instinct of a soil scientist when doing a field sampling might be to take 30 core samples across a field, pool them into a bucket, mix and subsample. The remainder of the material would be returned to the field. In a contaminated field the procedure might be the same, however the last step may be illegal. The breaking of the soil surface contact and lifting the soil into the bucket created a "hazardous waste." The contaminated soil (in place) is not legally defined as a waste. By returning the unsampled material to the field you are improperly disposing of a hazardous waste by land application and may be subject to significant penalties. Many regulations can be confusing to soil scientists doing field work in contaminated sites and appropriate legal guidance is usually provided by the site owners. The classically sited example of confusion is the legal definition of a "hazardous waste." In many cases the legal ramifications of the classification can be significant, expensive, and drive technology selection.

Flexibility. Many remediation projects now underway are operating under a legally imposed agreement on what will be done. Some of these "Consent Orders" or "Record of Decisions" have very little flexibility. Often soil scientist find out about the remediation problem or proposed technique too late to have a significant impact on a specific site. The time for interactive discussions, research, and involvement is prior to the legal decisions. Soil scientists just entering the field have more to offer in the initial stages of technology decisions and risk assessment than they have to offer in the actual cleanup process. As such it is imperative that they insert themselves into the discussion early in the process.

Technical Barriers

Soil scientists receive excellent training with soils that can support active plant growth. The lists of courses taught to them include the physics, chemistry, mineralogy, morphology and microbiology of soils. To a large degree, this expertise is centered on agricultural soils with surprising little variation (e.g. most soils studied can support plant growth). If soil scientists are to make significant contributions in the remediation area, they should be encouraged to expand to "aberrant" soils as well. As a general rule, we believe that in every course taught in the curricula, the question should be asked "if the student does not get a job in an agricultural field, what will they have learned that is useful." Each course should be re-evaluated to decide if it is appropriate to devote a fraction of the knowledge to industrial, urban, mining, wetland, and hazardous soils as well as aquifer materials. These soils are significantly different and this should be acknowledged in course work and research. Graduates with such training have vastly increased job possibilities.

Our lab has received "soils" for remediation studies that have widely differing physical, chemical and microbial parameters (pH of 2 to 11, explosive, plastic, etc.). Many soil matrices should be more technically referred to as "soil-like materials" as they often lack certain criteria of a true soil. It is these soils and their resulting chemical and biological characteristics that require study and remediation technology development. It is in this area that the abilities of soil science may expand to fulfill a new niche and market.

Another general example is that many soils and aquifers contaminated with organics quickly become anaerobic. Changing these soils, sludges, and aquifers to aerobic zones is often technically or economically impractical. Anaerobic training is increasingly playing a role in new remediation developments, especially in deeper zones, river beds and wetlands. Few agricultural based schools produce soil scientists that are functional in this area, yet most institutions could offer this expertise to their students and the research communities.

Most remediation techniques involve multiple skills, including engineering, toxicology, geology, hydrology, chemistry, microbiology, health and safety and air processes. Most treatment decisions and discussion are multi-disciplinary and "classically" trained soil scientist sometimes lack significant elements to enable them to interact. Lack of familiarity with equipment as well as physical and manpower limitations of pumps, screens, wells, and deep sampling operations is a hindrance. An introduction to engineering parameters and appreciations of techniques, including cost evaluations, is extremely useful to maximizing the contributions of soil scientists in the remediation field.

All of these technical limitations can be overcome as either portions of a relatively few courses taught for soil remediation scientists, or as an integral part of courses already offered. In any case, giving a student a chance for an intern or summer position in this field might drastically change their hirability and may also bring new problems or ideas back to the laboratory.

Soil scientists have a myriad of skills that could be used in the remediation business. There are many examples, but anyone who works with clays should be aware that they are used as liners, caps, and retaining walls in the contamination control business. Often soil scientists study the "sorption" of materials to clays.

This experience is invaluable in remediation and would save many remediation engineers from continually reinventing techniques.

Other Barriers

Another, less significant barrier, is title "agronomists" and the perception that if anything, agronomy (and agronomists) cause more environmental problems than it/they could solve. Nitrate pollution, pesticide runoff and leaching, soil erosion, and eutrophication of streams and ponds, are all common comments on "agronomists'" stewardship of the environment. Many remediation firms have remediated sites contaminated with such compounds as chlorinated pesticides. The National Pollutant Discharge Elimination System (NPDES) lists 18 pesticides as priority pollutants. Some of these are regulated in quantities as low as 10's of ppb as a sample leachate. The view of many in the environmental business is that past agriculturally based practices, sanctioned by both agronomists and extension workers, caused much harm. This perception exists to varying degrees in different remediation quarters. It is not however, by any means, an insurmountable obstacle.

There is substantial ignorance on the part of the remediation community as to what the soil science community has to offer. One would expect that this would be altered as more soil scientists move into the field and begin to make significant contributions. Most soil scientists, however, publish exclusively in soil science journals which few environmental remediation people read. Cross publishing into other media would raise the status of soil science in the eyes of the remediation engineering community. Soil scientists also should be labeled and function under the title "Soil Scientists." To a large degree, most soil scientists who currently work in the remediation field refer to themselves (or are billed to clients) as "geologists," "geochemists," or "bio-geochemists." This relabeling of academic training disguises the impact soil scientists have in the remediation community and diminishes job opportunities for more recent graduates.

HOW TO GET STARTED IN R & D

There are two basic approaches to beginning in this field. The first approach is to examine one's current skill base and then look for an environmental remediation problem it may solve. This "hammer in search of a nail" works well, but requires that the hammer holder be presented with a multitude of "nail like objects" in an efficient manner. Many labs have an enzyme/ catalyst/ microbe/ or soil amendment or technique that may be applicable to a particular remediation problem. Technology borrowed from one field has worked well in the remediation field previously, examples include: environmental engineers have begun to use horizontal wells (a petroleum technology) in remediation and white rot fungi are beginning to find a place in research and remediation of contaminants.

The down side of this "hammer in search of a nail" approach is a) most soil scientist don't have the opportunity to have all the remediation problems, techniques, market sizes and regulatory status arrayed in front of them so as to choose the appropriate "nail" on which to concentrate, and b) the remediation audiences may lack the sophistication to appreciate the possibilities inherent in a more conceptual proposal. The remediation community may wish to see results before funding opportunities might emerge.

Figure 16-4. presents one such opportunity for those soil scientists familiar with "an iron hammer" e.g., the ferric/ferrous cycling of seasonally flooded soils. The figure presented below is taken from a much larger engineering decision tree model involved with the bioremediation of contaminated groundwater. Inherent in the larger chart are questions concerning the specific nature of the site's geology, hydrogeology, microbiology, chemistry, and legal status. The portion represented in Fig. 16-4 reflects a problem that arises when pollutants that are easily biodegradable under aerobic conditions (e.g. many components of gasoline) are associated with groundwater high in dissolved (ferrous) iron. Unfortunately this is a somewhat common occurrence. Many polluted soils with high biological activity quickly become anaerobic with ferric iron forms reduced to the more soluble ferrous ion. Providing oxygen forms for aerobic metabolism to occur in these high ferrous iron backgrounds is a technically challenging and potentially extremely useful solution. This problem would seem to be of the type that might use the training and skill base of soil scientists, yet we know of no soil scientist looking at this question. The remediation field has hundreds of these opportunities for interdisciplinary research that soil scientists should be involved in answering.

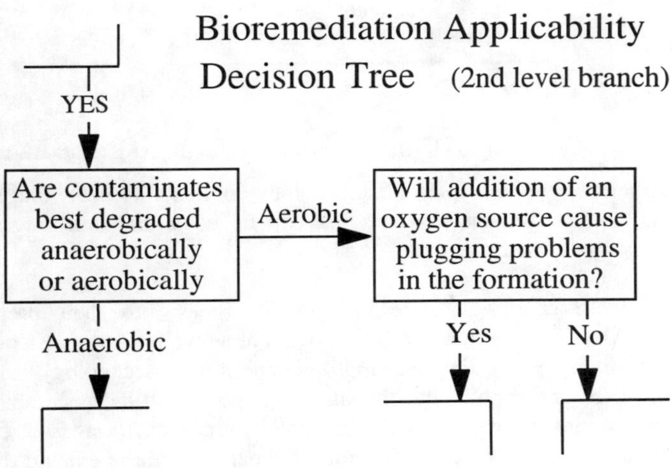

Fig. 16-4. A portion of a decision tree used in predicting the potential success of *in-situ* bioremediation in a contaminated aquifer.

The second common approach to choosing a research topic is to identify a significant, technical remediation problem and apply the full range of soil science offerings to that problem. This latter approach has the soil scientist sometimes working on somewhat more technically shaky ground but the need, economics and opportunities are often familiar to the remediation community. The results of moderately informed initial experiments are often easier to champion and funding follows more easily. The remediation community can offer a myriad of sites on which to explore and hone your skills. In general, the problem will be framed by the specific concerns of the site (matrix, volume of soil, and pollutant), but the judicious selection of sites can provide a useful basis for generalization to other sites and pollutants. Figure 16-5 presents an example of a generalized engineering chart that can be used to pinpoint and justify R & D projects. Each pollutant and matrix can be placed on a N-dimensional chart. Good remediation firms view possible treatment techniques in this manner (either explicitly or as a "gut feel"). Other dimensions include soil texture (sandy to clayey), concentration of pollutant (ppb to percent levels), years available for remediation (days to centuries), and willingness to incur cost (none to infinite).

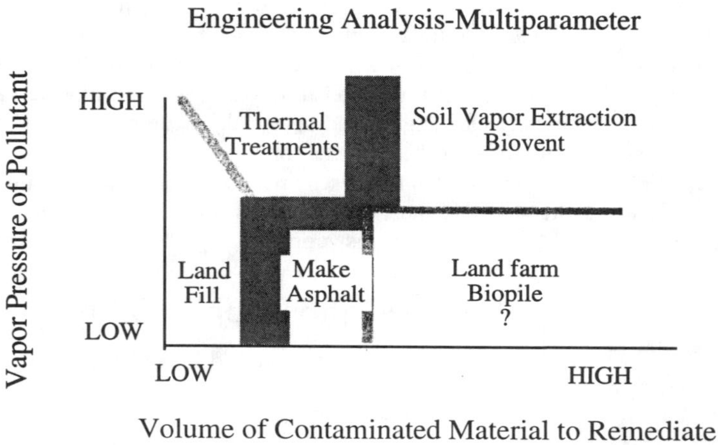

Fig. 16-5. A generalized example of a technology applicability chart that shows 2 dimensions of an *n*-dimensional set that can be used in identifying R & D opportunities for soil science.

By analyzing these charts, and the thought that goes into their preparation, researchers can identify opportunities for innovation, as well as inconsistencies that need critical testing. Evaluating the appropriateness of a technology relative to physical parameters of the pollutant, the site in question, and the generalized need (or market opportunity) for equivalent matrix-pollutant combinations is an excellent technical exercise for students, as well as professionals seeking to expend their R & D scope. Many engineering schools have chosen this approach and have received much commercial level support (both logistical and financial) for their efforts.

These two cases represent extremes in approach and one or the other community will be far out of their "comfort zone" for the initial interactions. The best cases represent true collaborative projects that are multidisciplinary in nature and tolerant of initial technical uncertainties. As an example of technical backgrounds that one can find in a single room is evident in the description of the "pollutant." The same material was referred to as "free product" by the plant site engineer, "contaminant" by the groundwater modeler, both "substrate" and "electron acceptor" by the microbiologist, and "the issue before us" by the legal representative.

The terms, experience, and approach were a direct reflection of the differing disciplines necessary to solve the problem.

Given the complexity of sites, regulations, and interwoven technologies, we recommend that the easiest way to become an integral part of remediation research is to establish a good working relationship with someone already in the field who can mentor the researcher, can champion and broker the technology to the appropriate source, and will act as a guide to keep the researcher out of legal trouble with sampling, handling, and personnel training issues.

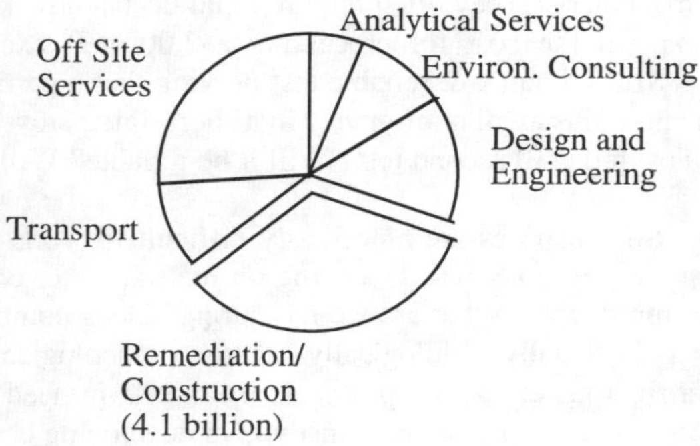

Fig. 16-6. 1991 U.S. hazardous waste spending (total spent $12.2 billion dollars) broken down by market (Lorenz et al. 1991).

PICKING A PROJECT/GETTING STARTED IN R & D

Soil science techniques, measurements and intuition are needed in the field in many areas. Figure 16-6 outlines the breakdown of spending allocation in the remediation business. Each sector represented presents significant opportunities for soil scientists. We have broken these sectors down into a short list of relatively practical needs in the remediation business. In addition to this list there are other lists of more basic questions that most of the participants will be able to generate themselves or that can be gleaned from reports on bioremediation by other groups e.g., *Scientific Foundations of Bioremediation-Current Status and Futures Needs* (Gibson and Sayler, 1992). Soil scientists are needed to help in the following areas:

Risk assessment and prioritization. Much of the science, regulatory testing protocols, and interpretations need significant scientific input. The decisions about regulatory limits and technologies to achieve them are multi-million dollar decisions that need excellent science and well reasoned regulations. The assumption that a fixed percentage of a toxicant in soil is available for mammalian uptake is wrong. A pinch of lead nitrate in a sandy soil is nearly 100% available to plants and to mammalian stomachs. A lead fishing weight or bullet in the same soil is not. Some analytical procedures discriminate between the two, others do not. Do we want to write soils regulations based on total Pb? Is there a better technique, measurement or rule that could be written? Discovering the species, flux, and toxicological properties of pollutants in soil should involve soil scientists. Modeling the fate and transport of these materials is often an important part of technical and regulatory considerations and at present, soil scientist generally are not involved. Broadening the scope beyond traditional soils- should a soil scientist have reviewed a decision by the NY public school system to remove asbestos (a mineral) from public schools? Could/should a soil scientist examine the need for this public expenditure if the scientist were the leading expert on dust flow and relative toxicology from contaminated matrices?

Quick tests for field screens. Field level decisions run in parallel with a full crew of men and machinery. They often require rapid decision making and error on the side of caution. If the cost for processing a 2,000 LB excavator/back hoe bucketful of soil is $100, a quick & reliable test proving that a particular scoop was clean can save an enormous amount of money in a short time. How do you sample a bucket? Can you develop a 30 second test? Will it be reliable? Will it be acceptable to regulators?

Lab tests. Soil matrices are notoriously difficult to work with and certain EPA protocols can vary up to 800% depending on matrix, water content etc. How can these tests be improved?. What are we measuring? Does nonmeasurement of a pollutant mean it is legally, biologically and/or toxicologically safe? Does measurement by a rigorous extraction protocols mean it is in need of remediation? Do "sorbed" materials count for you or against you in developing cleanup techniques and regulations? Soil scientists have developed plant and root bioassays to determine the effect of herbicides residue on subsequent crops. Are these tests appropriate for evaluating the inherent risk posed by contaminated soils? Are earth worm or microbial based tests better? Again soil scientists at all levels need to make their presence and skill known.

New techniques. New techniques to both clean up or better stabilize pollutants in soil. The goal of such research projects should be the development of a cost effective treatment for a contaminant that is wide spread and difficult or expensive to treat using current methods. Current techniques and their champions, however, may resist their entry into the marketplace. Many novices in the area often fail to consider the need for regulatory acceptance of the technique and an understanding of a complete life cycle of the process. The latter includes fate, mass balances, and energy inputs.

New techniques. New techniques to move bacteria, nutrients, oxygen, salts, cofactors, and metals in soil. This need applies to surface soils as well as down bore holes. In certain cases, like high iron groundwater, we know how to add nutrients and oxygen to remediate a gasoline spill, yet oxygen oxidizes the soluble ferrous iron and may plug the formation. Sometimes microbial biomass, in and of itself, plugs the formation preventing further mass transfer and metabolism. The remediation community needs microbiologists, engineers, plumbers, and regulators all sitting at the same table to solve both theoretical and practical application questions. Novel solutions are being tried (a biological curtain through which contaminated groundwater flows); however, all of these new techniques need better chemical, physical, and microbiological understandings for optimization. At a very basic level, soil scientists can contribute directly to the remediation engineers' understanding of the bonding and binding of contaminants, bacteria, and nutrients in the soil environment. Soil scientists have a very specific niche to fill in the remediation environment.

Expertise. The remediation industry has need of generalized experience in chemistry and biology. Many opportunities in the field are for "project managers" not either scientists or engineers. These managers need a wide spectrum of experience which enables them to make informed decisions. Remediation managers that are experts in only one technique (their "hammer") tend to view all problems as a variation on a "nail." Managers should draw comfortably from a wide variety of disciplines and techniques to choose the most appropriate alternative. Again few soil scientists have been hired into these positions.

Site investigations. Site investigations ranging from hazardous waste sites to simple property transfers (also called "due diligence" work) are required in most large property transactions. Performing this service would be a great place for young soil scientists to start out. Currently, however, environmental engineers are designing field evaluations.

Ombudsmen. Mediating the technical interactions between site operators, consulting firms, regulators, and general public is difficult. A university based group of scientists who understand the processes involved are needed to act as independent advocates and arbiters in the environmental debate. The debate is often highly emotional, political, and well publicized. An independent analysis of an environmental situation is often needed to calm the fears of the public, corporations, and regulators.

This list represents only a small fraction of where soil scientists can contribute to environmental problems outside of classical agronomy. To some degree the soil science community has begun to address environmental issues. These include incinerator ash and sludge disposal, land use questions, and beneficial use of industrial byproducts. In other areas such as product life-cycle analysis and fate and effect testing, soil scientists largely remain on the sidelines. As a general rule, the soil science community needs to show much greater creativity in wedging themselves into nonagronomic issues. Soil microbiologists study processes and isolate microbes. The polluted soils of the world will undoubtedly select for microbes with unusual metabolic, synthetic, or degradative products worthy of study. Can microbes be isolated from toxic waste dumps or pesticidally active soils ("problem soils") that have a use? Can soil microbiologists find bacteria, proteins, and genes that can help address the waste minimization questions that all industry is currently facing. It is well known that microbes can be isolated that make or destroy stereo specific chemical bonds. In some chemical processes only one isomer is the desired active product. Chemical synthesis may provide roughly even quantities of both isomers. Chiral specific synthesis can reduce the waste streams from these processes by 50%. As waste treatments costs are built into synthesis costs, the applicability and desirability of chiral synthesis becomes more and more appealing. These are only a brief list of where soil scientists can contribute outside of traditional soil foci. The contributions that excellent soil scientists can make are limited only by their lack of exposure to the breadth of environmental problems that exist.

FINDING MONEY

Targeting Federal/State funds

This is an area in which many academicians are familiar, hence little time will be spent here. There are, however, additional agencies and surrogates in the environmental arena of which most universities will be aware. In our experience a call to the engineering staff/or faculty of the university will produce more and varied funding options than a similar call to the dean of the Agricultural School. These include Departments of Energy, Defense (all branches), HUD, and the EPA. Certain EPA regions have active programs in soil remediation and checking with your state EPA coordinators will identify the Hazardous Substance Research Center in your area. Many state Environmental, Transportation and Parks and Recreation Departments also provide funding and opportunities for research.

Targeting Industrial Funds

Industry funds research in three ways.

1) Direct grants given to an individual, program, or university are common. DuPont's Committee on Education spends many millions/yr in these type of grants. These grants fund basic research on a wide variety of topics including environmental issues. These funds generally have few strings attached and may request only infrequent updates on the research being supported.

2) Corporations and industrial consortia fund both basic and applied environmental research. Such consortia include: PERF- Petroleum Environmental Research Forum, EPRI (Electric Power Research Institute), CGI- Coal Gasification Institute, RTDF - Remediation Technology Development Forum and others. These consortia between companies (and companies and universities) are becoming increasingly popular as resources shrink. At the moment, this is particularly true in the environmental field where a sense of shared industrial problems, better understanding of the antitrust laws regulating collaborations, and federal agency involvement is gaining ground. Many US governmental labs have an increased environmental focus, and are actively seeking collaborative agreements.

3) Certain large sites with huge economic liabilities also attract proposals and generate funds for academic labs. From an industrial perspective the general goal of such funding is to save the company money by demonstrating that: a) the site does not pose a particular environmental hazard due to the nature of the chemical, location, or supporting matrix, b) the pollutant is being naturally attenuated by an ongoing biological or chemical process and needs no further remedial activity; or c) the remediation method the regulatory agency is promoting is inappropriate or unnecessary at this site for site specific reasons.

In all cases, the chances of industrial funding improve significantly if the research proposal is reviewed by someone familiar with remediation technology. The goal of this review should be to review the grant for consistency with current technology needs and jargon, as well as adding additional statements of relevance and need. The authors have heard leading soil microbiologists make statement that although technically true, are naive from an engineering standpoint as the problem would never be addressed in that manner and can be solved with a minor change in the treatment train. In the search for industrial funding, grant writers should assume that engineers will often serve on the review panel for the proposed work.

SUMMARY

The environmental and remediation field need soil scientists to provide creative, novel solutions to a myriad of problems. There are significant, but not insurmountable, barriers to soil scientists contributing in this area: 1) No one is waiting with open arms for their input; 2) The jargon, rules, and regulations in the field are cumbersome; and 3) Universities continue to pump out "agronomic" soil scientists instead of "industrial, urban and environmental" soil scientists. Despite this, the soil science profession understands the soil matrix. The legal, political, and economic environment is ripe for soil scientists to enter into the field. Institutions that don't prepare themselves to meet this challenge and soil scientists that don't consider it, lose an opportunity for research, an outlet for graduates, and deprive the remediation industry of needed innovation/expertise.

REFERENCES

Bishop, J. 1992. 1992 Salary Survey. Hazmat World. 5:18-28.

Dzomback, D.A., P.A. Labieniec, and R.L. Siegrist. 1993. The need for uniform soil clean up goals. Environ. Sci. Tech. 27:766.

Gibson, D.T. and G.S. Sayler. 1992. Scientific Foundations of Bioremediation-Current Status and Futures Needs. American Academy of Microbiology. Washington DC.

Russell, M., E.W. Colglazier, and M.R. English. 1991. Hazardous waste remediation: the task ahead. Waste Management Research and Education Institute, University of Tenn, Knoxville, TN.

Staff, 1993a. VISITT bulletin. Solid Waste and Emergency Response. U.S. Gov. Print. Office, Washington D.C. EPA 542-N-93-004.

Staff, 1993b. Hazardous Waste Spending. Waste Tech News. 14:1.

Staff, 1993c. U.S. Bioremediation Demand. Waste Tech News. 13:1.

17 Science Needs for Implementation of Bioremediation

J.T. Wilson and M.D. Jawson
U.S.EPA, Robert S. Kerr
Environmental Research Laboratory
Ada, Oklahoma

Bioremediation has become a popular option in the cleanup of contaminated soils and aquifers. Nevertheless, bioremediation does not have a fully credible reputation within the regulatory community (MacDonald & Rittmann, 1993; NRC, 1993). Proponents of the technology have been described as modern day snake oil salesmen, despite the substantial cost savings generally associated with the use of bioremediation as compared to other remediation technologies (Ward et al., 1992; Norris et al., 1994). *Why is bioremediation suspect?* The capability of microorganisms to degrade contaminants in soil and groundwaters is well established. The difficulty, and perhaps the credibility failure, occurs during the transfer of the capability that microorganisms demonstrate in the laboratory into procedures that can be reliably applied in the field. There is a poor understanding, particularly at the field scale, of both the potential and limitations associated with bioremediation.

Transfer of bioremediation research and technology developed in the laboratory to the field is often a frustrating and unsatisfying activity. As researchers actively investigating bioremediation at contaminated field sites, we recognize the need for new information and insights that will allow the widespread and successful transfer of this technology to the field. Part of the problem reflects the levels of inquiry that typically occur in the laboratory as compared to what can occur in the field. Laboratory studies in support of bioremediation usually focus on biochemical or physiological processes. Appropriate controls are used to limit the number of mechanisms occurring during the study. In contrast, during field scale implementation and research on bioremediation, several processes can operate concurrently. These processes are chemical and physical as well as biological. Dilution, volatilization, partitioning and sorption all can play a major role in contaminant transport and fate. As a result, bioremediation requires the knowledge and skills of a variety of disciplines.

Which areas require research to enable bioremediation to be successfully implemented in the field? The National Research Council of the National Academy of Science (NRC, 1993) identified three primary needs. These are the development of: (i) innovative site characterization techniques; (ii) evaluation protocols, and (iii) improved mathematical models. Some of these needs are not directed at areas considered within the usual realm of "biological scientists." This only emphasizes that implementation of bioremediation will require a broad based understanding of the total environment in which the biological processes occur. Our goal for this chapter is to use a number of examples of actual field scale bioremediation projects as case studies to illustrate the kind of information needed for bioremediation to be implemented successfully in the field.

SITE CHARACTERIZATION

Applying the appropriate solution to a problem is most easily accomplished when the characteristics that define the problem are delineated and well understood. Consequently, the application of the appropriate bioremediation technology is most feasible when the nature and extent of the contamination is known. Current practices used for contaminated site characterization often do not adequately define the amount and distribution of contamination. As a result requirements for electron acceptors, supplemental mineral nutrients and projections of time to complete remediation, developed in the laboratory and based on inadequate information, will have much uncertainty when applied in to the field. The properties of the physical environment in which the contamination occurs must also be known because of their effects on contaminant and remedial chemical transport and microbial activity.

Site characterization techniques that are quick, reliable, and affordable would greatly aid all remediation efforts, because they would make it easier to determine the extent of the problem and the constraints the environment places on solutions. Assessment of the extent of contamination involves chemical, hydrological, and physical measurements. Biological measurements are predominately necessary to establish the feasibility of bioremediation. Examples of relevant site measurements include thorough characterization of the geological features, distribution of hydraulic conductivities, and determination of contaminant concentrations dissolved in solution and associated with solid or other nonaqueous phases.

Most organic contamination in the subsurface originates from spills of refined petroleum hydrocarbons, such as gasoline, or chlorinated solvents, such as trichloroethylene. These substances enter the subsurface as nonaqueous-phase oily liquids traveling separately from the soil and groundwater. Only a small proportion of these nonaqueous phase liquids (NAPLs) dissolve into the water. Most of it remains as a separate nonaqueous phase or sorbs to the soil solids. The amount of dissolution is specific for each NAPL and depends upon its solubility and sorption properties. As long as oily-phase liquid and sorbed material are present in the subsurface they will act as a continuing source of contamination, because contaminants contained in the nonaqueous phase will dissolve into the water. Site characterization techniques are required that can accurately estimate the total mass of contaminants that must be subjected to bioremediation.

Traditionally, monitoring wells have been installed and used to define the extent of contamination in the subsurface environment. However, well water cannot determine the extent of contamination by oily-phase materials because the water contains only a small proportion of the contaminant. If monitoring well data are the only data available, it is difficult to estimate, within an order of magnitude, the total contaminant mass that must be subjected to remediation (Case study I). Estimates of the total contaminant mass in the subsurface are required to predict the demand for nutrients and electron acceptor needed for bioremediation. If the total demand for nutrients and electron acceptor has been estimated, the rate of supply of the limiting

Case Study I

Limitation of Monitoring Well Data for Determination of the Extent of Contamination.

A spill of refined petroleum products occurred in south central Kansas as a result of a pipeline failure. The contaminated area was roughly circular with a diameter of 150 m. Thirteen monitoring stations were established uniformly across the spill. At each station a series of continuous cores were taken extending from clean material above the spill, through the spill, to clean material below it. Then monitoring wells were installed in the boreholes used to acquire the cores.

The cores were extracted and analyzed for the content of BTEX (benzene, toluene, ethylbenzene and xylene) and total petroleum hydrocarbons. Groundwater was collected and analyzed for the same parameters. Data from the 13 stations were subjected to geostatistical analysis to estimate the total contaminant mass in the aquifer and the total mass dissolved in the groundwater (Kennedy & Hutchins, 1992). Monitoring well data grossly underestimated the extent of contamination (Table 17-1). Less than 10% of the benzene, 1% of the total BTEX compounds and 0.01% of the total petroleum hydrocarbons were recovered in the monitoring well water samples.

Table 17-1. Comparison of dissolved phase hydrocarbons in a spill to the total quantity of petroleum hydrocarbons.

Samples‡	Contaminant: Core Samples†	Contaminant: Well Water
	(kg)	(kg)
Benzene	320	22
BTEX compounds	8,800	82
Total Petroleum Hydrocarbons	390,000	115

†Collected from soil mterials in the plume.
‡Collected from well water in the plume.

requirement can be used to estimate the time required for engineered or intrinsic bioremediation to take place.

It is also not appropriate to assume, because of sorption onto soil solids, that the elevation of oily phase hydrocarbon floating on the water in a well corresponds to the interval occupied by hydrocarbon in the aquifer. Nor is it appropriate to assume for the same reason that the absence of floating oil corresponds to the absence of oily phase hydrocarbon in an aquifer. Currently, core samples must be recovered and analyzed to obtain reliable information on the presence and distribution of oily phase hydrocarbons. Core samples can be acquired during advancement of the borehole used to install a permanent monitoring well. The cost of acquiring core material is no greater than the cost of installing a permanent monitoring well.

Analyzing many cores, however, may be too expensive to be practical in general use. The data set in Table 17-1 in the first case study contained more than 400 analyses. Screening cores in the field to determine whether the depth interval represented by a core is contaminated can eliminate some of the cost involved. There are several techniques such as analysis with a field gas chromatograph, organic vapor analyzer, or explosimeter that can reduce the subsequent laboratory analytical burden (Robbins et al., 1989; Kampbell & Cook, 1992). These techniques are inexpensive and generate data in real time, which allows the screening information to be used to guide decisions about depth and location of succeeding cores.

Detailed site characterizations are also necessary to determine the effects of the bioremediation process. Successful *in situ* bioremediation is more difficult to prove than *ex situ* or engineered bioremediation because experimental controls are usually unavailable at contaminated field sites. Performance monitoring at field sites is often limited to evaluation of the concentration of contaminants in groundwater and soil over time, and perhaps the concentrations of nutrients and electron acceptors. These data cannot insure that the biological process developed in the laboratory is responsible for contaminant removal at the field scale. The appropriate equivalent of experimental controls is a detailed site characterization including detailed characterization of the flow of remedial fluids and flux of amendments. This characterization allows an assessment of the influence of partitioning, sorption, dilution, and volatilization. Thus a detailed site characterization, provides a basis for evaluating the contribution and success of bioremediation.

Geophysical tools will provide much of the innovations needed for the required detailed site characterizations. With existing technology, cone penetrometers can be used to map the presence of sand and clay layers, to take water samples from discrete intervals, to estimate the vertical distribution of hydraulic head, to determine changes in electrical resistivity, and to estimate local pneumatic or hydraulic conductivity. Systems are under development and evaluation that use Laser Induced Fluorescence through fiber optics to map the vertical distribution of oily phase hydrocarbons. Other sensors to be used in conjunction with cone penetrometry are also being developed.

In addition, the use of non-evasive techniques needs to be researched at remediation sites. These techniques include ground penetrating radar, seismology, time domain reflectance tomography, and electrical resistivity. No one technique can be used universally at all sites because of limitation imposed by geological heterogeneity. Ward (1990) provides a detailed discussion of the uses and limitations of these various techniques.

Innovative site characterization techniques are needed to determine the extent of contamination, the progress of the remediation and to reduce the cost associated with extensive coring and analyses. The scientific needs for bioremediation in site characterization are not principally biological, but rather geophysical. Tools are needed that allow the full extent of the contamination and the properties of the site to be ascertained. Environmental microbiologists and chemists should be knowledgeable of site characterization procedures as well as the limitations of the data obtained by the various techniques. Their primary role is to project how the physical environment will effect the bioremediation processes so that appropriate solutions will be used at a given location.

Evaluation Protocols

Means and methods to clearly establish the biological contribution to contaminant dissipation are needed, because, as we have already stated, a variety of processes can reduce contaminant concentrations in the environment. These protocols need to prove the extent to which microorganisms can and will contribute to remediation. Laboratory studies have a role to play in this area since they are useful in determining the biodegradation possibilities and elucidating biochemical mechanisms at work. However, the laboratory data needs to be interpreted in a manner which will enable the potential for bioremediation in the field to be honestly assessed. Evaluation protocols are needed that can assess the true possibilities for bioremediation as well as the actual amount of biodegradation naturally or intrinsically occurring at a given location.

Case study 1 demonstrated the amount of material which can be partitioned into a nonaqueous phase and/or sorbed onto solid surfaces. Besides biodegradation-dilution, sorption, volatilization, and abiotic chemical reactions can also reduce contaminant concentrations in the subsurface. Therefore, to evaluate the benefit of bioremediation at a site, one must estimate the relative contribution of the microbes to contaminant degradation. Case study II demonstrates two methods which can be used to determine the amount of biodegradation at field locations and "prove the microbes did it." These and other methods (see for example National Research Council, 1993 and U.S. EPA, 1994) need to be widely tested and standardized.

Biological properties which should be measured as part of the site characterization include biodegradation potential and the composition and population of the microorganisms. Biodegradation potential measurements, often referred to as treatability studies, ascertain the susceptibility of the contaminants to biodegradation. This is usually best accomplished by obtaining actual field core samples and studying contaminant compound disappearance using microcosms in the laboratory where conditions such as oxygen status can be controlled. Also it is easier to measure rates, microbial adaptation, by-product formation, intermediary metabolites, and test the effects of stimulating materials such as various electron acceptors in laboratory studies. Development of innovative techniques is not the primary scientific advance needed, rather laboratory studies need to be conducted to obtain realistic biodegradation rate estimates. Realistic estimates allow the designs options necessary for successful remediation to be fully considered. A complete assessment must also consider the limitations imposed by the conditions present at the actual field site.

A major factor controlling biodegradation potential is the availability of the contaminants to the microorganisms. There is a pressing need to better understand bioavailability. Not only is bioavailability important for the assessment of biodegradation, but this information is also critical for evaluating the extent to which biodegradation can cleanup contamination. Bioavailability information, furthermore, is necessary for obtaining reasonable toxicity estimates and for determining "how clean is clean." Use of materials from actual contaminated field sites once again is advised, since the properties of weathered contaminants are different than those of fresh reagent grade materials in contact with soil for only a short time period. This information is ultimately needed for well considered risk assessments to be made and to subsequently enable best risk management decisions to be made.

In general, more is known about the biodegradation of fuel hydrocarbons than halogenated solvents such as trichloroethene. For this reason laboratory studies on reductive dehalogenation processes will provide more new basic information. For instance the microbial physiology of the reductive dehalogenated process is not completely understood. Mohn and Tiedje (1992) provide an extensive review of this process and McCarty and Semprini (1994) examine it from a field treatment perspective. The microbial ecology of the processes is also not well documented although in general it appears that the formation of mutualistic anaerobic microbial communities or consortia are critical for the reductive dehalogenation of many compounds.

Gaining a better understanding of the microbial ecology involved in bioremediation is needed to best mange this process. There is a strong interaction between the microorganisms and their environment. New techniques are needed to help better assess the microbial populations affecting biodegradation. These techniques will need to be able to determine community and consortia composition. Traditional techniques such as plate counts, most-probable-number (MPN) determination, and direct counts are almost as well known for their deficiencies as their utility. Techniques have been developed recently, primarily in the molecular biology area, which show promise in being adopted for use with environmental samples. These include DNA and RNA probes and fatty acid (e.g., FAME) analyses. Continued development of these tools and in particular adaptations which would enable them to be user-friendly so they could be routinely used is needed. Furthermore, almost all microbial assessments have been directed towards prokaryotes. Tools which help assess fungi and protozoan populations and activities are also needed.

Genetically modified organisms (GEMs) may be of potential use, particularly with recalcitrant contaminants such as PCBs. The ability of GEMs, however, to survive in the environment is as important as their ability to degrade contaminants. The greatest research needs at this time are in the areas of inoculation procedures and understanding microbial survival and activity. A genetically modified organism, or any microorganism for that matter, that is to be introduced into a "foreign" environment must first be delivered in high enough populations; secondly, must survive; and finally, must actively carry out the transformation of interest to be of any benefit. Research on ecological interactions, such as those just stated, is more critical than developing a new microorganism with novel degradative capability which cannot, however, function outside of the laboratory.

Evaluation protocol scientific needs are those which will enable us to determine the contribution from biodegradation to contaminant dissipation in the field. We need to prove the microbes did it, how much of it they did and how they did it. This will require fundamental research on the physiology of the reductive

Case Study II

Proving Bioremediation's Contribution at a Field Site.

Leaking distribution pipes from an underground storage tank released JP-4 jet fuel to the water table aquifer under the petroleum, oil, and lubricants storage depot (POL) at Eglin Air Force Base (AFB) in Florida. The POL is located over soils composed of sands and silty peats characteristic of most of the barrier island complex. At the time of the study, a plume of contaminated groundwater moved away from the residual JP-4 and discharged to a small creek approximately 300 ft downgradient. Oxygen has been depleted along the contaminant flow path and this groundwater was therefore anaerobic. Ground-water samples were acquired with a geoprobe, using an 18-inch screen.

Table 17-2 shows the attenuation of the compounds of regulatory concern--benzene, toluene, ethylbenzene, and the xylenes (BTEX)--compared with the attenuation of a tracer, a trimethylbenzene, along the groundwater flow path at this site (Wilson et al., 1994). Trimethylbenzenes are also components of the fuel. They can be used as tracers because they are recalcitrant to biodegradation. Such tracers enable attenuation due to dilution and sorption to be distinguished from attenuation due to biological destruction of the contaminants .

Benzene was not degraded along this anaerobic portion of the flow path. The concentrations of toluene, ethylbenzene, and m-xylene were reduced hundreds to thousands-fold along the flow path while the tracer was reduced only 3.5 times. Therefore, apparently concentrations of these compounds were reduced from one-half to one-fourth of the initial concentration due to dilution. The further reduction in concentrations can reasonably attributed to biological activity.

Table 17-2. Concentration of contaminants and an internal tracer along the groundwater flow path at the Eglin AFB POL site.

Compound	Concentration along Groundwater Flow Path		
	At edge of spill	50 m Downgradient	100 m Downgradient
	($\mu g\ L^{-1}$)		
Benzene	100	153	198
Toluene	5,150	18	1
Ethylbenzene	1,700	227	1
m-xylene	6,750	1,270	14
1,2,3-trimethylbenzene	406	182	115

Another way biodegradation can be evaluated in the field under anaerobic conditions is by analysis of the stoichiometry of electron acceptors. This type of analysis was also done for the jet fuel spill at Eglin AFB. Table 17-3 indicates there was very little oxygen, nitrate, or sulfate available for respiration of the BTEX compounds in the plume of contaminated groundwater. There was, however, an increase in the concentration of methane and iron and a corresponding decrease in aromatic fuel hydrocarbons along the flow path. Assuming the following stoichiometry for methanogenesis from BTEX compounds,

$$2CH + 3/2\ H_2O \rightarrow 5/4\ CH_4 + 3/4\ CO_2,$$

approximately 1.0 mg of methane is produced for each 1.3 mg of BTEX destroyed. The decrease in aromatic petroleum-derived hydrocarbons after corrections for dilution would be expected to produce amounts of methane in the order of magnitude observed along the flow path. The accumulation of methane is sufficient to rationalize the destruction of the aromatic fuel hydrocarbons along the flow path by methanogenic activity.

Table 17-3. Concentration of contaminants and potential electron acceptors along the groundwater flow path at the Eglin AFB POL site.

	Concentration along Groundwater Flow Path		
Compound	At edge of spill	50 m Downgradient	100 m Downgradient
		($\mu g\ L^{-1}$)	
BTEX	24	3	0.6
Oxygen	0.4	0.2	0.4
Nitrate	0.1	0.1	< 0.05
Sulfate	2.3	0.6	3.7
Iron II	2.3	7.8	3.0
Methane	3.7	16.8	13.3

BTEX - total concentration of benzene, toluene, ethylbenzene, and xylenes.

dehalogenation processes and on the microbial ecology of all bioremediation processes. Knowledge of these processes then needs to be translated into procedures that can be used to evaluate the amount of bioremediation occurring in the field. Research on the bioavailability of weathered contaminants is needed to enable the development of defensible clean up standards and performance of meaningful risk assessments and management.

Model Development

Models are potentially powerful tools for assessing bioremediation. If (and this currently is a very big "if" because of the lack of model validation) a model has been shown capable of describing contaminant fate in the field, it can provide weight to the limited number of observations one usually obtains in the field and it will aid in predicting the outcome of the bioremediation process. Since we are not modelers, the following discussion focuses not on the details of the modeling process but rather on what we perceive models should help us do better. Models should synthesize and integrate the information described above and as a result lead to more informed decisions.

A variety of models including geochemical, reaction-rate, multiphase-flow, etc. have been developed to describe subsurface processes. This variety is a reflection of the heterogenous nature of the subsurface and the complexity of the processes that occur there. Often model types are combined to describe bioremediation. For example, a model describing fluid flow will be linked to another defining geochemical reactions which may be linked to another delineating microbial responses.

Bioremediation can be evaluated using two types of approaches. The simpler approach is to calculate the difference between predicted mass balance and that observed. The difference, after abiotic processes are accounted for, is assumed to be due to biodegradation. This approach is most successful when there is substantial biodegradation. A simple example of this approach was used in the first part of the second case study.

Another generally more complex method, because it requires added details, uses models that describe microbial processes. The contaminant losses described by microbial activity is then compared to contaminant losses measured in the field. This latter use of models provides a more direct means of assessing bioremediation. More research is needed in this area since less effort has been spent on modeling microbial processes than has been spent on abiotic processes. This approach has an intuitive appeal because of its explicit representation of microbial processes. Information on processes and parameters needed for direct modeling often is available from laboratory studies or literature. A major difficulty in the use of models is in simulating the effects of site conditions on the processes.

Modeling, although useful, will never in itself be sufficient to prove biodegradation will occur at a site. It will never take the place of actual site data. Furthermore, every time a model is used it will need to be calibrated for the specific site. Many models still need to be tested and validated using real world data. Nevertheless, even if not validated models often are useful in enabling new insights to be developed. Models test our conceptual understanding.

As previously stated, the greatest practical use of models in fostering bioremediation is their ability to extend our inferences when only limited amounts of data are available. Models and/or decision support systems that can be used for on-site management are needed as field sampling and analysis becomes possible in "real time." Under these conditions decisions are more limited by a proper conceptual understanding and representation of the critical processes than by the time it takes for sampling and analysis.

Advancement in modeling of microbial processes during bioremediation is a major scientific need. Integrated into these biological models must also be the chemical, physical, and hydrogeological processes affecting the bioremediation process. Complex models will need to be translated into useable decision support systems for use in the field and in risk management.

SUMMARY

Determining both the potential and limitations of bioremediation requires an understanding of the dynamic interactions occurring in the subsurface.

Consequently, the scientific needs for implementing these technologies are not necessarily only those which fall in the realm of a biological scientist. The contaminated site needs to be fully described and the complete extent of contamination determined so that the most knowledgeable bioremediation strategies can be implemented. The primary needs in site characterization is for tools which will better describe the geophysical environment. As site characteristics control the remediation process, the biologist and chemist must be willing to develop an understanding of these methods. Cleanup is only possible when all parties are able to correctly use these data to evaluate the potential for bioremediation.

Protocols that evaluate the contribution of the biological processes to contaminant dissipation in the field need to be developed and standardized. Internal tracers would appear to be very useful in ascertaining the extent of bioremediation in the field. Analysis of electron acceptors under anaerobic conditions can also be useful. Additional procedures would be beneficial. A better understanding of contaminant bioavailability is necessary not only for evaluating bioremediation processes but also in making regulatory decisions. Fundamental microbial physiological information is needed for some processes such as reductive dehalogenation. Microbial ecological interactions need to be better understood. Tools for assessing subsurface microbial communities are lacking and advancements in this area would significantly increase our ability to manage bioremediation.

Models that can synthesize and integrate all factors involved in bioremediation need to be developed. The integration of microbial processes into transport models is a major scientific need. Finally, complex models will have to be made user friendly or translated into decision support systems to enable them to be used by non modelers for making real time decisions in the field and in risk management analyses.

REFERENCES

Kampbell, D.H., and M.L. Cook. 1992. Core assay method for fuel contamination during drilling operations. p.139-140. *In* Subsurface restoration conference proceedings. National Center for Groundwater Research, Rice University, Houston, Texas.

Kennedy, L.G., and S.R. Hutchins. 1992. Applied geologic, microbiological, and engineering constraints of *in situ* BTEX bioremediation. Remediation 3:83-108.

MacDonald, J.A., and B.E. Rittmann. 1993. Performance standards for *in situ* bioremediation. Environ. Scien. Tech. 27:1974-1979.

McCarty, P.L., and L. Semprini. 1994. Ground-water treatment for chlorinated solvents. p. 87-116. *In* R.D. Norris, R.E. Hinchee, R. Brown, P.C. McCarty, L. Semprini, J.T. Wilson, D.H. Kampbell, M. Reinhard, E.J. Bouwer, R.C. Borden, T.M. Vogel, J.M. Thomas, and C.H. Ward (ed.) Handbook of bioremediation. Lewis Publishers, Boca Raton, Florida.

Mohn, W.W., and J.M. Tiedje. 1992. Microbial reductive dehalogenation. Microbial Rev. 56:482-507.

NRC, 1993. National Research Council Report. p.2-11. *In Situ* Bioremediation: When Does It Work? National Academy Press, Washington, DC.

Nelson, C., R.J. Hicks, and S.D. Andrews. 1993. *In situ* bioremediation: an integrated system approach. *In* P.E. Flathman, D.E. Jerger, and J.H. Exner (ed.) Bioremediation: field experiences. Lewis Publishing, Chelsea, Michigan.

Robbins, G.A., R.D. Bristol, and V.D. Roe. 1989. A field screening method for gasoline contamination using a polyethylene bag sampling system. Groundwater Monit. Rev. 9:87-97.

Siegrist, R.L., and P.D. Jenssen. 1990. Evaluation of sampling method effects on volatile organic compound measurements in contaminated soils. Environ. Scien. Tech. 24:1387-1392.

U.S. Environmental Protection Agency. 1994. Symposium on intrinsic bioremediation of groundwater, Denver, CO. 30 Aug.-1 Sept. 1994. EPA/540/R-94/515. U.S. Gov. Print. Office, Washington D.C.

Ward, C.H., J.T. Wilson, D.H. Kampbell, and S.R. Hutchins. 1992. Performance and cost evaluation of bioremediation techniques for fuel spills. p.15-21. *In* S. Lesage, (ed.) Proceedings, *In situ* bioremediation symposium. Wastewater Technology Centre, Ontario, Canada.

Ward, S.H. 1990. Geotechnical and environmental geophysics. Investigations in Geophysics No. 5. Society of Exploration Geophysicists, Tulsa, Oklahoma.

Wilson, J.T., F.M. Pfeffer, J.W. Weaver, D.H. Kampbell, T.H. Wiedemeier, J.H. Hansen, and R.N. Miller. 1994. Intrinsic bioremediation of JP-4 jet fuel. p. 60-72. *In* U.S. Environmental Protection Agency. 1994. Symposium on intrinsic bioremediation of groundwater, Denver, CO. 30 Aug.-1 Sept. 1994. U.S. Gov. Print. Office, Washington D.C. EPA/540/R-94/515.

DISCLAIMER

This article has not been subjected to United States Environmental Protection Agency review and therefore does not necessarily reflect the views of the Agency and no official endorsement should be inferred.

18 Educational Needs in Bioremediation

C. P. L. Grady Jr.
Department of Environmental Systems Engineering
Clemson University
Clemson, South Carolina

H. D. Skipper
Department of Agronomy and Soils
Clemson University
Clemson, South Carolina

W. T. Frankenberger
Department of Soil and Environmental Science
University of California
Riverside, California

As made evident by the papers presented at the symposium and this book, bioremediation is a diverse and rapidly expanding field that encompasses a broad range of experimental approaches, but is not yet recognized as an academic discipline. Furthermore, a cursory glance at the affiliations of the speakers reveals that their backgrounds are diverse, attesting to the broad knowledge base that must be brought to bear on bioremediation problems. Reliance on chemical and physical means for remediating contaminated soils, sediments, and waters, which has been the tendency to date, is likely to be both ineffective and wasteful of scarce resources. Bioremediation, on the other hand, has much to offer, as demonstrated by its successful application in the Valdez oil spill (see Chapter 12), which increased interest in it by government agencies and private companies.

One impediment to broader implementation of bioremediation is a lack of professionals educated specifically to work in the area. This deficiency is a direct result of the experimental nature of the field and the breadth of skills required. This situation presents an interesting challenge to academics wishing to prepare new professionals. If there is no single recognized body of knowledge, then what should be taught from the fields of chemistry, engineering, geology, hydrogeology, microbiology, soil science, statistics and toxicology? Although there are differences from university to university, when a student tells someone that he/she is studying microbiology or chemistry, the listener has a pretty good picture of the knowledge required and the courses by which it is conveyed. No such picture exits for bioremediation. Furthermore, if a picture does exist, it is likely to be different for people from different traditional academic disciplines. Nevertheless, the potential associated with bioremediation suggests that we should be attempting to define the body of knowledge that should be instilled in future professionals in the field.

The purpose of this chapter is to present the thoughts of the authors on this topic. We do not anticipate that there will be broad agreement about what we have to say. In fact, it was difficult to reach agreement even among ourselves. Nevertheless, we hope that our thoughts and comments will stir debate that will assist educators in defining curricula in this important area. Before proceeding, it should be noted that the senior author participated in an education needs workshop sponsored by the Bioremediation Education Subcommittee of the U.S. EPA Bioremediation Action Committee in November 1992. Some of the concepts presented in this chapter were influenced by the discussions at that workshop and that influence should be acknowledged. However, it should be emphasized that the views expressed herein are those of the authors alone, and have not been reviewed or endorsed by the members of the Bioremediation Education Subcommittee.

MULTIDISCIPLINARY NATURE OF BIOREMEDIATION

In its broadest sense, bioremediation encompasses the use of living systems to destroy or alter pollutants found in water, soil, or air. In common usage, however, the term bioremediation primarily reflects the use of living systems to destroy or alter contaminants associated with soils, sediments, and groundwaters, whether that remediation is accomplished *in situ* or *ex situ* in engineered systems. That is the way in which it is used herein. As we explore educational needs we will not consider traditional applications of biological treatment for wastewaters, sludges, and agricultural wastes because educational programs for them are more well established. Rather, we will focus on the new needs.

Although we will not consider wastewater treatment herein, there is an important lesson to be gained by reviewing its history. Microorganisms have been the major means of removing pollutants from municipal and industrial wastewaters for almost a century. Using principles from chemical engineering and microbiology, environmental engineers developed techniques for designing reactor-based systems for removing organic materials, nitrogen and phosphorus from wastewaters, and those techniques have generally worked well for biogenic pollutants. However, when they used the same techniques to design bioreactors for *ex situ* treatment of contaminated groundwaters and for treatment of many industrial wastewaters, they found that the nature of the contaminants required a greater knowledge of microbial biochemistry and ecology. No longer was it possible for an engineer with a basic knowledge of microbiology to design a successful system. Rather, solution of the problem required a truly **inter**disciplinary team employing engineers and microbiologists working in concert. The lesson learned was that the complexity of modern environmental contamination problems requires a team approach to their solution. No single discipline is capable of solving the problems alone.

The need for a team approach is even greater when we turn to the problems of contaminated soils, sediments, and groundwaters because their solution requires knowledge of a solid matrix. No longer can the design team consider only microorganisms growing in aqueous media. Rather, they must consider the interactions with the solid matrix, as well as the impact of the matrix on fluid flow. Soils and sediments are complex matrices, from chemical, physical, and biological perspectives. One consequence of this is that a contaminant can react chemically and/or physically with them, thereby impairing the ability of microorganisms to attack it. Furthermore, they are porous media, which complicates the movement of water, nutrients and microorganisms through them. Finally, they are highly heterogeneous, which means that stochastic concepts often must be employed in their description and that high degrees of uncertainty are often associated with predictions of physical, chemical, and biological events occurring within them. The complexities introduced by the requirement to deal with a solid matrix means that the team must now be **multi**disciplinary. In addition to environmental engineers and microbiologists, we also require the skills of soil scientists to deal with contaminant/soil/sediment interactions and the modeling abilities of geohydrologists to help us predict the movement of materials in the subsurface. The question that we wish to address here is how to educate students from these four disciplines to work on teams addressing the remediation of contaminated soils, sediments, and groundwaters.

APPROPRIATE EDUCATIONAL LEVEL

As one thinks about the knowledge base underlying the undergraduate education of engineers, geohydrologists, microbiologists, and soil scientists in the context of what a person needs to know to function effectively as a bioremediation specialist, it becomes apparent that it would be difficult to establish a meaningful undergraduate curriculum in the area. There is simply too much knowledge required. Rather, it is apparent that in the broad sense, bioremediation must be carried out by multidisciplinary teams comprised of specialists from the various disciplines and led by a generalist who has sufficient knowledge to integrate the recommendations provided by the specialists. This suggests that the appropriate approach to education for bioremediation is to take specialists from the various disciplines and add sufficient knowledge to allow them to function in that field. That, in turn, raises the question of the appropriate level for conveying the additional knowledge. To meet the immediate need and demand for bioremediation, we propose to begin in graduate school at the M.S. level. The need to produce Ph.D.s in the various disciplines with emphasis in bioremediation will vary by discipline and is beyond the scope of this chapter. Each discipline should address the specific need for Ph.D.s in detail.

There are several reasons to select the graduate level as the appropriate level at which to educate engineers and scientists to work in bioremediation. First and foremost, students will be required to integrate knowledge from a number of disciplines, even though most will remain specialists in their traditional discipline. Completion of an undergraduate degree in one of those disciplines provides the intellectual maturity required to do that. Environmental engineering education has used this model successfully for decades, taking students from any undergraduate engineering discipline and giving them a common base of knowledge that allows them to function as environmental engineers. Second, an undergraduate degree in a classical discipline provides the breadth of knowledge required for career mobility. This is important in a field like bioremediation. No matter what we may think about the potential of the technology, we owe our students the maximum potential for career advancement and to educate them too narrowly too soon will be to their detriment. Third, the maturity associated with graduate degree holders will allow them to be productive rapidly in the field. Finally, properly constructed M.S. programs will be able to meet human resource needs in an economical manner.

Given the research orientation of current bioremediation practice, one might ask why we are emphasizing the M.S. level of education. There are several reasons for this. Research for the Ph.D. degree is more specialized than the M.S. The Ph.D. requires more time and investment. The Ph.D. is viewed as less conducive for a team approach than the M.S. degree. Certainly, environmental engineering practice has been very successful with the M.S. degree as the major degree. In fact, most engineering project managers hold M.S. degrees. However, M.S. programs have not been as successful for other disciplines, where the Ph.D. is considered more essential for advancement. Nevertheless, it is our opinion that an integrated, team approach at the M.S. level should be the major focus of bioremediation education for the immediate future, although Ph.D. level education will always play an important role.

Another question to arise is whether all human resource requirements can be met with M.S. level graduates. The answer to that is no. There will certainly be a place for technicians, both in the field and in the laboratory. Persons to fill such positions can be trained best at technical and community colleges, probably through Associate Degree programs. It should be quite possible to devise programs that would give them the skills required to take field samples, run routine analytical tests, etc. However, since none of the authors is associated with technical education, we have not attempted to develop an educational program for them. Nevertheless, we see a strong need for such programs and encourage linkages with technical and community colleges to produce bioremediation technicians with the required skills.

SPECIALIST VERSUS GENERALIST

Having selected the M.S. level as an appropriate level at which to educate engineers and scientists to work in bioremediation, it is now appropriate to address the question of how the program should be structured. Should there be a "Department of Bioremediation" which admits students with undergraduate degrees in engineering, geohydrology, microbiology, and soil science and puts them through a single curriculum in bioremediation, thereby producing generalists? We think the answer to that question is clearly no. The multidisciplinary nature of the field suggests that a more successful approach would be for students to pursue M.S. degrees in their respective disciplines, while taking a core of courses that will give them a sufficient common knowledge of bioremediation to allow them to work effectively in a team environment. In other words, they should be specialists in the application of their disciplines to bioremediation. A key component of the core, which would form a coherent minor in bioremediation, would be a "capstone" design course that requires the students from the various disciplines to work together in multidisciplinary teams. One objective of that course would be to develop skills in multidisciplinary problem solving.

Although the majority of students will follow the specialist route, there may well be some who desire a general education in bioremediation, allowing them to function as project managers and team leaders. We think that such students are best handled through a multidisciplinary program in which the student would take courses in all of the various departments. Such a program would be administered outside normal departmental lines since it would cut across college boundaries.

EDUCATION OF THE SPECIALIST

Let us now examine the knowledge required for specialists in environmental engineering, geohydrology, microbiology, and soil science to work effectively as members of a bioremediation team. Such an examination will serve two purposes. It will help establish the common knowledge that all must have, thereby defining the content of the coherent bioremediation minor, and it will establish an appreciation of what each discipline brings to the team. In the interest of brevity, we have organized the material into typical curricula containing eight courses of three semester credit hours each for each discipline. This allows time for a typical thesis project during a 21 month residency. Most of these courses already exist at many universities. Our intent is to draw them together in a way that will benefit bioremediation education.

The Bioremediation Minor

Three courses are common to all of the curricula and constitute the coherent minor that would be followed by all students seeking to become competent to work in bioremediation while remaining specialists within their own disciplines. They are Environmental Organic Chemistry, Biodegradation and Bioremediation, and Design of Bioremediation Systems. The knowledge imparted through each course is listed in Table 18-1. All students minoring in bioremediation would be enrolled in the same sections of these courses, which means that each discipline must prepare its students to benefit fully from them.

It is anticipated that each of the core courses would be cross-listed among two or more of the departments within which students could minor in bioremediation and that in many cases, the courses would be team taught. In addition to the common skills imparted, there is another important reason why all students pursuing the bioremediation minor should be enrolled in common courses, regardless of background. The success of the team approach to problem solving requires a mutual appreciation of the skills and knowledge inherent in each discipline. The development of such an appreciation is hampered by the traditional disciplinary approach to education. By bringing students and professors from four disciplines together into a shared environment, barriers to communication will be removed and development of the mutual appreciation will be facilitated.

Since many environmental contaminants of interest in bioremediation projects are organic compounds, it is important for students to understand the types of reactions that they undergo and the manner in which those reactions are influenced by the physical and chemical environment surrounding the contaminants. Furthermore, because the list of organic contaminants is potentially almost endless, knowledge of how to deduce the reactivity of a particular chemical from its structure is a powerful tool. Such understanding and knowledge would be developed in Environmental Organic Chemistry.

Most courses on microbial metabolism provide information only about the common catabolic routes, such as those associated with carbohydrates, amino acids, aromatics, and other biogenic compounds. However, many organic environmental

Table 18-1. Courses constituting a coherent minor in bioremediation that could befollowed by students in environmental engineering, geohydrology, soil science or microbiology. Each course represents three semester credit hours of effort.

Environmental Organic Chemistry
Molecular structure
Reactivity (hydrolysis, oxidation, reduction, sorption)
Relationship between molecular structure and reactivity
Effect of soil and water environment on reactivity
Quantification of reactive processes
Quantitative structure/activity relationships (QSARs)
Biodegradation and Bioremediation
Biodegradation mechanisms (primary & secondary substrates, cometabolism)
Microbial consortia
Effects of multicomponent substrates
Biotransformation of inorganic pollutants
Biodegradation testing (both stoichiometry and kinetics)
Aqueous phase bioreactors
Soil slurry bioreactors
Land treatment
In situ bioremediation
Design of Bioremediation Systems
Regulatory requirements
Multidisciplinary aspects of bioremediation
Team approach to problem solving
Quantitative and nonquantitative decision making techniques
Statistics
Site characterization
Treatability testing
Case studies
Field exercises
Team design projects

contaminants are biodegraded by metabolic pathways that are quite different from those. Furthermore, in most biodegradation situations, both substrate interactions and microbial consortia are involved. If the bioremediation specialist is to design systems to facilitate those reactions, he/she must have a basic understanding of them. Furthermore, he/she must understand how to assess the biodegradability of the contaminants in quantitative terms (i.e., kinetics). Finally, he/she must understand the functioning and the design approach to bioremediation reactors, whether they be *ex situ* or *in situ*. All of these topics would be covered in Biodegradation and Bioremediation.

The final course, Design of Bioremediation Systems, is a capstone course that has several elements in it that are crucial to the ability of the graduate to function effectively as a member of a bioremediation team. A significant portion of such a course should be devoted to building the skills required for effective participation in multidisciplinary teams. One way to accomplish this is through emphasis on the team approach to problem solving by pursuing group projects which require input from persons from each of the individual disciplines. In addition, case studies will illustrate the multidisciplinary approach to problem solving while providing important background on those situations in which bioremediation has been successful. Finally, field exercises will give the students experience with some of the procedures they have learned about.

One point needs emphasis. Because of the team approach required in the development of bioremediation systems, it is important that "people skills" be developed in the students. While this is an explicit goal in Design of Bioremediation Systems, it should also be an implicit goal across the curricula. Group projects and written and oral presentations should be utilized whenever possible.

Environmental Engineer

The engineer on the team designing a bioremediation system has two primary responsibilities. First, since bioremediation systems generally are part of larger overall remediation projects, he/she must ensure that the bioremediation system forms an integral part of the overall system. This requires environmental process engineering skills as well as knowledge of physicochemical unit operations. The latter is provided by the course of the same name listed in the curriculum in Table 18-2.

The engineer must also design the system(s) for delivering nutrients, oxygen, microorganisms, etc. to the site of bioremediation. This must be done in concert with the geohydrologist and soil scientist, and requires familiarity with contaminant transport, including its modeling. Consequently, a course on this subject is also provided. Students enter environmental engineering graduate programs with undergraduate degrees in any field of engineering. Unfortunately, most have had no chemistry beyond freshman chemistry and few have had any courses in the biological sciences. Consequently, most are unprepared to take the courses in the coherent minor. Environmental Chemistry and Microbiological Principles are designed to overcome those deficiencies by focusing on the aspects of chemistry and biology that are particularly relevant to environmental engineering practice. It is apparent that the inclusion of more biology and chemistry in undergraduate engineering programs would make it possible to design a stronger curriculum for environmental engineers interested in bioremediation. With the exception of chemical engineers, most beginning engineering graduate students have not been exposed to heat and mass transfer (transport phenomena), reaction kinetics, and simulation techniques, which are fundamental to development of the environmental process engineering skills that the engineer brings to the bioremdiation team. Thus, the course in Environmental Engineering Principles.

Geohydrologist

The major roles of the geohydrologist on a bioremediation team are to provide information about the extent and movement of the contaminants in the

Table 18-2. Masters degree curriculum for an environmental engineer working in bioremediation. Each course represents three semester credit hours of effort.

Environmental Engineering Principles	**Microbiological Principles**
Transport phenomena	Organic nomenclature
Reaction kinetics	Microbial physiology
Simulation techniques	Central metabolism
	Microbial nutrition
	Microbial energetics
	Microbial kinetics
Physicochemical Unit Operations	**Environmental Chemistry**
Mixing	Thermodynamics
Gas transfer	Acid-base equilibria
Oxidation	Solubility equilibria
Adsorption	Electrochemistry
Ion exchange	Complex formation
Coagulation	Colloid chemistry
Precipitation	Adsorption
Contaminant Transport	**Bioremediation Minor**
Surface water hydrology	Three courses
Groundwater hydrology	See Table 18-1
Wells	
Steady and transient flow	
Soil & sediment characteristics	
Contaminant movement	
Soil & sediment/contaminant interactions	
Contaminant transformation	
Modeling contaminant transport	

subsurface and to provide the basic data for design of a pumping system to either bring water-borne contaminants to the surface for *ex situ* treatment or to deliver nutrients, electron acceptors, microorganisms, etc. to the site of contamination for *in situ* treatment. This requires advanced modeling skills which incorporate physical/chemical impacts on the contaminant as evidenced by the courses Groundwater Modeling and Subsurface Remediation Modeling listed in Table 18-3. In order to properly model the groundwater environment the geohydrologist must also understand geochemistry, which is related to, but distinct from the environmental chemistry taken by environmental engineers. Hence, the course in Groundwater Geochemistry. Finally, since it will also be the responsibility of the geohydrologist to design and supervise the studies required to characterize the extent of subsurface contamination, he/she must have experience with field testing procedures. The course Hydrogeological Techniques will provide it. Although most students entering a graduate program in geohydrology will be well prepared in the geosciences, most will have had little exposure to microbial physiology and metabolism. Consequently, before they can take the bioremediation minor they

Table 18-3. Masters degree curriculum for a geohydrologist working in bioremediation. Each course represents three semester credit hours of effort.

Groundwater Geochemistry	**Microbiological Principles**
Solution equilibria	Organic nomenclature
Chemical weathering	Microbial physiology
Oxidation/reduction	Central metabolism
Complex formation	Microbial nutrition
Ion exchange	Microbial energetics
Use of tracers	Microbial kinetics
Groundwater Modeling	**Subsurface Remediation Modeling**
Modeling groundwater flow	Groundwater extraction
Modeling solute transport	Vapor extraction
Conservation equations	Steam flooding
Analytical techniques	Modeling flow of multiphase and multicomponent mixtures
Numerical techniques	Bioremediation modeling
Stochastic modeling	
Hydrogeological Techniques	**Bioremediation Minor**
Analysis of cores	Three courses
Geophysical well logs	See Table 18-1
Subsurface mapping	
Pumping tests	
Tracer tests	

must have basic preparation in microbial physiology and metabolism. The course in Microbiological Principles, which is also taken by the environmental engineers, can provide the required knowledge.

Microbiologist

The microbiologist on the bioremediation team brings in-depth knowledge of microbial processes. His/Her role will be to assess the biodegradative potential of the available microorganisms, to determine factors limiting expression of that potential, to develop strategies for alleviating limitations, and to isolate and/or develop strains or consortia that can be introduced to solve a particular biodegradative problem. In addition to a sound fundamental understanding of microbial biochemistry and physiology, this role requires that the individual be well versed in the use of modern molecular microbiology techniques for both strain identification and genetic manipulation, Furthermore, since most bioremediation strategies capitalize on the robust degradative character of microbial consortia, the individual should also be knowledgeable of microbial ecology. Finally, since soil is often the contaminated medium, the microbiologist must have knowledge of soil microbiology, including mycology, as well as soil chemistry and physics. The courses listed in Table 18-4 provide this required knowledge. The courses in Microbial Metabolism and Soil Microbiology are the same as the ones taken by the

Table 18-4. Masters degree curriculum for a microbiologist working in bioremediation. Each course represents three semester credit hours of effort.

Microbial Ecology	**Soil Microbiology**
Microbial interactions	Mycology
Population dynamics	Bacterial nitrogen fixation
Subsurface microbiology	Transformation of organic, nitrogen, and mineral substances in soil
Molecular microbiology techniques	Microbial/plant/soil interactions
Biodegradative capabilities of microbial consortia	
Genetic exchange	
Soil Chemistry & Physics	**Microbial Metabolism**
Soil structure	Major anabolic, amphibolic, and catabolic pathways
Soil mineral composition	Fermentation
Colloid chemistry	Aerobic respiration
Ion exchange	Anaerobic respiration
Adsorption	Energetics
Flow through porous media	
Solute transport	
Bacterial Cytology and Physiology	**Bioremediation Minor**
Structure and chemistry of cell components	Three courses
Physiology of bacterial growth, reproduction, and death	See Table 18-1
Regulation of enzyme and nucleic acid synthesis	

soil scientist and are more advanced than the course in Microbiological Principles taken by the engineer and geohydrologist, reflecting the specialized skills brought to the team by the scientists. The course in Soil Chemistry and Physics, on the other hand, is provided to allow the microbiologist to understand the limitations imposed by the environment where the bioremediation must be performed and to interact more effectively with the soil scientist and geohydrologist.

Soil Scientist

The soil scientist brings essential knowledge of the complex chemical, physical, and biological properties of soils and sediments to the team. Responsibilities of the soil scientist would include characterization of those properties relative to the site and use of the resultant knowledge in the design of the bioremediation system in cooperation with other team members. The courses that would be pursued by the graduate student in soil science minoring in bioremediation are listed in Table 18-5. Two courses are common with those taken by the microbiologist, Microbial Metabolism and Soil Microbiology, reflecting the concern of both disciplines with the functioning of microbes in the soil environment. In

Table 18-5. Masters degree curriculum for a soil scientist working in bioremediation. Each course represents three semester credit hours of effort.

Soil Chemistry
- Ion exchange
- Surface phenomena
- Chemical equilibria
- Acid-base equilibria
- Oxidation/reduction
- Structural and chemical properties of soil collids

Soil Physics
- Unsaturated flow phenomena
- Steady and transient flow
- Solute transport

Soil Fertility
- Soil types and plant stress
- Nutrient availability
- Behavior of essential elements in soil
- Microbial/plant/soil interactions

Soil Microbiology
- Mycology
- Bacterial nitrogen fixation
- Transformation of organic, nitrogen, and mineral substances in soil
- Microbial/plant/soil interactions

Microbial Metabolism
- Major anabolic, amphibolic and catabolic pathways
- Fermentation
- Aerobic respiration
- Anaerobic respiration
- Energetics

Bioremediation Minor
- Three courses
- See Table 18-1

addition, one course, Soil Chemistry, is similar to the course in Groundwater Geochemistry taken by the geohydrologist, although the emphasis and details are different, reflecting the unique applications associated with soil science. The course in soil physics provides basic knowledge of liquid and solute transport in soils. Since plant growth and the rhizosphere can impact bioremediation of surface contamination and movement of pollutants to groundwater, the course in Soil Fertility will prepare the soil scientist to consider these effects during design of remediation systems. It will also allow the team to meet regulations that require restoration of the landscape to its precontamination vegetative state. In addition, the factors that influence nutrient availability to plants are the same as those that limit their availability to microorganisms carrying out biodegradation.

EDUCATION OF THE GENERALIST

As stated earlier, there may be some students who desire a general education in bioremediation, allowing them to function as project managers and team leaders, rather than as technical specialists. As argued previously, the knowledge base required for bioremediation is so broad that it cannot be confined to a single discipline. Furthermore, examination of the curricula presented in Tables 18-2

Table 18-6. Masters degree curriculum for a generalist in bioremediation. Each course represents three semester credit hours of effort.

Courses in Bioremediation Minor	Physicochemical Unit Operations
Environmental Chemistry	Microbiological Principles
or	or
Groundwater Geochemistry	Microbial Metabolism
or	or
Soil Chemistry	Soil Microbiology
	Subsurface Remediation Modeling
Groundwater and Contaminant Transport	
or	
Groundwater Modeling	
or	
Soil Physics	

through 18-5 makes it clear that each discipline brings a unique perspective and collection of skills to the problem. This suggests that it would be unwise to develop a degree program in a single department for generalists in bioremediation, although some students may desire one. Rather, their needs can best be met by a multidisciplinary program administered outside normal departmental boundaries. Such a program would have to be flexible in order to accommodate the various backgrounds that the students would bring to it.

The proposed curriculum for the generalist is outlined in Table 18-6. The contents of the various courses are given in the previous tables, and thus are not listed. The most obvious inclusion is the coherent minor. Since those three courses are fundamental to bioremediation, the generalist must be expert in their content. In addition, because transport processes, kinetics and modeling are basic to the design of bioremediation systems, it would be necessary for all generalists to be familiar with Environmental Engineering Principles. The other four courses in the curriculum fall in the areas of chemistry, microbiology, groundwater and contaminant transport, and physical operations, giving the generalist an appreciation of the activities of each of the specialists. However, the particular course taken in each area will depend on the student's background. For example, a soil scientist might take Soil Chemistry, Soil Microbiology, Soil Physics, and Physicochemical Unit Operations whereas a geohydrologist might take Groundwater Geochemistry, Microbiological Principles, Groundwater Modeling, and Subsurface Remediation Modeling. It is this flexibility that requires that the program be administered outside of normal departmental lines.

THESIS OR NON-THESIS PROGRAM?

Before closing, one other issue should be addressed. Throughout this chapter it has been assumed that all students would perform research and write a thesis as part of their degree requirement. The advantage to this is that it prepares the graduate to work in an experimental setting, which bioremediation is likely to remain for the next several years. The disadvantage is that such a requirement is time consuming and expensive. As a consequence, most students require financial aid to complete their degrees. An alternative approach would be to develop a non-thesis option in which students take six to nine additional credit hours of course work in lieu of a thesis. The obvious advantage is the savings in time and money and the ability to get larger numbers of educated professionals in the field faster.

The obvious disadvantage is the loss of professional maturity gained by the student as he/she goes through the process of developing and completing a project of significant scope. This loss can be mitigated somewhat by requiring an internship or practicum as part of the nonthesis program. Some would even argue that such real world experience would be preferable to a thesis in preparing students to work in the bioremediation field. The factor limiting application of such a concept at this time is the paucity of trained professionals to serve as off-campus mentors for the students. As bioremediation matures as a remediation option, this limitation will disappear. In preparation for that time, educators should begin to seriously debate the issue of whether the thesis or nonthesis route to education of bioremediation professionals is the best approach.

SUMMARY AND CONCLUSIONS

Design and implementation of bioremediation facilities is a multidisciplinary activity that demands an educational approach which prepares professionals to work in teams. Consequently, most students should be educated at the M.S. degree level to be specialists in their particular disciplines while developing an understanding of the fundamentals of bioremediation and an ability to work in teams. Curricula are presented that will allow specialists in environmental engineering, geohydrology, microbiology, and soil science to work on bioremediation projects. Common to all curricula is a coherent minor that includes the basics of bioremediation plus a capstone design course requiring a team approach to problem solving. Since there is also a need for a limited number of generalists to act as team leaders, a multidepartmental curriculum is also presented that draws upon courses in the various specialty areas, as well as in the coherent minor.

APPENDIX

Environmentally Important Chemicals

Common name : Chemical name

ABS : alkylbenzene sulfonate

Acenaphthene : 1,8-ethylenenaphthalene

Alachlor : 2-chloro-*N*-(2,6-diethylphenyl)-*N*-(methoxymethyl) acetamide

Amitrole : 3-amino-1,2,4-triazole

Anthracene : paranaphthalene

Anthraquinone : 9,10-anthraquinone

Aroclor 1254 (PCB's) : 11% tetrachlorobiphenyls; 49% pentachlorobiphenyls

Arsanilic acid : p-aminobenzenearsonic acid

Arsenic acid : orthoarsenic acid

Arsine : arsenic trihydride

Atrazine : 6-chloro-*N*-ethyl-*N'*-(1-methylethyl)-1,3,5 triazine-2,4-diamine

Benz[a]anthracene : 1,2-benzanthracene

Benzene : cyclohexatriene

Benzo[a]pyrene : 3,4-benzopyrene

BHC : hexachlorocyclohexane

Bibenzyl : 1,2-diphenylethane

Bromacil : 5-bromo-6-methyl-3-(1-methylpropyl)-2,4(1H,3H)pyrimidinedione

BTEX : benzene; toluene; ethylbenzene; xylene

Captafol : *N*-[(1,1,2,2-tetrachloroethyl)thio]-4-cyclohexene-1,2-dicarboximide

Carbofuran : 2,3-dihydro-2,2-dimethyl-7-benzofuranyl methylcarbamate

Carbon tetrachloride : tetrachloromethane

Chloramphenicol : D(-)-threo-2,2-dichloro-*N*-[B-hydroxy-alpha-(hydroxymethyl)-p-nitrophenethyl] acetamide

Chloroform : trichloromethane

Chloropyrifos : O,O-diethyl O-(3,5,6-trichloro-2-pyridyl)phosphorothioate

Chrysene : 1,2-benzophenanthrene

Creosote : Creosote is a complex mixture of 100's of individual poly-nuclear aromatic hydrocarbon (PAH) compounds, plus a few phenolics.

Cysteine : 2-amino-3-mercaptopropanoic acid

2,4-D : (2,4-dichlorophenoxy) acetic acid

DDT : 1,1,1-trichloro-2,2-bis(p-chlorophenyl)ethane

Diazinon : O,O-diethyl O-(6-methyl-2-(1-methylethyl) 4-pyrimidinyl phosphorothioate

Common name : Chemical name

Dicamba : 3,6-dichloro-2-methoxybenzoic acid

Dieldrin : 1,2,3,4,10,10-hexachloro-6,7-epoxy-1,4,4a,5,6,7,8,8a-; octahydro endo,exo-1,4:5,8-dimethanonaphthalene

Diquat : 6,7-dihydrodipyrido[1,2-alpha:2',1'-*c*]pyrazinediium ion

EDB : ethylene dibromide

Fluoranthene : 1,2-benzacenaphthene

Homocysteine : 2-amino-4-mercaptobutyric acid

Lindane : 1,2,3,4,5,6-hexachlorocyclohexane

Malic acid : hydroxysuccinic acid

MCPA : (4-chloro-2-methylphenoxy) acetic acid

Mecoprop : 2-(4-chloro-2-methylphenoxy)propanoic acid

Methyl bromide : monobromomethane

Methyl isobutyl ketone : 4-methyl-2-pentanone

Metolachlor : 2-chloro-*N*-(2-ethyl-6-methylphenyl)-*N*-(2-methoxy-1-methylethyl) acetamide

Monuron : 3(p-chlorophenyl)-1,1-dimethylurea

Napropamide : *N,N*-diethyl-2-(1-naphthalenyloxy)propanamide

Oleic acid : cis-9-octadecenoic acid

Orcinol : 3,5-dihydroxytoluene

Oxalic acid : ethanedioic acid

PAH : polycyclic aromatic hydrocarbon

Paraquat : 1,1'-dimethyl-4,4'-bipyridinium ion

Parathion : O,O-diethyl-O-(4-nitrophenyl)phosphorothioate

PCBs : Mixture of polychlorinated biphenyls

PCE : perchloroethylene

PCP : pentachlorophenol

Phenol : hydroxybenzene

Picloram : 4-amino-3,5,6-trichloro-2-pyridinecarboxylic acid

Prometryn : *N,N'*-bis(1-methylethyl)-6-(methylthio)-1,3,5-triazine-2,4-diamine

Propazine : 6-chloro-*N,N'*-bis(1-methylethyl)-1,3,5-triazine-2,4-diamine

Pyrene : benzo(def)phenanthrene

Quinoline : 1-benzazine

Common name : Chemical name

Quinone : 1,4-cyclohexadienedione

Simazine : 6-chloro-*N,N'*-diethyl-1,3,5,-triazine-2,4-diamine

Syringic acid : 4-hydroxy-3,5-dimethoxybenzoic acid

TCE : trichloroethylene

Tetrachloroethylene : see PCE

TNT : 2,4,6-trinitrotoluene

Toluene : methylbenzene

o-toluidine : 2-aminotoluene

Toxaphene : chlorinated camphene

Trifluralin : 2,6-dinitro-*N,N*-dipropyl-4-(trifluoromethyl) benzenamine

Vanillic acid : 4-hydroxy-3-methoxybenzoic acid

Vinyl chloride : chloroethylene

Xylene : dimethylbenzene